Microbial Aggregation

Author

G. B. Calleja, Ph.D.

National Research Council
Ottawa, Canada

CRC Press
Taylor & Francis Group
Boca Raton London New York

CRC Press is an imprint of the
Taylor & Francis Group, an **informa** business

First published 1984 by CRC Press
Taylor & Francis Group
6000 Broken Sound Parkway NW, Suite 300
Boca Raton, FL 33487-2742

Reissued 2018 by CRC Press

Library of Congress Cataloging-in-Publication Data

Calleja, Gode B., 1937-
Microbial aggregation.

Bibliography: p.
Includes index.
1. Microbial aggregation. I. Title.
QR73.6.C34 1984 576'.11 83-20883
ISBN 0-8493-5708-X

A Library of Congress record exists under LC control number: 83020883

Publisher's Note
The publisher has gone to great lengths to ensure the quality of this reprint but points out that some imperfections in the original copies may be apparent.

Disclaimer
The publisher has made every effort to trace copyright holders and welcomes correspondence from those they have been unable to contact.

ISBN 13: 978-1-315-89542-0 (hbk)
ISBN 13: 978-1-351-07452-0 (ebk)

Visit the Taylor & Francis Web site at http://www.taylorandfrancis.com and the
CRC Press Web site at http://www.crcpress.com

PREFACE

Somewhere I have read that an author's prefatory notes are usually written last, if at all written, and that they are usually read last, if at all read. (It must have been in somebody's preface.) And yet they come, by definition, before the main text. First shall be last and last first. This one is no exception.

A preface has a number of uses. Here I will use the device for the most part as a means of acknowledging my debt to those who helped me assemble this book. For this is nothing more than an aggregation, conscious and unconscious, of the words and ideas of others and a bit of mine, chiefly the string to bind them together.

I originally attempted to cover all microbial aggregation systems known and studied to date, but halfway in the writing process, it appeared that to do just that, with justice, required at least two volumes. Thus, the present volume covers in detail bacteria and yeasts. A second volume will comprise the cellular slime molds, the filamentous fungi, algae, and protozoa. Nonetheless, an overall perspective of microbial aggregation as fundamental form and function is presented here to include systems still to be treated in detail. Such an overview, I feel, is demanded by the subject matter.

There is a long list of people I have to thank. First among them is Byron F. Johnson of the National Research Council of Canada in Ottawa. A good many of his suggestions, I have unabashedly appropriated. Without him around to read patiently the drafts as the typewriter spewed them, this little book would have been doubly difficult to write and shape. Among those I have worked with in the laboratory, Bong Yul Yoo of the University of New Brunswick in Fredericton has helped me much by way of many insightful discussions. Both Dr. Johnson and Professor Yoo have been most generous collaborators in other projects.

With deep appreciation, I like to bare my indebtedness to Teena Walker and Susan Levy-Rick. Without their help, intellectual and manual, I would have given up gathering the materials for this volume; instead, I would have reached for the nearest waste basket. Writing may be physically accomplished with one hand, left or right, but the making of a book of this kind takes more than a pair of hands. It requires friends, who do not just watch as the author bleeds. The more difficult references were collected with the kind assistance of Margaret Schade and Noreen Brady.

In addition, the book and I profited much from corrections, comments, and criticisms (some caustic, others kind, never with a hint of malice) from other colleagues, who took some of their precious time to read segments of the manuscript: Isabelle Boisclair, Jack Christ, J. R. Colvin, Patricia Douglas, Allen P. James, C. V. Lusena, Brian L. A. Miki, F. Moranelli, Alain Vaisius, as well as others who would prefer to remain anonymous for fear that their contribution might be only minor. I disagree that their contribution is small. Anne Daley, Chris Gobey, and Lynda Boucher took turns to type the final script for the printer. The original illustrations were drawn by Celia Clyde and Denise Ladoucer. The photographic plates were prepared by Harry Turner. I would like to thank them all.

As well, I would like to thank all the authors who have consented to the re-use of their original materials. They are individually acknowledged at the appropriate places in the text. Here I will just list their names in alphabetical order: M. Achtman, C. E. Ballou, J. T. Bonner, T. D. Brock, M. Crandall, J. P. Duguid, G. M. Dunny, G. Gerisch, R. J. Gibbons, R. B. Gilliland, R. N. Greenshields, B. D. Hartong, E. Helm, W. Heumann, J. Hodgkin, J. Kohli, R. P. Levine, V. L. MacKay, S. A. MacKay, D. Malchow, T. R. Manney, B. L. A. Miki, W. L. Orton, P. C. Newell, N. H. Poon, A. Pühler, H. Reichenbach, G. G. Stewart, N. W. Taylor, R. S. W. Thorne, A. Tomasz, Y. Tsubo, H. van den Ende, M. J. Vold, L. J. Wickerham, N. Yanagishima, B. Y. Yoo, and M. Yusa.

In spite of the many forms of assistance generously given by associates, the book is far from perfect. Of course, my friends are not responsible for the imperfections. That responsibility is greedily reserved by the author for himself. The fool and his book are never parted.

This book is not for my wife, Mia, and the children, Maria and Raoul — even as they suffered as I wrote. There is world enough, I hope, to write another book just for you. You are most welcome to read this one — or even to show it to friends, yours, mine, and ours — provided they are not minors or prudes.

Gode B. Calleja

TABLE OF CONTENTS

Chapter 1

GATHERING THINGS TOGETHER

I. INTRODUCTION

A first book on an ill-defined subject must begin with definitions. For, in spite of the literature it has generated, microbial aggregation remains an ill-defined subject. This book and a subsequent volume are an attempt to gather in two bundles the scattered literature on the subject.

II. AGGREGATION DEFINED

Aggregation is the gathering together of units to make a larger unit (Figure 1). The resultant larger unit is also called an aggregation. In addition, the state or condition of being gathered together is also known as aggregation. The clumping of cells in liquid culture is aggregation. The resultant clump is an aggregation, or an aggregate, of cells. In the first sense, aggregation is a process, a function, the gathering together. In the second, it is a product, a structure, the larger unit. Such a melding of function and structure, despite the mutual disrespect of the physiologist and the morphologist for each other, appears to be a common feature of many languages.

The gathering together and the units may be concrete or purely conceptual. A compilation such as this book is therefore an aggregation. A cell is an aggregation of organelles, an organelle of molecules, a molecule of atoms. Society is an aggregation of individuals. A forest is an aggregation of trees. A sentence is an aggregation of words, a word of letters. The mathematical operation of addition is aggregation. The resultant sum is an aggregation, too, and so is any collective noun or pronoun. However, we shall concern ourselves here not merely with either grammar or numerical manipulation, but primarily with the concrete and the palpable, in particular, the aggregation of microbial cells.

The Milky Way is an aggregation of stars, but only structurally and conceptually. Its origins are not likely due to an aggregation process. Concatenated paper dolls make an aggregation (Figure 2). They are the product of a process not of aggregation but of paper cutting. A chain of bacilli, another concrete example, may be thought of as an aggregation, conceptually, but even in the conceptual sense, the process of chain formation among bacilli is not aggregation. Instead, it is properly called cell division or even cell multiplication, which is outside our present concern. Aggregation is addition, mathematically and rhetorically.

III. CELL AGGREGATION DEFINED

With that in mind and with the unabashed assumption that aggregation is a fundamental condition, function, and structure in biology, we shall define "cell aggregation" as the gathering together of cells to form fairly stable, contiguous, multicellular associations under physiological conditions (Figure 3). To various investigators, it is also known as adhesion, adherence, agglomeration, agglutination, association, autoagglutination, clumping, coagulation, coherence, cohesion, flocculation, flocculence, flotation, isoagglutination, sedimentation, stickiness. The resultant structure is also called agglomerate, agglomeration, aggregate, aggregation, agglutinate, agglutination, clump, cluster, coagulum, coremium, film, floc, flock, grex, head, pellet, pellicle, plasmodium, rhizomorph, ring, sclerotium, slime, slug, strand, stroma, synnema.

FIGURE 1. Aggregation, nonmicrobial. (Reprinted by permission of the *Bulletin of the Atomic Scientists*, a magazine of science and public affairs. Copyright (c) by the Educational Foundation for Nuclear Science, Chicago.)

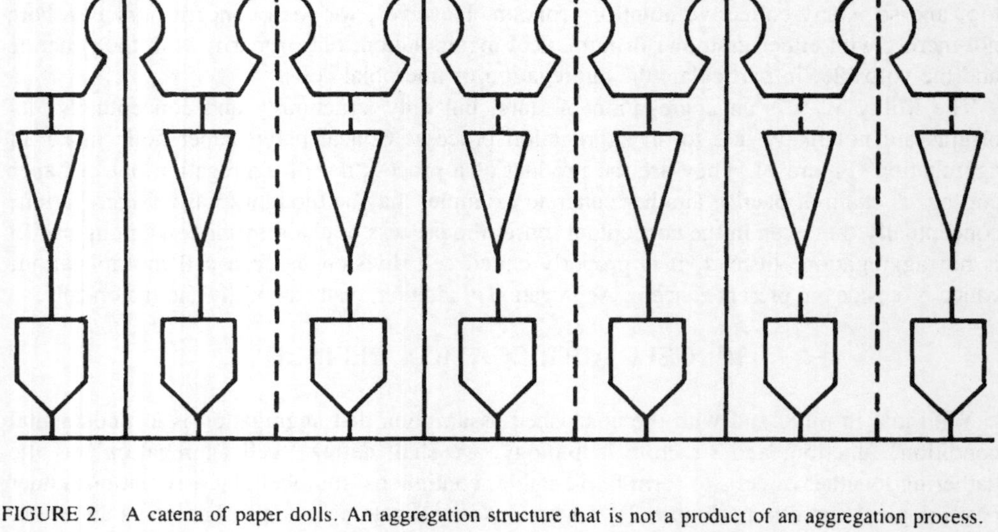

FIGURE 2. A catena of paper dolls. An aggregation structure that is not a product of an aggregation process.

The abundance of names, although a source of minor confusion, must not be viewed as a hindrance to the unification of the field of study. Rather, it is in keeping with the richness of the English language, especially with regard to words describing collectives or aggregations of animals. A few of these more colorful names are a bale of turtles, a charm of finches, a clutter of cats, a congregation of plovers, a cowardice of curs, a cry of hounds,

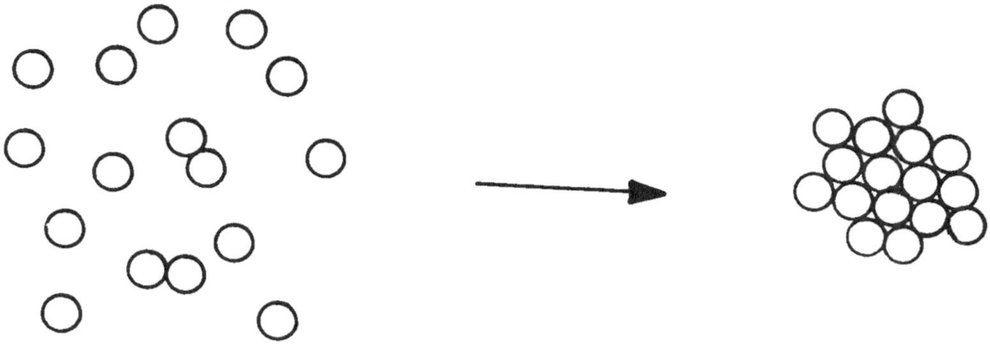

FIGURE 3. A generalized schema for cell aggregation. Cells are represented as spheres.

a drift of hogs, an exaltation of larks, a gaggle of geese, a leap of leopards, a murder of crows, an ostentation of peacocks, a parliament of owls, a pod of whales, a pride of lions, a shrewdness of apes, a skulk of foxes, a sleuth of bears, a spring of teal, an unkindness of ravens, a watch of nightingales.[1,2] In like manner, we may speak of an aggregation of myxobacteria, a clump of cells, a floc of yeast, a slug of amoebae, a strand of hyphae.

The definition contains two principal elements: physical movement and stable multicellular contacts. Both need to be present in any cell aggregation system.

Movement may be either directed (active), like chemotaxis in the cellular slime molds, or stochastic (passive), like random collision in a shaken liquid culture of bacteria (Figure 4). However it is achieved, there must be movement to allow cells to come together. This condition presupposes a state, prior to aggregation, in which the cells are disperse.

There must be actual physical contacts among the aggregated units, rather than merely conceptual grouping or lumping. Mere proximity of units, or even juxtaposition, is not good enough. Contacts must be more intimately close than close encounters of any kind. Moreover, they must be intercellular. Adsorption of cells onto inert surfaces fails to satisfy the definition. Furthermore, contacts must be multicellular. The minimally plural condition is not sufficient. Nor is the number of cells prior to aggregation strictly singular. The reason is more of convenience than of grammar: if we were to include pair formation (mechanistically, but not formally by our definition, the minimum aggregation), then all conjugation systems would have to be included, surely, not an easily managed lot. A lower limit of ten cells is a convenient, albeit largely arbitrary, size for a minimum group. It appears reasonable for certain systems, sex-directed flocculation in fission yeast, for example.

The multicellular condition connotes multivalency of the involved cells. It is clear that the final aggregated structure is determined by the combining power, or valence, of the individual components (Figure 5). Monovalent cells can only form pairs, divalent cells only chains (open or closed).

As in gametic agglutination of unicellular algae, contacts may be transient, but must last long enough and be strong enough to be experimentally describable. An aggregation that does not result in conjugation or cell fusion must be able to withstand Brownian buffeting under physiological conditions. A sediment of nonflocculent yeast cells, for instance, when gently agitated, disperses as a fairly homogeneous suspension in wort, but that of flocculent cells disperses as discrete flocs.

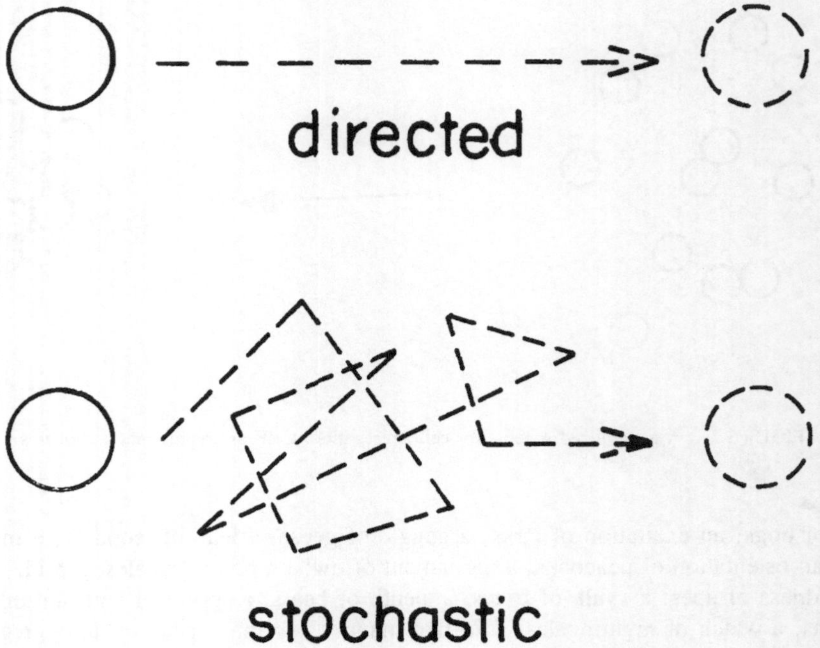

FIGURE 4. Movement of cells from one site to another.

IV. EXCLUDED PHENOMENA

Listed below are phenomena that may be, and have been in the literature, confused with cell aggregation. Some of them are graphically illustrated in Figure 6. They are things with which we shall not be concerned. For convenience, they are grouped as follows:

1. Noncellular aggregation: colloidal suspension, crystallization of viruses, gel formation, macromolecular self-assembly, polymerization, precipitation of molecules
2. Multicellular or multinuclear condition due to failure of progeny to effect complete separation after cell division or due to failure of cells to divide after nuclear division: chain, coenocyte, colonial form, filament, fruiting body, mycelium, packet, plasmodium, pseudomycelium, syncytium
3. Grouping (due to growth, buoyancy, or gravitation) under nondispersive conditions: bloom, colony on agar, focus of infection, pellet, pellicle, sediment
4. Association with noncellular objects: adherence, adhesion to glassware, adsorption onto solids, phage adsorption, stickiness
5. Interaction short of the multicellular condition: anastomosis, cell fusion, conjugation, copulation, heterokaryosis, mating, pair formation, palmella formation, zygote formation
6. Tropic response without subsequent stable contact: aerotaxis, chemotaxis, phototaxis, pellicle, swarm
7. Aggregation provoked by substances not usually found as components in the culture medium nor produced by the organism: agglutination, flocculation, and sedimentation provoked by antibodies, exotic lectins (e.g., concanavalin A, ricin, wheat germ agglutinin), native or denatured enzymes (e.g., lysozyme, ribonuclease), serum proteins (e.g., bovine serum albumin), artificial flocculants (e.g., bentonite, borate, undefined clays), and synthetic polymers (e.g., polyethyleneimine, polylysine)

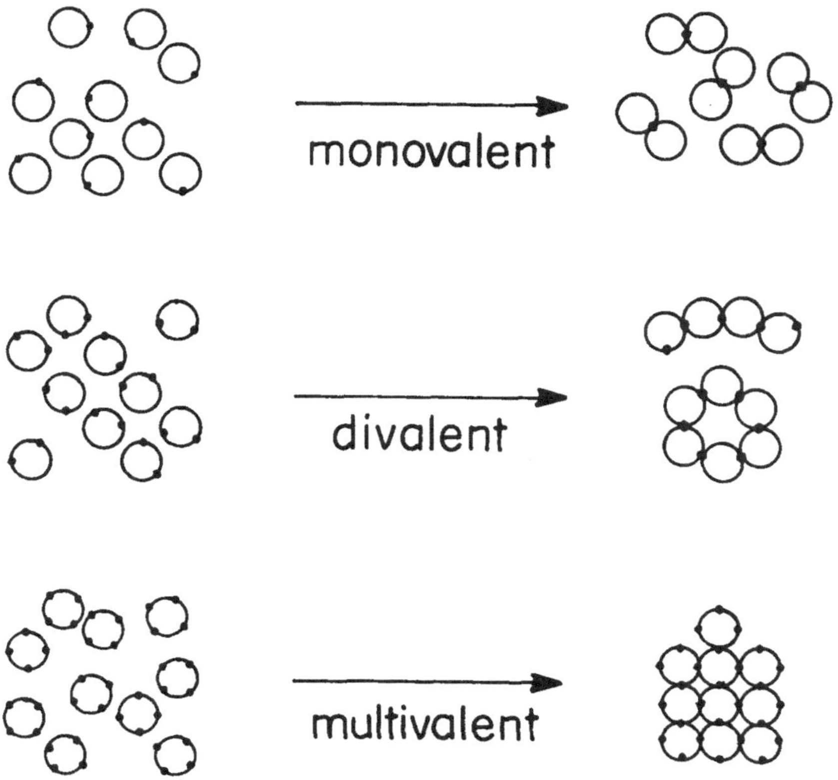

FIGURE 5. Valency of cells determines final aggregated structure. Dots represent potential and actual sites of union between cells.

8. Aggregation provoked under obviously nonphysiological conditions: agglutination, coagulation, denaturation, precipitation, and sedimentation due to strong acids and alkalis, organic solvents, centrifugation, dehydration, desiccation, heat, and lyophilization
9. Phenomena directly or indirectly associated with cell aggregation: anastomosis, cell fusion, competence, conjugation, copulation, fruiting-body formation, gametogenesis, growth, hyphal elongation, mating, meiosis, sexuality, sporulation, taxis, zygote formation

V. CRITERIA FOR INCLUSION

The following criteria are derived from the definition and the enumerated exclusions. To be included in a list of cell aggregation phenomena, a system must be (1) active (there is a change from a disperse condition to an aggregated condition), (2) inducible (there exists a physiological situation in which aggregation is absent), (3) stable (as an aqueous suspension), (4) intercellular (adhesion to noncellular surfaces is excluded), (5) multicomponent (a minimum group of ten cells), and (6) spontaneous (compatible with the life cycle of the organism and not due to obviously nonphysiological perturbations).

Notwithstanding the expedient limitations of the definition and the exclusion of many situations, cell aggregation remains a widespread condition in the microbial world. It is

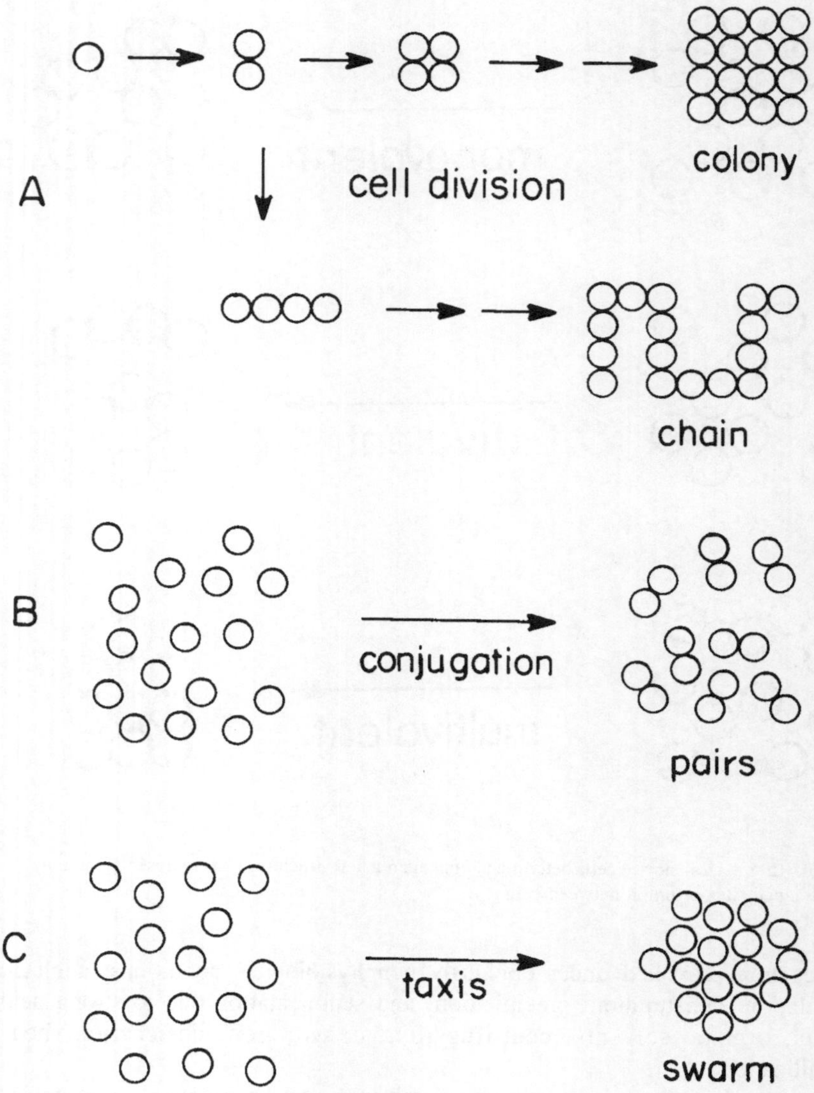

FIGURE 6. Phenomena that are related to, but excluded from, cell aggregation as defined in the text. (A) is excluded because the multicellular condition is achieved by a process that is more properly termed cell division; (B) because the final structures fall short of multicellularity; (C) because the final structures are units not in intimate material contact; (D) because the contacts are not intercellular; (E) because the process of achieving the multicellular condition is not physiological; (F) because the multicellular condition is elicited by exotic polymers; (G) because the final structure is achieved only under nondispersive conditions.

found among bacteria, yeasts, cellular slime molds, filamentous fungi, algae, and protozoa. In addition to well-known phenomena such as flocculation of brewers' yeasts, aggregation of cellular slime molds, and agglutination of gametes in chlamydomonads, the definition includes mating-aggregate formation during bacterial conjugation, bacterial star formation, agglutination associated with competence for bacterial transformation, dental plaque formation, aggregation in myxobacteria, sex-directed flocculation in fission yeast, and sexual agglutination in *Hansenula*, *Saccharomyces*, and other yeasts. It also includes mating re-

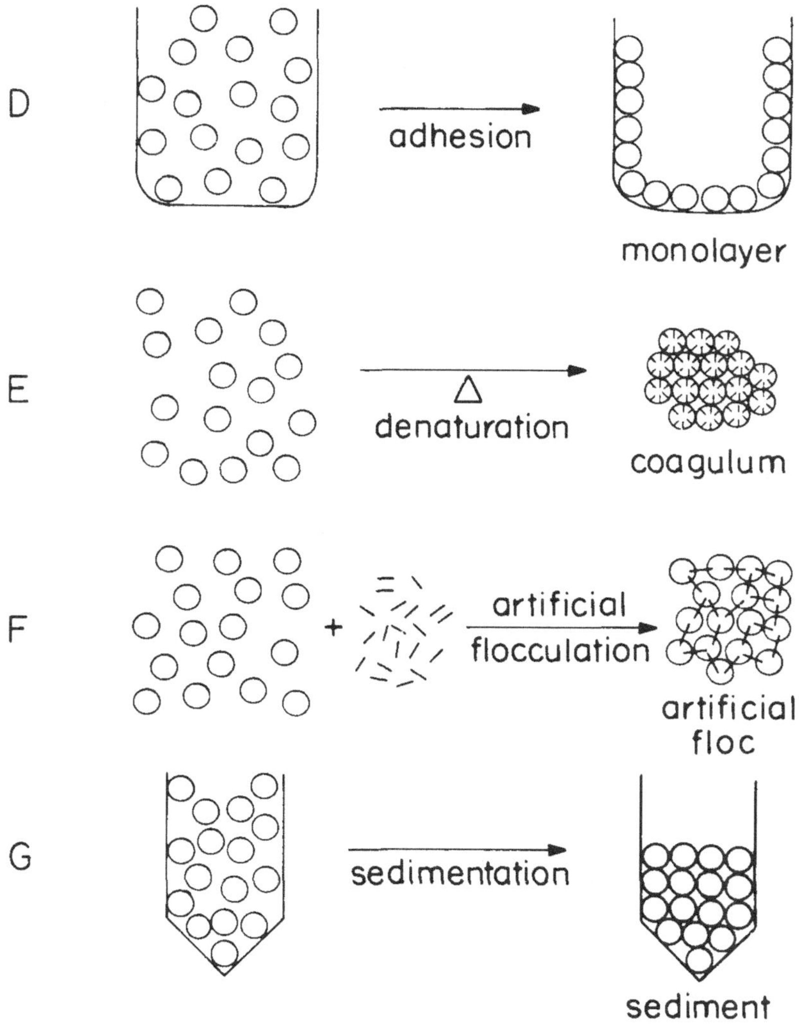

FIGURE 6.

actions in protozoa and the formation of strands, synnemata, rhizomorphs, sclerotia, and coremia in the filamentous fungi. As you well see, we have not run out of subject matter.

VI. REVIEWS REVIEWED

Of cell aggregation that is nonmicrobial, many reviews of general coverage have been written. A few of these must be mentioned, if for no other reason than to direct the reader to them.[3-7] Microbial systems, in contrast, have been less dealt with in broad terms, perhaps because of their apparent diversity.

Microbial aggregation has been reviewed from various vantage points. Hoffman[8] in 1964 treated bacterial aggregations as models of morphogenesis. His concept of aggregation, however, remains exactly that, that which we have dismissed as purely conceptual. The bacterial aggregations he reviewed are cell division, failure of cells to completely separate after cell division, chain formation, and colonies on agar, items excluded from the definition. Indeed, the review virtually confines itself to these phenomena.

Morris's treatment,[9] in 1966, of aggregation in yeasts likewise includes colony formation in addition to flocculation and sexual aggregation. Strangely enough, the formation of pseudomycelia is also entertained as aggregation and labeled asexual conjugation.

The literature on the role of polymers in microbial aggregation was ably summarized in 1973 by Harris and Mitchell.[10] Systems discussed include bacterial aggregation by synthetic and natural polymers, formation of dental plaque, yeast aggregation during conjugation and fermentation, and aggregation in heterogeneous microbial communities, with emphasis on soil ecology and waste treatment. The ecological and practical import of microbial aggregation is underscored.

A different approach was made in 1974 by Reissig[11] in viewing the cell surface as a seat of cellular regulation and recognition. Touching on many different aspects of ektobiology,[12] his admirable synthesis covers the whole range of the microbial spectrum and, among other things, includes bacterial conjugation, sexual agglutination in yeasts, mating reactions in protozoa and algae, chemotaxis, and bacterial transformation. The emphasis is information content of the cell surface, but the significance of microbial aggregation comes out very clearly.

The survey by Atkinson and Daoud[13] in 1976 is from the standpoint of fermentation process engineering. It is quite useful because of its detailed handling of methodology, mechanisms, and process applications. However, a great majority of the flocculation systems described are outside the scope of our present concern.

Ottow's review[14] in 1975 on fimbrae and pili peripherally touches on bacterial aggregation, and so does Smith's review[15] in 1977 on microbial surfaces in relation to pathogenicity.

The subject of mating-type interactions in microorganisms was extensively reviewed in 1977 by Crandall[16] for the new series *Receptors and Recognition*. Most of the systems covered fall within bounds of our definition. The coverage, however, is limited to sexual systems.

In the same new series is an anthology, edited by Reissig[17] in 1977, of contributions from various laboratories on microbial cell-cell interactions. It includes aggregation in the cellular slime molds, bacterial chemotaxis, bacterial transformation, mating-aggregate formation in bacteria, and mating reactions in *Saccharomyces, Chlamydomonas*, and the ciliates. A summary by the editor gives the reader a broad overview of the subject.

Fungal aggregation in its various guises has been dealt with adequately by Burnett[18,19] in his book on mycology. Other reviews on fungal aggregation from various biased viewpoints are those by Hawker,[20] Garrett,[21] Butler,[22] Willetts,[23] and Chet and Innis.[24] A review by Carlile and Gooday[25] in 1978 on cell fusion in fungi and myxomycetes embraces a good number of fungal aggregation systems.

By far, the most adequately covered systems are those of brewers' yeast and the cellular slime molds. It is no accident that it is in these systems where there is most intensive research activity. Reviews exclusively concerning these systems will be dealt with in chapters devoted to them. For the moment, it is sufficient to note that, in Bonner's book on the cellular slime molds,[26] cell aggregation is treated as if it were a virtual monopoly of Acrasiales, among the eukaryotes, and of myxobacteria, among the prokaryotes.

VII. THE STRATEGY OF THE MONOGRAPH

The recent flurry of reviews, general or specialized, on microbial aggregation is a welcome sign of a growing interest in the subject among researchers of diverse prejudices. Still, there is the need to gather together the many scattered bits of information concerning the subject. The need to gather the bits into a monographic whole becomes more pressing as the clutter and the confusion intensify. Because many associated and more complicated phenomena, including unidentified objects, have been mistaken for microbial aggregation, it has become imperative that a unifying treatment of the subject in rather broad and comprehensive terms be written, if only to put into proper perspective the extent and the meaning of a general biological phenomenon that has been too long neglected.

This monograph, an expansion of an earlier compilation on the nutritional aspects of microbial aggregation,[27] limits itself to microbes and to cell aggregation as defined above and as defined in the earlier review. The word ''microbes'' will be taken to mean organisms that are commonly accepted as cellular microorganisms by students of microbiology. Much space has already been spent on quibbling over definitions. There will be no further attempt to quibble in a taxonomic manner. Instead, the book will attempt to approach microbial aggregation as a unified subject and as a fundamental structure and a fundamental function in biology. As a way of emphasizing diversity, a number of chapters will be devoted in some detail to the better studied systems. Cellular slime molds, filamentous fungi, algae, and protozoa will be dealt with in a subsequent volume. In this volume, bacterial and yeast aggregation systems will be examined in detail. In a final chapter of this volume, these diverse systems will be gathered once again into an aggregated whole. The objective is to permit us, so to speak, to see the forest for the trees, the exaltation for the larks.

Chapter 2

CELL AGGREGATION IN THE MICROBIAL WORLD

I. CLASSIFICATION ATTEMPTED

The different forms of cell aggregation may be classified into ''homotypic'' and ''heterotypic'' systems (Figure 1). A homotypic system is made up of similar units, whereas a heterotypic system is made up of dissimilar units. Flocculation of brewers' yeast during fermentation is homotypic. Co-flocculation of yeast with bacteria is heterotypic. These examples seem to demonstrate the usefulness of the classification scheme. However, all is not that clear-cut.

If intergeneric interactions are obviously heterotypic, interspecific ones must no less be so. Homotypic systems are then limited to intraspecific interactions, but how does one classify interactions of mating types? Certainly, mating reactions involve cells of the same species, but cells that must differ in some way as to make them distinguishable from one another.

At the level of molecular structures, homotypic interactions may be achieved in two ways: (1) the interaction is mediated by a functionally symmetrical molecule and (2) the interaction is mediated in a lock-and-key fashion, both lock and key being found on the surface of the same cell (Figure 2). At this level, all systems involving mating types will have to be considered as heterotypic.

In the strict sense of the attempted classification, there is scarcely any cell aggregation system that has been shown to be truly homotypic. Perhaps, all interactions are among units that are dissimilar, in one way or the other, but more specifically at the level of the interlocking structures. To be truly homotypic, a system must be shown to be mediated by functionally symmetrical molecules or by interlocking structures that are found on the same cell.

The term ''heterotypic'' will be confined to systems that are obviously heterotypic and ''homotypic'' to those that do not formally require that the components be different. Many systems that will be described in detail are not obviously heterotypic and therefore, for the present purpose, classified as homotypic. Sex-directed flocculation or sexual agglutination, although among units within a species, is classified as heterotypic because it requires two types of complementary cells, however difficult it might be to distinguish them morphologically. Sexual agglutination aside, there are not many heterotypic systems that have been investigated carefully.

Many different ways of classification are conceivably useful. For instance, teleologic, according to the reason of a system for being, or mechanistic, according to how the interaction is mediated. Perhaps it is wiser to resist the temptation to classify, when the present state of the science does not warrant or even require classification, or failing that, to merely classify the systems taxonomically, according to the organisms involved.

Convenience, then, rather than logical insight, prompts us to group the various microbial aggregation systems according to taxa: bacteria, yeasts, cellular slime molds, filamentous fungi, algae, and protozoa. There will be a minimum reference to heterotypic systems, except to mean intergeneric and interspecific systems and interactions of mating types.

II. DISTRIBUTION AMONG MICROBIAL TAXA

Microbial aggregation systems are distributed among many microbial taxa. They are found in both prokaryotes and eukaryotes. A list of systems that will be discussed in detail is found in Table 1.

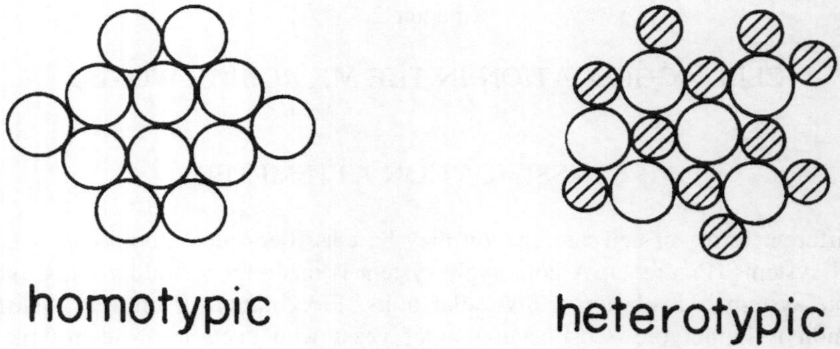

FIGURE 1. Homotypic and heterotypic aggregations. Cells are represented as spheres.

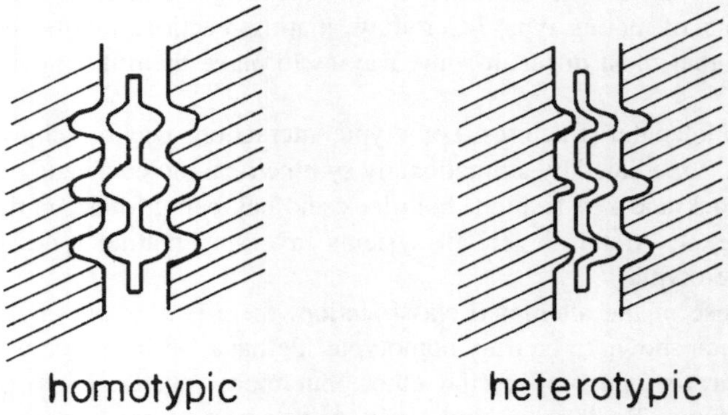

FIGURE 2. Cell-cell interactions. Hatched areas represent cell walls or cell membranes or wall appendages. Top: interactions mediated by a discrete molecule. Bottom: interactions mediated by lock-and-key structures which are integral parts of the cell wall or the cell membrane or the wall appendage.

The terminologies for aggregation as used by the authors of primary references are retained where possible. Such an arrangement, it is hoped, does not become a source of inconvenience to the reader. Rather, it lends some quaintness to the minor confusion, making the list look more like an oriental bazaar than a department store. Where there is lack of agreement as to the appropriate terminology among several laboratories working on the same system, the better known nomenclature, not necessarily the earliest, is chosen. Where there is no specific

Table 1
A LIST OF MICROBIAL AGGREGATION SYSTEMS

Bacteria
- Mating-aggregate formation during conjugation in *E. coli*
- Star formation in *Pseudomonas, Rhizobium, Agrobacterium*
- Pellicle formation in fimbriate bacteria
- Agglutination associated with transformation competence in bacteria
  - *Streptococcus*
  - *Pneumococcus*
  - *Bacillus*
- Plaque formation and agglutination in oral bacteria
- Aggregation in *Streptococcus faecalis*
- Aggregation in myxobacteria
- Other bacterial aggregation systems

Yeasts
- Sexual agglutination in *Hansenula wingei*
- Sex-directed flocculation in *Schizosaccharomyces pombe*
- Sexual agglutination in *Saccharomyces cerevisiae*
- Flocculation of brewers' yeast
- Other yeast aggregation systems

Mycetozoa
- Aggregate growth in *Physarum*
- Aggregation in cellular slime molds
  - *Dictyostelium*
  - *Polysphondylium*
  - *Acytostelium*

Filamentous fungi
- Strand formation: *Merulius, Helicobasidium, Agaricus*
- Rhizomorph formation: *Armillaria, Marasmius*
- Sclerotium formation
  - *Sclerotium*
  - *Sclerotinia*

Microsclerotium formation in *Verticillium*
Synnema formation: *Hirsutella, Isaria, Graphium*
Coremium formation: *Penicillium, Sphaerocybe*

Algae
- Mating reaction or sexual agglutination in *Chlamydomonas*
- Agglutination reaction in *Dunaliella salina*
- Isoagglutination of gametes in *Pandorina*
- Agglutination of gametes in *Ulva mutabilis*
- Other algal aggregation systems

Protozoa
- Mating reaction in *Blepharisma*
- Mating reaction in *Euplotes*
- Mating reaction in *Paramecium*
- Mating-aggregate formation during chemically induced conjugation in *Paramecium*
- Clumping in *Mayorella palistinensis*
- Other protozoan aggregation systems

terminology proposed, the author takes the liberty of inventing a descriptive title as a means of identifying a system and preventing its being mistaken for similar or related phenomena.

The making of this list is like making an all-star selection for a national hockey team. The choice and the order would be quite different if the list were made by somebody else. It would look different again, if it were to be submitted to a book of lists.

In spite of its innumerable contributions to general biology, the bacterium thus far has not found much favor among workers in the field of cell aggregation. Perhaps this is due to the fact that the field is barely past its phenomenological stage. Consequently, prokaryotes, being considerably smaller, carry with them a built-in disadvantage. Two bacterial systems, nevertheless, have attracted a loyal following. These are dental plaque formation caused by streptococci and aggregation in myxobacteria. Studies on dental plaque formation show signs of being in the forefront even as the myxobacteria may prove to be as engaging as their eukaryotic analogs, the cellular slime molds.

The yeasts, especially brewers' yeast, have been the standard models for microbial aggregation. Ever since fermentation was proved in the last century to be due to cellular activity, yeast flocculation has earned the serious attention of the brewer. So far there is no evidence of it being sexual. Sexual agglutination in the same organism, however, has drawn much attention since the discovery of mating factors, and sex-directed flocculation in fission yeast has been advanced as a model for development. Among the yeasts, it is *Hansenula wingei* which has the best claim to being the organism of choice in the study of sexual agglutination as mediated by glycoproteins on the cell surface.

Very interesting findings have been made with the cellular slime molds, the best studied of the microbial aggregation systems, at least with regard to the chemotactic signals involved, and easily the most popular. Various approaches, theoretical and experimental, have been brought to bear on this intriguingly beautiful process. The morphology and the mechanics of the movement, the generation of cyclic-AMP pulses, and the reception of these signals have occupied the time of many laboratories. These, however, form a small part of the investigations that treat these organisms as models for morphogenesis and differentiation.

Despite their potential role in plant pathogenesis, the aggregation systems of filamentous fungi have not been exploited to the fullest. Except for studies on sclerotium development, there has been no sustained effort to unravel the intricacies of the aggregation process in filamentous fungi, a process known to the early mycologists. There are some inherent difficulties, not the least of which is the difficulty in separating the process of aggregation from that of growth. The coalescence of hyphae to form hyphal aggregates is normally achieved concomitant with hyphal elongation.

Research on algal aggregation has been centered on the mating reaction or sexual agglutination of species of *Chlamydomonas*. Gametogenesis precedes the mating reaction, mediated by gametangial flagella. The morphological description of this reaction has been pursued with intensity together with attempts to isolate the mating factors. There is hardly any other alga that has been carefully investigated with regard to aggregation. And, of course, the colonial forms are excluded from our definition.

The mating reactions of a number of ciliates have been investigated, but species of *Paramecium* remain the mainstay. Upon mixing, cells of complementary mating types agglutinate by their ciliary tips, a process analogous to flagellar agglutination in chlamydomonads. The ciliary reaction is not the only way agglutination may be elicited. Mating-aggregate formation is also observed when conjugation is induced chemically.

A few of the systems found in the list are somewhat unexpected and require justification. These are formation of mating aggregates during conjugation in *Escherichia coli*, agglutination associated with the development of transformation competence in bacteria, and the mating reactions in *Blepharisma* and *Euplotes*. Since its discovery, bacterial conjugation has been assumed to be uncomplicated pair formation. Recently, however, it has been convincingly shown that the mating structure is an aggregate of many cells.[28] Genetic transformation in bacteria eventually gave us molecular biology. It has been reported lately that associated with competence for transformation is an agglutination reaction.[29-31] The mating reactions in *Blepharisma* and *Euplotes* have been known for many years. Unlike those in *Paramecium* and in chlamydomonads, these reactions are so weak that they have been assumed to be mere pairing of gametes. Ways of bringing about the agglutination reactions have been found not too long ago.[32,33] However, there has not been any report as to the capability of *Tetrahymena* for multicellular aggregation, and hence its absence from the list.

The list, of course, is representative rather than comprehensive. It is safe to assume that a large majority of microbial taxa still await investigative attention with regard to their aggregative capabilities. When the preferred orientations of other secretively sensuous microbes are unmasked, the list will undoubtedly become much longer.

Too often, cell aggregation is not the most conspicuous (nor the most interesting) phenomenon in a complex series of events or structures. This has occasionally led to it being just about ignored. When it gets more than the perfunctory attention, what often is scored is the event that comes before (e.g., gametogenesis, chemotaxis) or the event that comes after (e.g., conjugation, fruiting body, sporulation) or the associated event (e.g., competence for transformation, competence for conjugation), rather than aggregation as such. These events and others will be discussed, but only insofar as they help clarify the principal subject matter.

III. THE IMPORTANCE OF BEING AGGREGATED

A. Practical Considerations

Many a culture in the laboratory gets thrown out because what has been expected to be a pure culture turns out to be seemingly mutated, or worse, contaminated: an erstwhile homogeneous suspension of cells has become a collection of clumps. Instead of being sidetracked into finding out the reason for clumping, the all-too-careful investigator decides to throw away the errant culture and start afresh, this time being more careful not to encounter again the clumping phenomenon. Of course, if it occurs again and again, the nuisance is then taken seriously. Then the investigator is hooked on studying microbial aggregation. If you can't lick it, you might as well get a research grant for it.

To the investigator, however, who has to grow a homogeneous suspension of cells, because he has to count the cells or to sample the culture for analysis, clumping is no more interesting than a nuisance that must be wished away.[34] If he loses sleep on it, it is merely to find ways to avoid it.

Microbial aggregation has more than just negative value. In the case of yeast, it is a case of beer. To the brewer, yeast flocculation makes or unmakes a beer. Flocculation influences the fermentation process to a very large extent. Its onset signifies the end of primary fermentation. When the cells fall in flocs to the bottom of the brewing vessel, their access to the nutrients in the wort becomes restricted. At racking, the beer from a flocculent strain is poorly attenuated, sweeter, and brighter. Furthermore, flocculation influences the maturing of beer and the processes after the primary fermentation. Because it effectively reduces the yeast population in contact with the wort, it leaves very few cells for secondary fermentation, and the yet unborn beer becomes prone to spoilage by unwanted contaminants which may outfeed the yeast on the rich medium. Thus, the choice of yeast strains is critical. The quality of the final product depends on the flocculation characteristics of the chosen strains.

The flocculation characteristics of a strain are exploited not only in the production of beer but in the production of metabolites and biomass as well. The recovery, by filtration or by centrifugation, of yeast cells after fermentation is better accomplished in flocculent strains because flocs are fluffier than a sediment of "powdery" cells. Flocculent strains are also required in tower-fermentation systems to maintain adequate cell concentration and prevent washout.

In general, flocculation and the flocculating capacity of an organism may affect the design, performance, and yield of a fermentation set-up by facilitating the recovery of biomass, improving the flow properties of the brew at high concentrations of biomass, increasing microbial hold-up, and thus, presenting opportunities for novel fermentation design.[11] On the other hand, flocs hinder substrate diffusion and thus reduce substrate uptake and retard the metabolic processes.

Much of the potential usefulness of flocculation to the fermentation industry, for instance, tower-fermentation design in the brewery, the production of single-cell protein, and the production of biochemicals, remains to be realized. One organism that has commercial possibilities is *Saccharomyces kluyveri*.[35] A fast and strong fermenter, it may be grown for food or feed in place of *Candida utilis*. When the two complementary mating types are combined in water, a dramatic sexual agglutination reaction occurs (Figure 3). The mating types may be grown in separate containers and, at the end of growth, allowed to mix in a collecting tank. The agglutinated mass may then be recovered at much reduced time and expense.

One may suspect that cell aggregation plays a decisive role in the pathogenicity or invasiveness of a pathogen, or at least in its ability to colonize tissues. The best studied model on which one may base this suspicion is dental plaque formation. The possible relation of the aggregative capacity of *Streptococcus mutans* to cariogenesis and periodontal diseases has been the principal incentive in the surge of research activity on this system. The evidence has been summarized by Gibbons and van Houte.[37]

The ability of aggregated hyphal structures to withstand adverse conditions aggravates the hardships in eradicating plant pathogens. The destruction of these resistant structures and the prevention of their development are targets of measures to control the outbreaks of plant diseases.

Flocculation by natural and synthetic polymers is part of waste treatment by the activated sludge process. It is of minor significance in the trickling filter process.

Heterogeneous microbial communities may include bacteria, fungi, and protozoa in an aggregated mass. It is known that soil aggregation is enhanced by its biotic component, but as to whether cell aggregation, as contrasted to mere microbial growth, is of significance is open to question. However, aggregative microorganisms, including the cellular slime molds, are a significant component of the flora and fauna of the soil.

The ecological importance of microbial aggregation appears to lie in its contribution to the microbiota of a habitat. The environment at large selects for organisms that can adequately respond to its whimsical demands and selects against those that cannot. As an alternative state of a microbial population, cell aggregation may add to the viability of a biotic community beleaguered by unpredictable stresses.

B. Fundamental Function and Structure in Biology

Regardless of whether microbial aggregation is of any concern to the experimenter, the brewer, the biochemical engineer, the dentist, the plant pathologist, the sanitary engineer, the agronomist, or the ecologist, the microbes have their own reasons for coming together. They do not socialize for our pleasure or our concern or our consternation. The biological meaning of microbial aggregation clearly goes beyond its contribution to man and his daily needs. Its widespread distribution and the variety of its forms emanate from an intrinsic realization of the irrepressible living plasm. The force that drives cells to come together springs from an eternal design no less profound than man's search for purpose and far more profound than man's search for profit.

Microbial aggregation systems may be grouped, according to their apparent *raison d'être*, into three teleological categories: (1) a prelude to morphogenesis and differentiation, (2) a prelude to sexuality and parasexuality, and (3) a mechanism for survival under trying circumstances. Most of the better studied microbial aggregation systems may be grouped into these three teleological categories. The cellular slime molds may be included in the first category, the many sexual systems in the second, and the filamentous fungi in the third. The categories are not mutually exclusive. Some systems, or possibly most, may embrace all three, for instance, the fission yeast or the myxobacteria. However classified, a microbial aggregation system is a developmental system, simply because it is an inducible component of the life cycle of an organism.

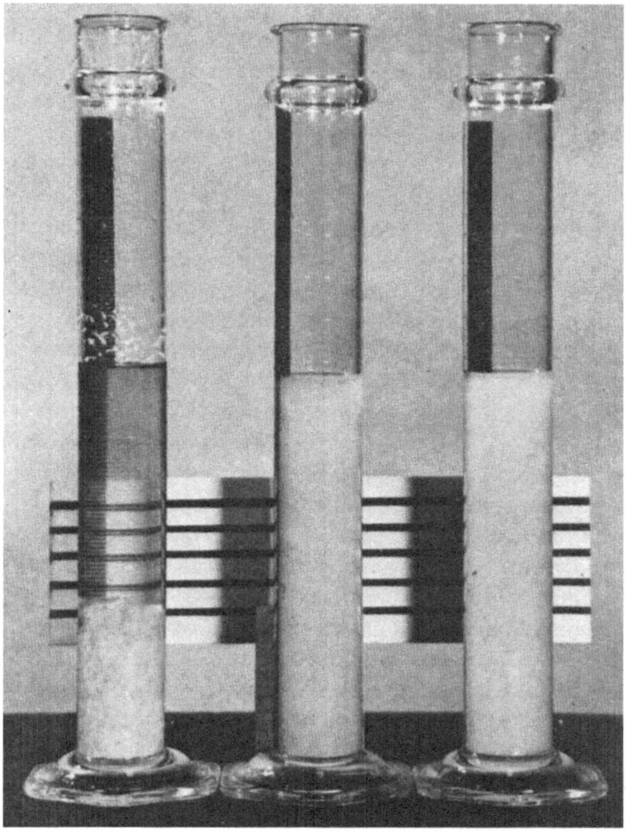

FIGURE 3. Sexual agglutination in *Saccharomyces kluyveri*. The left-most graduate contains a 3-min-old mixture of complementary mating types. Separate cultures of mating types are shown in the other two graduates. (From Wickerham, L. J., *Antonie van Leeuwenhoek J. Microbiol. Serol.*, 35(Suppl. Yeast Symp.), 31, 1969. With permission.)

That aggregate formation is a prerequisite to sex has been shown in many sexual systems. In instances where sex does not obviously follow, one might still argue that some parasexual function is accomplished during the process of aggregation, or that the competence for aggregation remained untouched even as mutation followed by selection had dropped the sexual function.

Sexual agglutination reactions in yeasts tend to become more intense as species evolve from haploid to a higher ploidy.[35,36,38,39] Wickerham surmises that haploid species developing strong agglutination reactions could hardly remain haploid for long. He believes that the evolutionary import of sexual agglutination is its contribution to the increase of ploidy in yeasts.

Microbial aggregation may be viewed as a reaching out by lower forms for the multicellular condition. In some systems, a truly multicellular condition is achieved, in that the aggregate, until now a collection of individuals, now behaves as an integrated organism made up of functionally differentiated cells, a myxobacterial fruiting structure, for instance. Like the proverbial girl who goes to Hollywood or the proverbial boy who goes to Broadway, a population of cells after aggregation is never the same again. The population is embarked on a new program. The events after aggregation normally follow a sequence inimical to individual cell proliferation. In this sense, the multicellular condition exists, however loose the integration and the coordination of functions might be.

Precisely because it is a structure derived from an erstwhile homogeneous collection of units, an aggregation seemingly defies the universal tendency toward an increase in entropy. It confers upon itself a higher degree of order, at the expense of its immediate environment, of course. Although it might not be bound by a new set of rules, it certainly has more information at its disposal than its components had.

In the filamentous fungi, the aggregation of hyphae leads to the production of a network which serves as nutritional pipelines, as alternate routes for communication and repair, and as an organized system for special functions.[40,41] In times of stress, a bundle of hyphae is better able to cope than a community of nonaggregated hyphae. The aggregate is a microenvironment less hostile and less variable than the larger environment in which it finds itself.

Alternatively, one may look at microbial aggregation as the mere consequence of some other events, instead of attempting to identify its survival value. Microbial aggregation is a surface phenomenon. The cell surface, be it membrane or wall or mural appendage, must undergo some change to allow and promote cell-cell interaction. Whether or not the change is in fact for the purpose of bringing the cells together is another matter. It is sufficient for the sake of argument that the change leads to a fairly stable aggregation structure. It may be solely for the purpose of facilitating genetic transformation, but the change in the structure of the cell surface confers on the cells a greater capacity to stick together.

How come sexual agglutination, not mere pair formation, not even just a *ménage à trois,* but a *mênage formidable*? The teleological doctrinaire will see in the agglutination structure the enhancement of mating efficiency. The enhancement is due to mass action, the result of the increase in collision frequency and in cell population density within the microcosm of the agglutination structure. Or it may be due to the increase in contact time, the result of mass immobilization of cells with respect to one another in a Brownian boiler. Or perhaps, the aggregate creates an ambiance, a microenvironment conducive to the mating game. The hedonist has a different explanation. He sees in the orgy a Maxwellian demon of sorts in the role of Eros, or better still, as group therapist whipping up the cell population into a frenzy. Passion uncontrolled precipitates into an exaggerated reaction: an agglutination. In yeast, the reaction is a veritable bacchanalia in ethanol. The mechano-materialist views sexual agglutination in a puritanic light. It is no more than a superficial change (in the literal sense) with more than superficial consequence (now both literally and figuratively): sexual agglutination is the consequence of a structural and functional change on the cell surface.

Whatever its reason for being, a motive force or a mere consequence, a microbial aggregate is a society of cells. The advantages that accrue to social organization may be thought of as applicable even to a society of cells. What cannot be attained by many individuals severally may be attained by the synergistic effort of a union of many individuals. It is the Marxist paradox: in uniting, individuals have nothing to lose but those that bind them, their common weaknesses as individuals. The union may be a means to a particular function, but it may be for no other reason than pure and naked survival.

C. Microbial Aggregation Systems as Models

Microbial aggregation systems make good models for cell aggregation of higher forms, for just the reasons that make microbes the organisms of choice for working out more complicated biological problems. Microbes are simpler, easier to grow, easier to handle. They require very little bench space. They have very short generation times. As a consequence, it is possible to work with a very large sample of cells. Mutations are easier to come by, genetic manipulations convenient, but more important, what is true for *E. coli*, according to Monod's act of faith, is even truer for the elephants.

Because of the fundamental nature of microbial aggregation and because of the diversity of its many guises, there is a microbe for everybody's needs. There will always be a microbe from which to learn the rules of sexuality, differentiation, morphogenesis, and development of higher organisms, including man. Fertilization in higher gametes, for example, has its counterparts in the conjugation systems of the algae, the protozoa, the yeasts, the filamentous fungi, and the bacteria. The problems of embryogenesis may be simulated in the cellular slime molds. One sees in the sexual agglutination of *Hansenula* the designs of an antigen-antibody reaction, an enzyme-substrate complex. In many microbial aggregation systems may be seen the patterns for cellular adhesion, sorting-out of cells, intercellular communication and recognition, the generation and reception of molecular signals, cell fusion, the emergence of cells into tissues. As a spin-off, several aggregation systems may serve as back-up systems for the study of genetic expression, induction of protein synthesis, catabolite repression, metabolic regulation, infection, mitochondrial development, allosteric interaction, karyogamy, meiosis, gene switching.

Problems of a higher order of complexity may be worked out by dissecting the more simple. It is in the best tradition of microbiology, that of pointing the way to seeing the simple in the complex, the unity in the diversity. For this reason alone, if for no other, the study of microbial aggregation may contribute to the clarification of the conundrums not only of cell aggregation in higher plants and animals but also of human society. Who knows, the study of microbial aggregation for its own sake may ultimately lead to the proper study of mankind.

Chapter 3

IDENTIFYING THE PROBLEMS: THEORETICAL AND EXPERIMENTAL APPROACHES

I. GROSS PHENOMENOLOGY

To get started on the experimental study of microbial aggregation, one has to simply observe an organism that exhibits the phenomenon or to look for it in an organism in which it has not been observed before, or, as in certain cases, to stumble upon it while looking for other things. Any of the three approaches pertains, because the phenomenon is grossly observable, at least in most of the systems.

A number of microbial aggregation systems are old phenomena. The aggregative nature of Acrasiales was known to Brefeld[42] in 1869 and to van Tieghem[43] in 1880, and that of the myxobacteria to Thaxter[44] in 1892. Algal aggregation was described as early as 1881 by Berthold.[45] Pasteur reported yeast flocculation in 1876 in his classical studies on beer.[46] At the unfolding of this century, the literature on yeast flocculation had begun to accumulate.[47-54] Mycologists of the last century, Brefeld,[55] de Bary,[56] and Reinhardt[57] among them, were well aware of hyphal aggregation in filamentous fungi. Although the impetus to its study came after the discovery of mating types in *Paramecium* by Sonneborn[58] in 1937, conjugation in ciliates was known to Bütschli[59] in 1876 and to Maupas[60] and Hertwig[61] in 1880. The rest of the better studied systems were identified fairly recently.

Some microbial aggregation systems are so obvious there is no question as to their being observable at all. A good example is the sexual agglutination of the yeast *Hansenula wingei* (Figure 1), but even such an impressive reaction had to await demonstration because the complementary mating types had to be discovered at the same time.[62] The reaction is observable only upon mixing of mating types, even as the mating types are ascertained experimentally by the agglutination reaction: like types being unable to bring it about, complementary types being able to do so. The resulting diplont is normally not agglutinative.

Wickerham's discovery in 1954 of sexual agglutination in *Hansenula* species prepared the way for the discovery of sexual agglutination reactions in other yeasts. Very vigorous reactions have been shown for *Saccharomyces kluyveri* (Figure 3, Chapter 2), *Citeromyces matritensis* (Figure 2), and its haploid form *Torulopsis globosa*,[35,36,63] but until recently, only these four genera were represented in a short list of sexual agglutination systems in yeasts.

So long as the fission yeast *Schizosaccharomyces pombe* was grown on agar, its ability to flocculate remained unreported, although the mating types had been known for some time.[64] Homothallic strains grown in malt-extract broth flocculate during stationary phase.[65] Flocculation is sex directed, being a prerequisite to conjugation.[66] The heterothallic strains, unlike those of *Hansenula*, do not flocculate immediately when mixed together. They require at least a round of cell division together.[67] Sex-directed flocculation is found also in other species of fission yeast.[68]

Aggregation in other sexual systems came to be recognized with hesitation. The demonstration of aggregate formation during the mating reactions in *Blepharisma*[32] and in *Euplotes*[33] has already been mentioned in the previous chapter. Sexual agglutination in *Saccharomyces cerevisiae* was noticed[69] only many years after the identification of mating types.[70] Cells of **a** and **α** mating types agglutinate when mixed under appropriate conditions, but this reaction had been regarded as not so worthy of study as the pheromonal control of the mating phenomenon.[71,72] Still other systems had to be proved with difficulty. The formation of mating aggregates during conjugation in *E. coli* is now convincingly demonstrated,[28] but the hard evidence came long after the initial demonstration of bacterial sex.[73] The occurrence of clumps during bacterial conjugation had been known in the literature but only as a minor inconvenience to be avoided or ignored by those who aspired to the rewards of molecular biology.

Other than in the well known or in the very obvious, where does one look for microbial aggregation? How does one go about looking for it, what is there to look for, and how does one know when he has found it? Because agglutination and sex tend to go together, one might look for aggregation in conjugation. One suspects that a good many sexual systems are indeed aggregation systems if only the right conditions are found.

In systems where mating types are concerned, the demonstration of agglutination calls for the mixing of cultures of complementary mating types. Where the complementary mating types are descended from a clone, as in homothallic strains of yeasts, the mating types are already present in one culture and it becomes a matter of inducing the culture to agglutinate under appropriate conditions. Potential aggregation systems might have been missed because the right conditions for induction or the opportune time when the cells are ready for mixing has not been found. Quite often, because it is a developmental system, cell aggregation is found at the end of or outside the cell cycle. In liquid culture, it is often found during stationary phase. It may be grossly visible in the form of clumps or flocs of cells. The primary flocs may coalesce to form even larger flocs. They may separate from the uninduced cells, fall out of suspension, and settle to the bottom of the culture vessel (Figures 1 and 2).

One has to show that aggregation is not merely pairing but the formation of an aggregate of many cells. Where it is glaringly a multicomponent association of cells, one has still to show that the aggregated structure is not the consequence of failure of cells to physically separate after cell division and that there are fairly stable contacts among them. Indeed, where aggregation is suspected, one has to show that it is not excluded by the definition and that it meets the criteria for inclusion among microbial aggregation systems.

One can screen for agglutinative strains of yeasts by exploiting sex and the events that occur after aggregation. An example is the heat-treatment method worked out by Wickerham and Burton.[74] Yeast spores are slightly more heat resistant than vegetative cells. A sporulated culture then may be subjected to high temperatures, say 58°C, and the survivors plated. Presuming that spores are the products of cell complementation of mating types, one may test the agglutinative properties of spore clones when mixed together in various combinations. The agglutinative strains of *Hansenula* were found this way. In this connection, the alcohol-treatment method of Leupold for the recovery of fission yeast spores is also worthy of mention.[75] Indeed, spores are a good starting material for the isolation of mating types.

Gross observations may reveal right away whether an aggregation is obviously heterotypic or not. If it is not obviously so, it is far from easy to prove it is homotypic. Initial observations may yield valuable information, such as when aggregation occurs with respect to the life cycle of the organism or with respect to the stages of growth of a cell population, cultural conditions of its occurrence, hydrodynamic properties of the aggregates, certain environmental factors, including temperature, ion concentration, various chemical effectors, etc. The primary objective of a gross phenomenological study is simply to demonstrate the phenomenon in such a way as to allow others, if they so desire, to demonstrate it to themselves with reliability and to their satisfaction.

II. QUANTITATION OF MICROBIAL AGGREGATION

Far too many reports concerning microbial aggregation have relied upon subjective judgment by eye. Yet the description of any phenomenon must eventually become quantitative,

FIGURE 1. Sexual agglutination in *Hansenula wingei*. The first two vials from the left show complementary mating types separately dispersed. The third vial is a mixture of the mating types. Right, the mixture has been rolled into a hard ball, which maintains its shape. (From Wickerham, L. J., *Antonie van Leeuwenhoek J. Microbiol.*, 35 (Suppl.), 31, 1969. With permission.)

FIGURE 2. Sexual agglutination in *Citeromyces matritensis*. The first two graduates from the left show complementary mating types separately dispersed. The rest contain mixtures of the mating types at 35 sec, 55 sec, and 4 min, respectively, from the left, after mixture. (From Wickerham, L. J., *Antonie van Leeuwenhoek J. Microbiol.*, 35 (Suppl.), 31, 1969. With permission.)

if its study is to achieve a reasonable rate of progress. For many systems, rapid and precise quantitative procedures for estimating aggregation structures and processes are yet to be devised.

Quantitation of systems that concern only a few cells is straightforward. It is no more complicated than counting marbles. The smallness of the cells is offset by the microscope or the electronic particle counter, but in systems that consist of very large aggregated masses, the estimation of the degree of cell aggregation is somewhat complicated. For one thing, counting cells in a three-dimensional array of more than 20 is hardly possible in the microscope. For another, obtaining a homogeneous and representative aliquot of a clumpy culture presents a formidable challenge. To go around this difficulty, one uses the whole culture in one container as one sample and sets up many simultaneous cultures as subsequent samples. Else, one may have to resort to drastic disruption of the culture by sonication or by enzymic dispersal of clumps.

Furthermore, it is rather difficult to decide beforehand just what parameter to measure. Is it how many aggregates there are? Or what percentage of a cell population is aggregated? Or is it how large the aggregates are? Or how rapidly do they form? Or even, how stable they are under a specified set of conditions? Clearly, the degree of aggregation of a culture is a matter of definition. The aggregation index of a culture graded by eye tends to a compromise among four criteria: number, size, stability, and the rate of formation of the aggregates. Each contributes in a different way to the estimate of how much biomass in a population of cells is in the aggregated condition. The critical reader will note in the literature the personal bias that gets into the description of an aggregation system: poorly, highly, weakly, strongly, or intensely agglutinative or flocculent. These adverbial shibboleths betray the dominant criterion employed by the observer: "poorly" and "highly" suggest number and size, "weakly," "strongly," "intensely" are a reference to stability and the rate of formation of aggregates.

All one needs to determine experimentally is the number of cells in the nonaggregated fraction, if the total cell population of an aggregated culture is known. The difference between the total population and the nonaggregated fraction is the aggregated mass. The nonaggregated fraction, however, must be delimited operationally. There is no problem with regard to single isolated cells nor with regard to macroscopic aggregates, but is a five-cell unit an aggregate or not? How about a pair? The cutoff number must be convenient and repeatably useful. It must be defined empirically for every system and it too must be decided beforehand.

As defined in Chapter 1, cell aggregation is limited to groups of no less than 10 cells each. Any cell in a group of less than that number we shall call a free cell, a convention based on convenience. In fission yeast, however, the definition finds a rational basis, the distribution of aggregation numbers of free-cell populations.[76] The term "aggregation number" simply refers to the number of cells in a group. Thus, an isolated cell has an aggregation number of 1, because it belongs to a group comprising one cell. Each of a pair of cells physically linked to each other for whatever reason, as well as the pair, has an aggregation number of 2. Each cell in a group of three, as well as the group, has an aggregation number of 3, and so forth. An aggregation number is a property of a group as well as of each cell in that group.

The distribution profiles of aggregation numbers in free-cell populations of fission yeast are shown in Figure 3. The profiles amply demonstrate that the aggregation numbers of free-cell configurations conveniently fall below 10. However, the modal aggregation number is not 1 (as would have been expected if the free-cell condition were confined to the singular number), but 2, the minimally plural condition. Besides, aggregation numbers greater than 2 are not infrequent. Note, on the other hand, that a floc of fission yeast cells on the average contains 10^5 cells, a number that must be imagined as extremely far to the right of the profiles.[76] Hence, the number 10, or some number thereabouts, is a convenient and reasonable

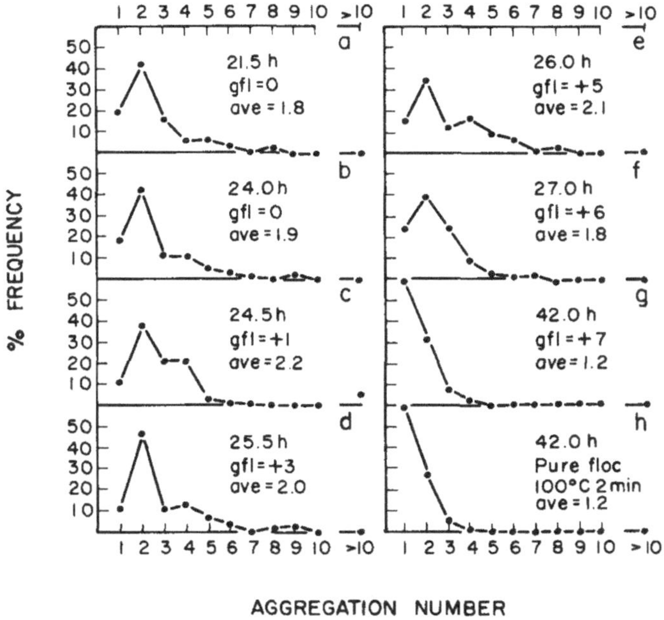

FIGURE 3. Distributions of aggregation numbers in free-cell populations of fission yeast. A flask culture was aerated from the time of inoculation and sampled for free cells at times indicated. Cells were counted in a hemocytometer. Whole culture (a to c), free-cell fraction only (d to g), purified flocs deflocculated by heat (h); gfl, gross flocculation as graded by eye; ave, average aggregation number. (From Calleja, G. B. and Johnson, B. F., *Can. J. Microbiol.*, 23, 68, 1977. With permission.)

cutoff separating free cells and aggregates. A cutoff number significantly less than 10 would be most difficult to apply in fission yeast, and, perhaps, in many other microbial aggregation systems as well. Incidentally, the profiles also serve as evidence in favor of an active process of aggregation, rather than of an aggregation structure that is the consequence of the failure of progeny to physically disengage after cell division.

The diverse methods for measuring microbial aggregation may be classified into five approaches: (1) visual grading of whole cultures in arbitrary units;[30,66,76-87] (2) particle counting and sizing of whole cultures;[28,88-90] (3) estimation, with or without dispersion of aggregates, of the number of free cells;[30,65,68,76,91-102] (4) estimation of the aggregated mass after its separation from the free cells;[66,69,76,85,103-107] and (5) measurement, by eye or by some optical arrangement, of the clearing rate of the supernatant during gravity sedimentation or the rate of fall of the sedimentation boundary.[76,92,95,107-115]

The first approach pertains when aggregates are large enough to be perceptible to the unaided eye or at least identifiable in the light microscope. However, visual grading is susceptible to deception by appearances. The aggregated fraction usually consists of many aggregates which vary in size, shape, and stability. The second approach is practicable only when the range of aggregate sizes is not too broad. A histogram of aggregation numbers may be derived from a microscopic (Figure 4) or an electronic particle count (Figure 5). In the third approach, the free-cell fraction may be estimated in a hemocytometer, an electronic particle counter, a spectrophotometer, or a nephelometer. Without the benefit of a procedure for dispersion, this method, like the first, is useful only in a comparative way. Absolute numbers may be obtained by scoring for free cells before and after dispersion of aggregates. The fourth approach separates the aggregated fraction by the difference in sedimentation

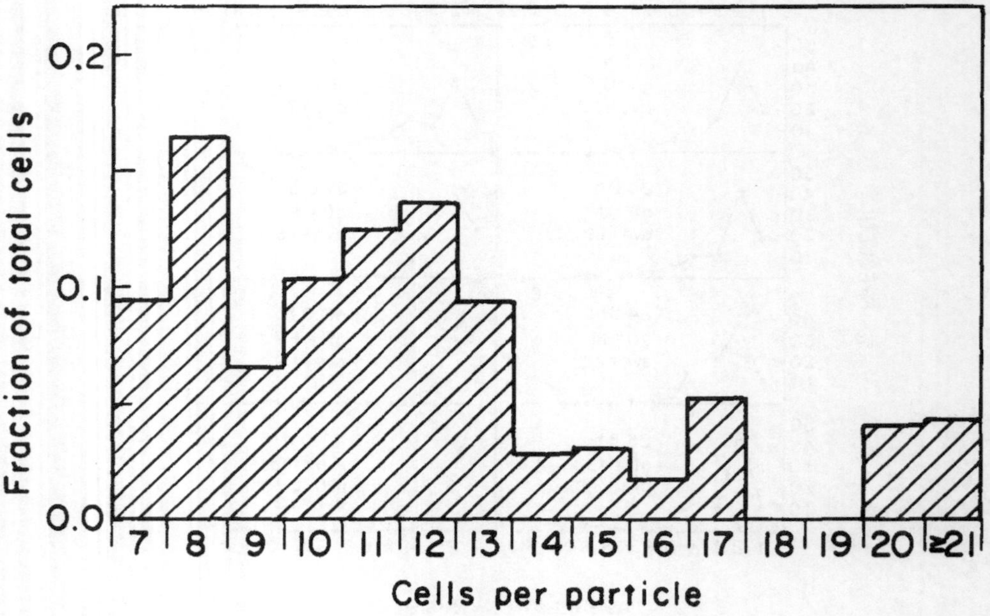

FIGURE 4. Distribution of aggregation numbers: mating-aggregate formation during conjugation in *Escherichia coli*. The histogram excludes aggregates containing less than seven cells. A total of 91 aggregates of 965 cells were scored by light microscopy. (From Achtman, M., *J. Bacteriol.*, 123, 505, 1975. With permission.)

rates between aggregates and free cells. The aggregated fraction may then be assayed for mass, volume, or cell number. The fifth approach estimates the size of the aggregates, rather than their number in a sample. The assumption is that there is a direct relationship between the size of the individual aggregates as reflected in their sedimentation rate and the intensity or the extent of the aggregation process.

Examples of methods for measuring aggregation in liquid cultures are described in detail below. These methods, based on differential sedimentation of aggregates and of free cells, have been applied to the study of the kinetics of sex-directed flocculation in fission yeast.[76] In the first method, one determines the number of cells that do not settle within a given period of time and then the total number of cells, free and in aggregates. It is necessary to allow the aggregates to settle before sampling for free cells because aggregates interfere with the sampling and the counting. Because of the difficulty in counting cells in aggregates, the aggregates are dispersed by proteolytic treatment before the total cell count is made. The difference between the counts is the number of cells in the aggregates. In the second method, the turbidity of a culture is read before and after aggregates settle during a short period (Figure 6). The assay in effect measures the decrease in turbidity and the difference in readings is equated to the turbidity component due to aggregated cells.

A. Hemocytometric Method

While gently swirled (Figure 7), a 10-mℓ culture in a flask is quantitatively transferred to a 15-mℓ conical centrifuge tube (Step 1) and let to stand undisturbed. Exactly 5 min later, when all the visible aggregates shall have settled (Step 2) while leaving the free cells in suspension, the supernatant is sampled about 2 cm below the meniscus and the 0.5-mℓ sample diluted tenfold for counting in a hemocytometer or an electronic particle counter. The tube is then spun in a table-top centrifuge to bring down the remaining cells, free and aggregated, to a pellet (Step 3). The culture medium is decanted and replaced with tap water

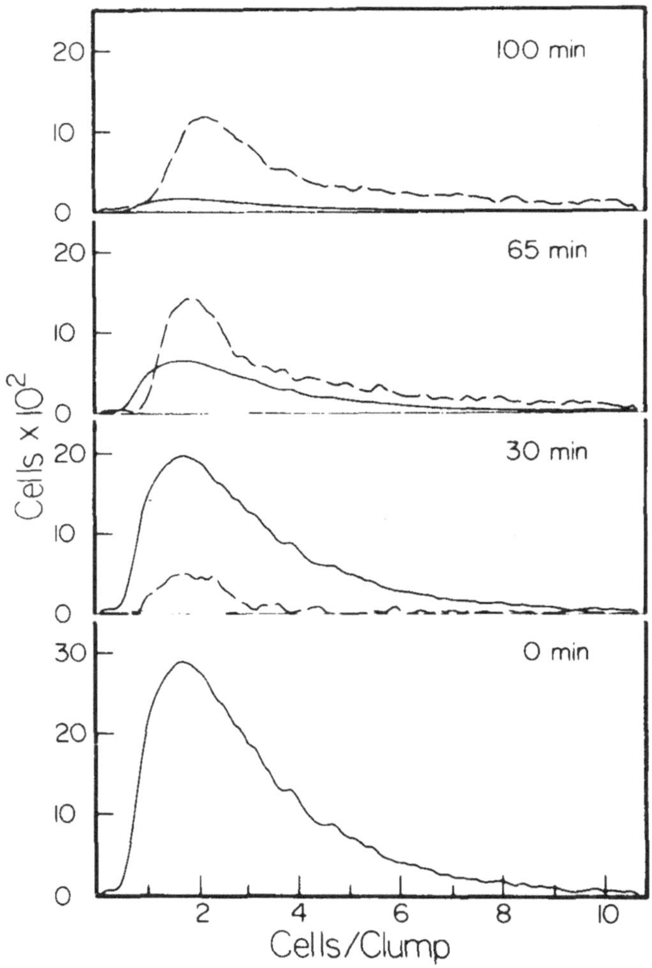

FIGURE 5. Distribution of aggregation numbers: sexual agglutination in *Saccharomyces cerevisiae*. Samples were withdrawn for counting in an electronic particle counter at times indicated after mixing of mating types **α** and **a**. Broken lines represent sex-specific aggregates, solid lines control cells. The distributions of sex-specific aggregates were derived by subtracting the normalized distribution of control cells from the distribution of the mixture of mating types. (From Manney, T. R. and Meade, J. H., in *Microbial Interactions*, Reissig, J. L., Ed., Chapman and Hall, London, 1977, Chap. 7. With permission.)

plus the unused remainder of the diluted sample (Step 4). The pellet is resuspended and Pronase (or trypsin or papain) in powder form added (Step 5). (The culture medium has to be removed because it inhibits the dispersing capacity of Pronase.) The suspension is stirred until all aggregates are dispersed (Step 6). Dispersion is complete within 5 min. The suspension of free cells is then sampled for total cell counts. The small error due to the initial sampling for free cells may be easily calculated or even disregarded. Note that the method is destructive in that the entire content of a flask is used for the assay. The results of an application of this method are shown in Figure 8.

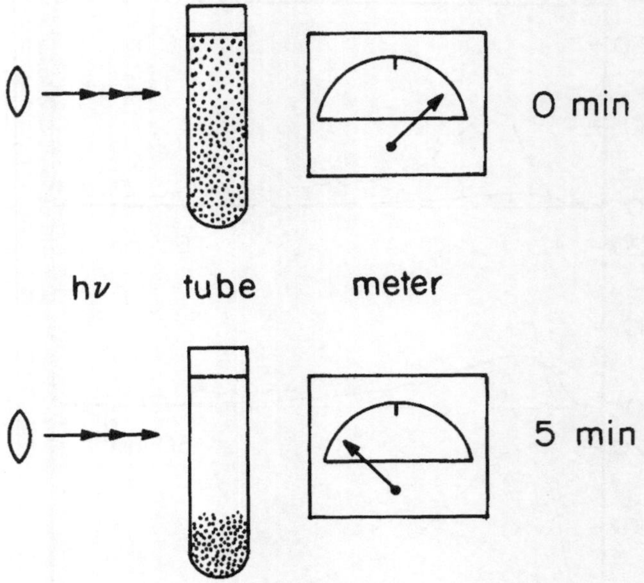

FIGURE 6. Gravity sedimentation of aggregates and consequent decrease in turbidity of a culture. Changes in turbidity of a suspension of cells may be detected by either transmitted or reflected light.

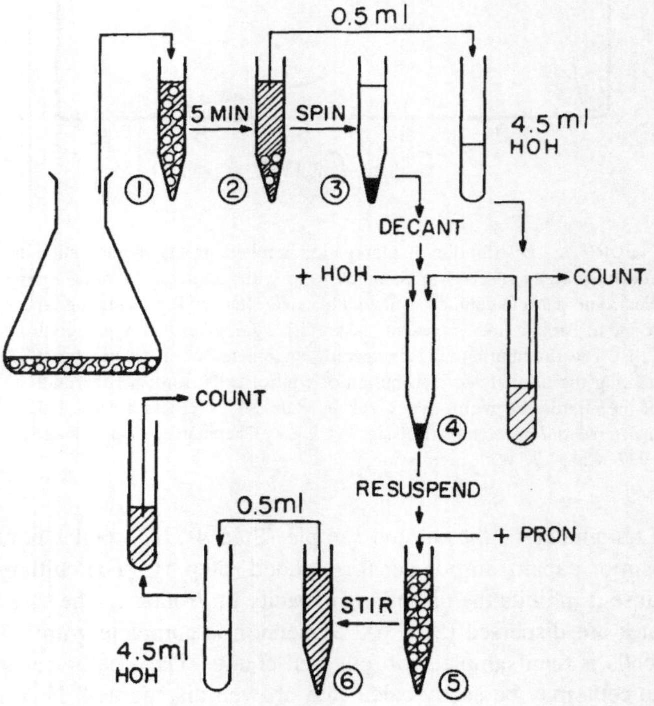

FIGURE 7. Hemocytometric method for measuring sex-directed flocculation in fission yeast. Circles represent flocs, stripes free cells. See text for explanation of procedure.

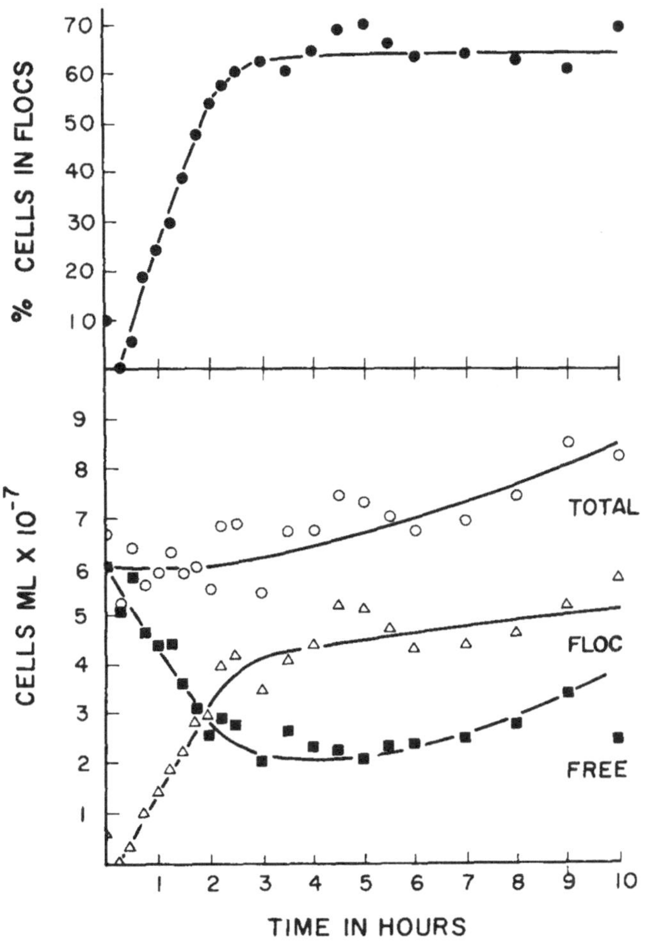

FIGURE 8. Development of sex-directed flocculation in fission yeast as measured by the hemocytometric method. The start of the induction procedure is 0 hr. Total cell population, ○; free-cell fraction, ■; cells in flocs, △; cells in flocs percent cells, ●. (From Calleja, G. B. and Johnson, B. F., *Can. J. Microbiol.*, 23, 68, 1977. With permission.)

B. Turbidimetric Method

A culture grown in a flask with side arm may be assayed periodically for aggregation *in situ*. At sampling time, the culture is gently swirled and tipped into the side arm, which is inserted quickly into the cuvette well of a Klett-Summerson colorimeter. (Any other nephelometric device should work, provided its light path and sensor are well above the sedimentation line.) As soon as the side arm is in place, the turbidity of the culture is read (Figure 6). Then the culture is left undisturbed. Exactly 5 min later, the turbidity is read once more. The difference between readings at 0 min and at 5 min is the aggregation index. The readings may be repeated several times for precision. Because the assay is nondestructive, the flask may be returned to the incubator after the assay. Results obtained by this method are shown in Figure 9.

C. Volumetric Method

An aggregated culture is left undisturbed for a short period of time in a graduated centrifuge tube. The volume occupied by the aggregates at the bottom of the tube is noted. This method

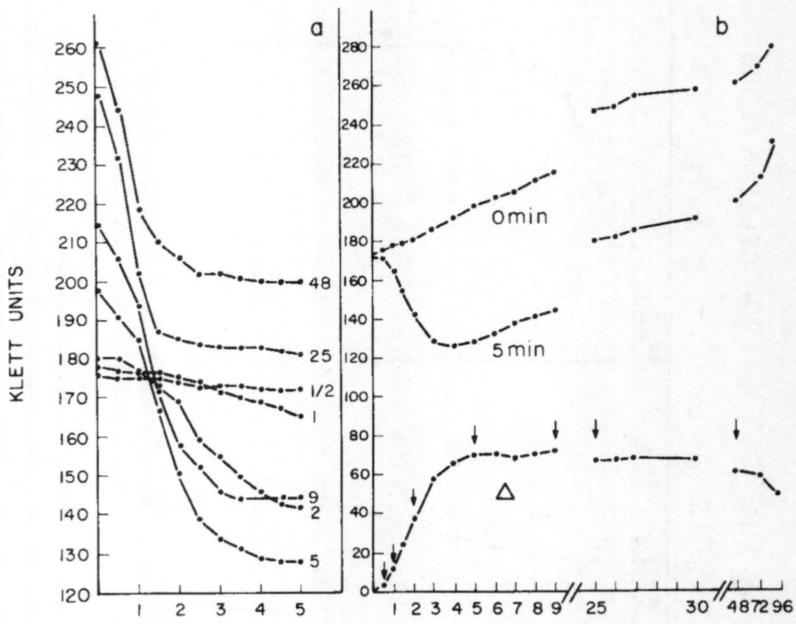

FIGURE 9. Turbidimetric method for measuring sex-directed flocculation in fission yeast. (a) Decrease in turbidity of a culture allowed to stand undisturbed for 5 min. The age of the culture in hours after the start of the induction procedure at 0 hr is shown as a number at the end of each curve. (b) Development of sex-directed flocculation in a culture. The turbidity readings at 0 min and 5 min generate the upper two curves. The bottom curve (marked Δ) is the difference between them. Points derived from curves on the left (a) are indicated by arrows. (From Calleja, G. B. and Johnson, B. F., *Can. J. Microbiol.*, 23, 68, 1977. With permission.)

may be used in the course of applying the hemocytometric method. Because of its lack of reproducibility and sensitivity, it is applicable only to large samples, when aggregation is voluminous, and in combination with other methods. Calibration is necessary because the volume of an aggregated mass comprising a given number of cells is considerably greater (about tenfold in fission yeast) than that of an equal number of free cells. Nonetheless, it has its worth as a rapid and nondestructive assay.

D. Kinetic Method

The three methods described so far estimate the total aggregated mass, without regard for the individual mass or size. The size distribution of the aggregates may be tediously determined in the microscope (Figure 4) or by means of an electronic particle counter (Figure 5). Alternatively, if the time-course of gravity sedimentation is followed (by the turbidimetric method, for example), the rate of clearing may be obtained. This would be a function of the average mass of the individual aggregates. By taking the slopes of the clearing curves (Figure 9a), one may monitor the increase in the average mass of the individual aggregates as aggregation progresses in a culture (Figure 10). Obviously, the kinetic method measures only how large the aggregates are, not how many there are.

E. Estimation by Eye

Finally, before using any of the objective methods, one may grade by eye the cultures for their degree of aggregation. There is no common prescription for doing this. Each observer has his own inimitable way of scoring and his own arbitrary unitage. Although variable and

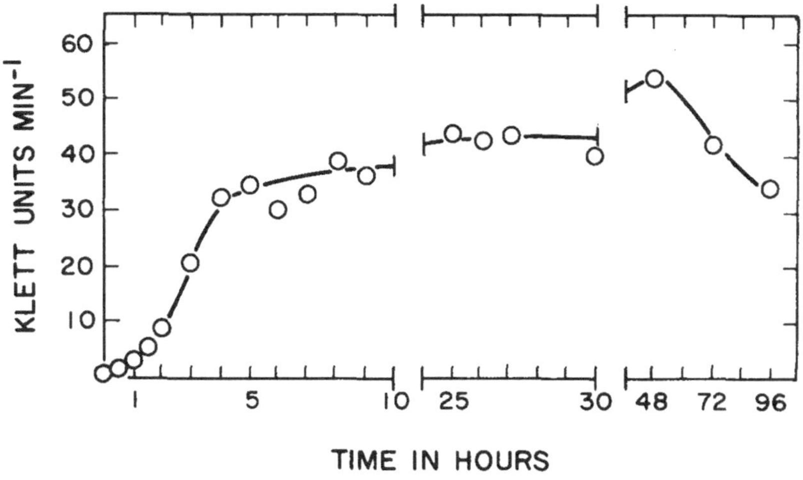

FIGURE 10. Increase in the sedimentation coefficients of aggregates during the development of sex-directed flocculation in fission yeast. The values are derived from the slopes of clearing curves, some of which are shown in Figure 9a.

imprecise, estimation by eye entails the least interference with cultures. More important, it is nearly instantaneous. Its principal usefulness lies in simultaneous comparison of many cultures.

The hemocytometric method is destructive and time consuming. However, it answers a little more directly the question of how many cells there are in the aggregated fraction of a culture. Because of its sensitivity, the method is most useful in ascertaining the extent of contamination of an isolated aggregated fraction with free cells. It may be combined with cytological observations concerning related events, such as copulation, conjugation, zygote formation, and sporulation. That advantage, of course, is lost when the cells are counted by an electronic particle counter.

The choice for routine monitoring of aggregation is the turbidimetric method. Because the method is nondestructive, as is estimation by eye, a sample may be assayed over and over again. It is markedly faster and more convenient than the hemocytometric method. Of the methods described, it is the most reproducible.

Just how many cells make up an individual aggregate? Clumps of a few cells such as those found in *E. coli* or in *Paramecium* pose no problem. Information regarding large masses, however, is hard to come by in the literature. The number of cells in a single aggregate may be determined by counting the number of cells in an aggregated fraction and dividing this number by the number of aggregates, except that the number of aggregates is often difficult to ascertain. Moreover, aggregates often coalesce to form even bigger aggregates.

The number of cells in a floc of *Schizosaccharomyces pombe* is determined in the following manner.[76] A culture in a flask is gently swirled and then broadcast in one jerk of the hand to make a thin spread on a very large surface of solidified agar. Isolated flocs on the agar surface are carefully picked up with a spatula and individually dispersed in a known volume of Pronase solution. The suspension of cells is then sampled for counting. Of course, the dimensions of an aggregate, instead of the number of cells it contains, may be determined. The sizes of aggregates in a few systems are compared in Table 1.

Table 1
SIZES OF AGGREGATES IN SELECTED MICROBIAL AGGREGATION SYSTEMS

System	Size of aggregates
Mating-aggregate formation during conjugation in *Escherichia coli*[28]	Up to 20 cells
Sex-directed flocculation in *Schizosaccharomyces pombe*[76]	Average: 10^5 cells
Flocculation of brewers' yeast[116]	Diameter: up to 1.3 cm
Aggregation in the cellular slime molds[88,117,118]	10^3 to 10^5 cells; minimum: 15 to 325 cells
Mating reaction in ciliates[119]	Up to hundreds of cells
Flocculation of activated sludge[89]	25 to 2000 μm

III. SEPARATION AND PURIFICATION OF AGGREGATES

If all cells in a culture were to aggregate, there would be no free cells from which aggregates must be separated. In due time, aggregates will have purified themselves. In most systems, however, a sizable fraction of the cell population are not competent to form aggregates among themselves. Hence, the need to separate the aggregates from free cells for various experimental reasons: for assay, for studying the physical properties of aggregates as compared to nonaggregates, for isolation of aggregation factors, for answering physiological, biochemical, and genetical questions.

Aggregates when large enough are easy to separate from free cells. In an aqueous suspension, they fall to the bottom of the container, or rise to the top, in a few systems. Provided they are stable enough, they may be mechanically separated from the free cells by decanting the contaminant free cells or by filtration. Stability cannot be overstressed; it is futile to isolate any material that does not stay together.

Flocs of fission yeast are fairly stable to dilution in an aqueous solution at room temperature and may be washed after their separation by gravity sedimentation.[99] A 10-mℓ flocculated culture transferred to a 15-mℓ conical tube is gently agitated by tipping the tube to a horizontal position five times. Then the tube is left to stand undisturbed in a perfectly vertical position. During this period, free cells remain in suspension while flocs settle to the bottom of the tube. At the end of 5 min, the supernatant of free cells is sampled, decanted, and replaced with phosphate buffer. The tube is once again tipped five times and left to stand undisturbed for 5 min. This regimen constitutes a washing cycle. At the end of five or six cycles, the recovered material attains a purity greater than 99.9% (Figure 11). A recovery of about 50% of all cells in the original culture may be achieved by this homely procedure. Note that the procedure is partially destructive in that repeated washing after the fifth does not quite bring down the level of free-cell contamination. Presumably, truly competent cells are dislodged from parent aggregates even as some aggregates are broken apart into smaller ones, small enough to remain in suspension. More gentle handling, on the other hand, would not provide for the removal of incompetent cells opportunistically ensconced in aggregates, in the same way that mediocrity gets ensconced in an authoritarian bureaucracy.

When aggregates are rather small, gravity sedimentation in the culture medium may not be sufficient. Samples may be layered on top of a solution of dextran.[90] When aggregates are even smaller, like those of conjugating *E. coli*, one may have to resort to sucrose-gradient centrifugation for enrichment.[28] Many reports in the literature are less than forthright in their assessment of the degree of contamination of isolated aggregates. In most instances, it really does not matter much.

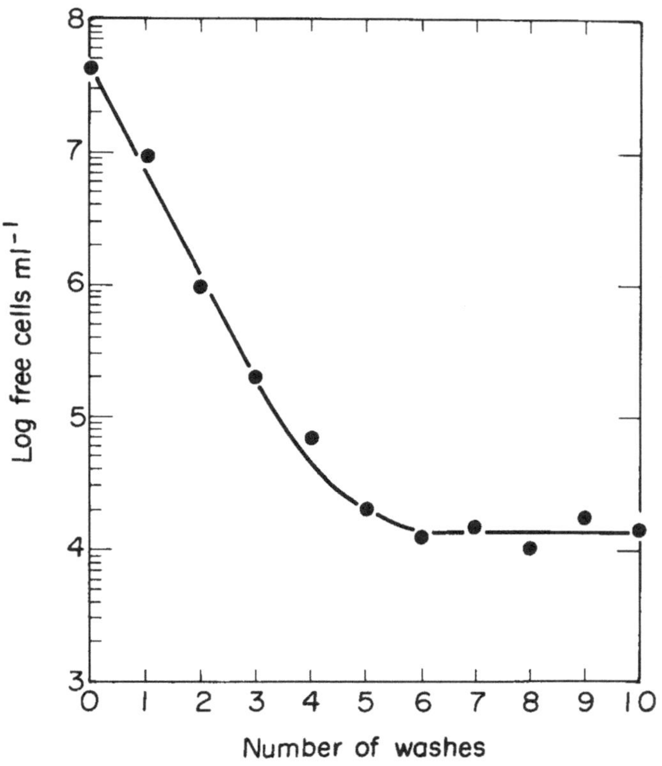

FIGURE 11. Purification of flocs of fission yeast. A 10-mℓ culture was washed ten times in phosphate buffer (*M*/100, pH 7.0) according to a procedure described in the text. Original culture comprised 7 × 10^7 cells $m\ell^{-1}$, about 70% of which were cells in flocs. Free cells were counted in a hemocytometer. The procedure described here is an improvement of a procedure described earlier.[99]

IV. IDENTIFICATION OF ASSOCIATED EVENTS

Many a microbial aggregation system has been investigated not as such but as component of a more complicated sequence of events: a prelude, a consequence, or a halfway house. When aggregation, rather than its associated events, is observed first, a primary objective is to discover its overt meaning in terms of the events that precede it or the events that come after it or other correlative measurable function.

Because it involves whole cells, microbial aggregation is a complicated affair. Movement and contact are required by definition, but aside from these, there are many gross morphological and physiological events associated with it. Not surprisingly, for by definition it is a component of the life cycle of an organism and being so, it is both the consequence and the cause of other events.

In the cellular slime molds, the aggregation phase comes at the end of growth and before morphogenetic development, which includes slug formation and migration and culmination into a fruiting structure (Figure 12). Aggregation in the myxobacteria follows essentially the same program, in spite of the fact that they are quite unrelated taxonomically, the cellular slime molds being eukaryotes, the myxobacteria prokaryotes.

In sexual systems, aggregation may be followed by copulation, conjugation, zygote formation, meiosis, and sporulation (Figure 13). But is aggregation, as contrasted to mere

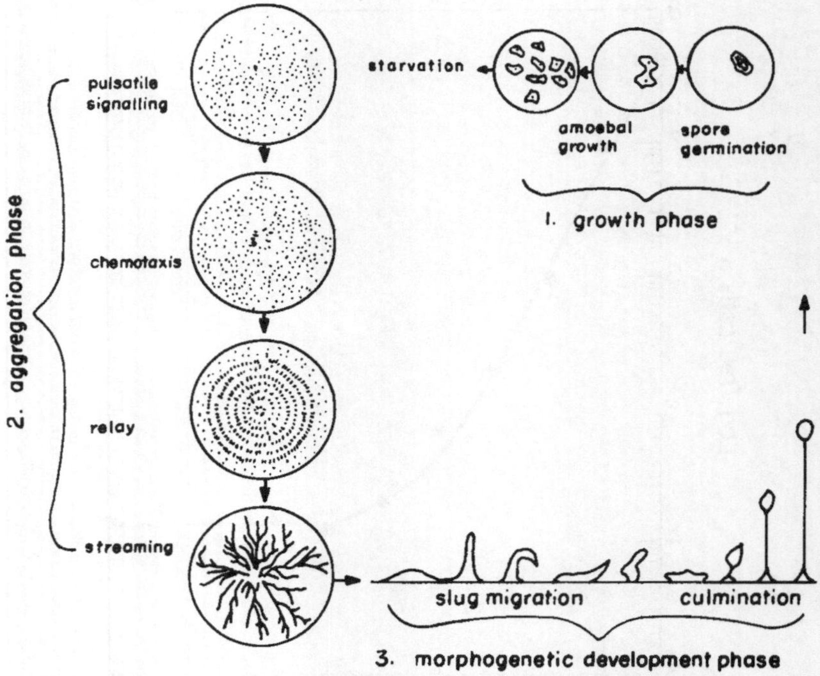

FIGURE 12. The life cycle of *Dictyostelium discoideum*. The fruiting structure at the end of the third phase bears the spores that will germinate to re-start the first phase. Adapted from Newell.[117]

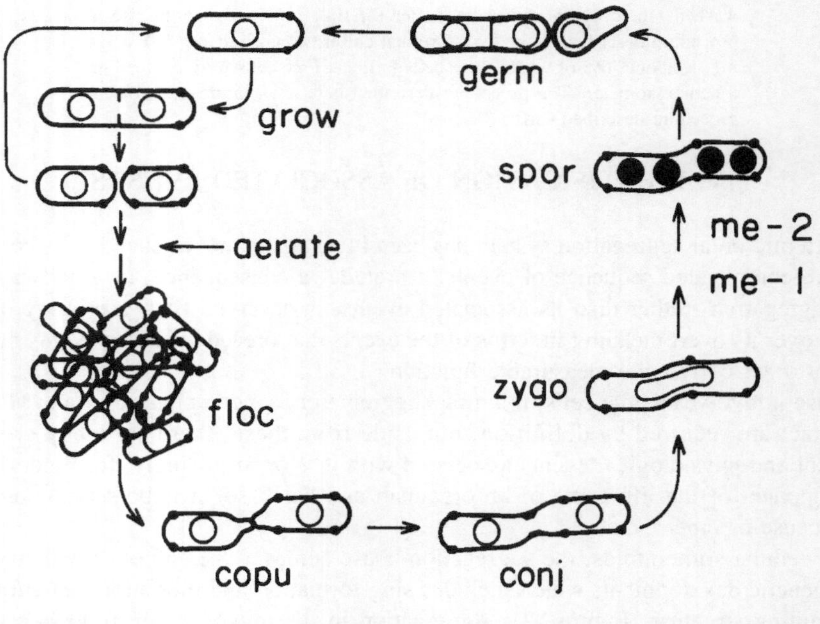

FIGURE 13. The life cycle of the homothallic *Schizosaccharomyces pombe* NCYC 132. Stages: grow, vegetative reproduction by fission; floc, sex-directed flocculation (hydrogen-bonded aggregate dispersible by urea or protease); copu, copulation (covalent union between cells); conj, conjugation (cytoplasmic fusion); zygo, zygote formation (nuclear fusion); me-1, first meiotic division; me-2, second meiotic division; spor, ascospore formation; germ, spore germination. Dots correspond to fuscannels.[120,121] (Reproduced with permission from Pergamon Press, Toronto.)

pairing, a prerequisite to conjugation? In the ciliates, the agglutination reaction precedes pair formation. It is grossly obvious that agglutination precedes conjugation in systems which involve mating types that agglutinate immediately when mixed together. It is not so in homothallic systems or when agglutination does not happen immediately upon mixing of the complementary mating types. It had to be shown, for instance, that sex-directed flocculation in fission yeast is a prerequisite to conjugation.[66] It still is not clear whether sex in any form follows flocculation of brewers' yeast during fermentation. The mating reaction of the same species of yeast, on the other hand, involves agglutination prior to conjugation.[69]

That aggregation is a requirement for any subsequent event is in many cases difficult to demonstrate. The problem especially arises because of the time dependence of both aggregation and the associated event and because of the lack of synchrony in a culture. When conjugation appears soon after aggregation or when there is an overlap of the events, the temporal sequence might be difficult to resolve. Fortuitousness must be ruled out and precedence must be demonstrated. A perturbation in the precedent must be reflected in the subsequent event. The use of aggregationless mutants may prove decisive.

The identification of gross events preceding aggregation is trivial in most cases. Gametogenesis precedes the mating reaction in unicellular algae. It is a prerequisite to the mating reaction because it is the gametes which are involved. Taxis of some sort must come before slug formation. How else can a slug of amoebae come to being? In general, aggregation comes after population growth, after nutrient exhaustion of the medium. In filamentous fungi, however, it is hardly possible to resolve the processes of growth and of aggregation.

Obviously, the development of competence for aggregation must come before the very process of aggregation. But does it precede competence for taxis? Whereas gross events are easy to identify, the events at the molecular level call for a more profound understanding of the mechanism of aggregation. As a beginning, it is sufficient to merely locate aggregation with respect to the life cycle of the organism. The initial objective is to relate the phenomenon to the more apparent morphological and physiological events in the cycle.

V. INDUCTION OF MICROBIAL AGGREGATION

For cells in a culture to aggregate, a number of conditions have to be fulfilled. These conditions are as many as one cares to enumerate. For the sake of discussion, they are sorted into four categories: (1) genetic, (2) physiological, (3) environmental, and (4) manipulative. The first two categories comprise conditions inherent to the cell, the other two external to it. The cells must first be genetically and physiologically competent. Then the environmental factors must allow the expression of that competence. Ultimately, the experimenter must be able to manipulate situations such that aggregation becomes demonstrable. These various conditions will be discussed in some detail below according to category.

A. Genetic Conditions

In the final analysis, the information for aggregation must reside in the genetic makeup of the organism involved. Despite its widespread distribution, the capacity for aggregation is not found in all microbes. Not all species are aggregative. Within a species, strains differ in their capacity for aggregation. Some are more aggregative, some less. Others are absolutely incompetent. Even among members of a large enough clone, there are deviants. Mutations followed by selection may convert a genetically competent strain to one that is not, an incompetent strain to one that is.

Mating types, which call for sexual complementation in heterothallic systems, must differ in the nucleotidal sequence of their genomes. Perhaps, homothallic systems are no different. Recent evidence in yeasts suggests that there might exist no such creature as a homothallic cell, homothallic in the sense that a single cell expresses simultaneously the information of both complementary mating types.[122-124] A homothallic clone of many cells is made up of two complementary mating types that frequently interconvert from one type to the other.

Ploidy differences among isolates may result in hereditary differences with respect to their aggregative properties. In yeast, the higher the ploidy, the more agglutinative an isolate tends to be.[36] Yet the immediate diplonts, derived from the union of complementary agglutinative haplonts, of *Hansenula wingei* and *Saccharomyces cerevisiae* are not normally agglutinative.[125,126]

Extrachromosomal inheritance may also affect profoundly the genetic competence of an organism. The presence of the fertility factor, F, determines the ability of *E. coli* to conjugate. Mitochondrial DNA in eukaryotes contains information for functions which might be necessary for the expression of nuclear information for aggregation. There is some evidence to show that mutation to respiratory deficiency ("petite" formation) results in complete or partial loss of flocculence in brewers' yeast.[87,127]

Finally, the structural genes for aggregation competence may be intact in a cell but are not expressed because of the repressive presence of regulatory genes. The disposition of these genes, structural and regulatory, may determine whether the aggregation function is constitutive or inducible.

B. Physiological Conditions

If the capacity for aggregation is constitutive, whatever state the organism is in with regard to its cell cycle or its life cycle, aggregation will always be present, provided the environment and the experimenter allow it to occur. If not, it is inducible and will be present only when the cells are physiologically competent, only at a specific portion of the population growth curve or only at a particular stage in the life cycle.

By definition, microbial aggregation is inducible. In instances where it appears to be constitutive, it is a matter of finding the physiological state when competence is lost. Constitutivity is, after all, useful only in a restricted sense here. A system is constitutive only for the duration the experimenter is willing to tolerate. It is difficult to show that the capacity for aggregation is always present in any system. The existence of aggregation requires that cells be competent. However, in the absence of aggregation, the cells must be shown to be nonetheless competent, to show that for the duration of the experiment, the function is constitutive.

At any rate, genetically competent organisms may be physiologically competent only during a stretch of time in the cell cycle or the life cycle. Conditions that favor cell proliferation do not make for good aggregation. The end of growth phase is the time for the development of aggregation competence in many systems, including ciliates and algae, the myxobacteria, the cellular slime molds, and the yeasts. A notable exception is *Saccharomyces cerevisiae*. Whereas flocculation of brewers' yeast occurs at the start of stationary growth phase, sexual agglutination in the same species is optimum during logarithmic growth.[125] However, in an asynchronous population, only nongrowing (unbudded) cells undergo the mating reaction. The unbudded cells are presumed to be in G1 stage of the cell cycle. Work with cell-cycle mutants of this yeast species points to a period very near the initiation of DNA replication as the time for sexual activity, of which sexual agglutination is an initial component.[128,129] During this period, the cell arrives at a point of decision either to initiate differentiation or to reinitiate another mitotic cycle. The mating reactions in *Hansenula wingei* and *Schizosaccharomyces pombe* are also presumed to occur in G1.[98,130-132] The G1 stage has also been implicated in gametogenesis, a prerequisite to sexual agglutination, in *Chlamydomonas*.[133]

Another aspect of the physiological state of the cell is the development of some other function upon which aggregation depends. The development of respiratory sufficiency is a necessary condition for the development of competence for sex-directed flocculation in fission yeast.[84] Mitochondrial function and development are repressed during cell proliferation. Consequently, flocculation occurs only after derepression. Other eukaryotic aggregation systems that depend on respiration for energy may behave in a similar manner. Catabolite repression, which affects the synthesis of many enzymes, may prevent the expression of competence for aggregation.[134-136]

There is no doubt that the metabolic state of a culture determines its immediate inducibility. Whether the requirement of the aggregation process is primarily for energy, or for protein synthesis in the cytoplasm or in the mitochondrion, or for a special enzyme or metabolite, the cell to be induced must be poised for a change in the developmental program. No amount of environmental manipulation can induce a culture that is not ready for induction.

C. Environmental Conditions

The observation that aggregation occurs at a given time in the population growth curve does not mean that the cells have not become physiologically competent long before the occurrence. The environment in which the cells find themselves must allow aggregation to occur. Equally important, the environment must allow the development of competence among cells physiologically poised to be induced. The environmental conditions include physical, mechanical, and chemical effectors. Principal among them are light, temperature, agitation, cell population density, aeration, ionic strength, pH, nutrition, energy source, pheromones, metabolic inhibitors, and stimulators. Whether their effects are direct or indirect, of course, is something to be carefully checked. The objective of studying these conditions is initially to maximize, minimize, or optimize the aggregation process and ultimately to use them as probes in the study of its mechanism and its regulation.

A few aggregation systems are sensitive to light. In photosynthetic systems, like the unicellular algae, light serves the energy requirements of the organisms. More specifically, it appears to be required for gametic differentiation in *Chlamydomonas*.[137,429] The diurnal cycle of light and dark influences the intensity of the mating reaction in *Euplotes crassus*.[33] The reaction is best observed at the start of dusk and is least intense at dawn. In *Paramecium bursaria*, the peak is observed halfway in the light cycle.[138,139] Fairly synchronous outbursts of aggregation have been observed in the cellular slime molds when transferred from dark to light.[140] Sexual flocculation in *Schizosaccharomyces japonicus* has been reported to also respond positively to light.[68]

Because it is dependent on a series of chemical reactions, presumably involving enzyme catalyses, aggregation is dependent on temperature. The higher the temperature, the faster the reactions, except that at elevated temperatures, stability of the aggregation structure is affected, for weak bonds required for the interaction begin to break. The development of competence would likewise be affected, but for a slightly different reason, namely, the denaturation of enzymes. Furthermore, there is some evidence that the development of competence for aggregation is more temperature sensitive than the growth processes in general.[141-143] Indeed, one may suspect that thermosensitivity is more widespread than recognized at present, if for no other reason than that developmental programs in many microorganisms, the fungi in particular, have been long known to be more thermosensitive than the mitotic cycle.[20,144-148]

A high population density helps increase the rate of fruitful collision, the aggregation of cells being a reaction of many bodies. Formation of aggregative centers in the cellular slime molds requires a threshold density.[118] Moreover, in the same system, there is a minimum number of amoebae that will form a fruiting structure.[88] Mechanical agitation likewise helps cells in liquid culture attain a high collision rate over and above the rate due to gravity, diffusion, convection, turbulent flow, and Brownian motion, but too much agitation interferes with the formation of aggregates and renders them prone to rupture and disaggregation.

The requirement of certain systems for aeration is probably due to the energy requirement of many other processes, such as protein synthesis, which are required for the development of aggregation competence. In catabolite-repressible systems, aeration serves not only the respiratory function but also the development of that function.

Ion concentration has been a controversial issue in a number of systems. Where it has been studied assiduously, results from different laboratories are nonetheless contradictory. Cells may be thought of as charged particles. For them to approach one another, they must break through a repulsive barrier. The effect of ions in increasing the rate of aggregation or in stabilizing the aggregates already formed is possibly due to neutralization of charges on the cell surface. Counter-ions merely decrease repulsive forces on the surfaces of already competent cells.

Hydrogen-ion concentration also affects the stability of cell aggregates. Flocs of fission yeast, however, exhibit stability over a very wide pH range, virtually ruling out the involvement of surface molecules ionizable between pH 2 and 11.[99] The early literature on flocculation of brewers' yeast records the efforts of investigators to explain yeast flocculation on the basis of ionizable groups.[149-160] The demonstration of agglutination associated with transformation competence in streptococci is accomplished by acidification.[29] On the other hand, dextran-induced agglutination of *Streptococcus mutans* is most intense at pH 8.5.[82]

A discussion of the other environmental conditions, for example, nutrition, pheromones, metabolic inhibitors, and stimulators, will be deferred for later sections. It suffices for the moment to remind the reader that they, like the other environmental conditions already discussed, may influence the development of competence or the aggregation process itself.

D. Manipulative Conditions

Finally, the experimenter must put all of the aforementioned conditions together to demonstrate the phenomenon of aggregation, thus making his action a condition in itself. It is not enough that the cells be competent and the environment right. A competent cell in isolation in the right medium needs other cells nevertheless. So long as aggregation is not demonstrated, the competence of the cell remains only potential. The experimenter becomes the efficient cause of the demonstrated phenomenon. Aside from selecting the strains, he must, for instance, adjust the rate of agitation of a liquid culture or the cell population density or the temperature or the pH. In heterothallic systems, which involve mating types, he must place the complementary mating types in the same container. In heterotypic systems in general, he must adjust, wittingly or unwittingly, the input ratio of the complementary components. Furthermore, he may choose to disrupt the aggregates and follow reaggregation instead.

The manipulative condition is not trivial, because in many systems, it constitutes the zero hour of a time-course experiment. Zero hour is the time of inoculation, the start of aeration, the time of mixing the mating types, the time of resuspension of cells, the time of a shift-up, the time of addition of an exogenous stimulant. In many reports, it is carelessly referred to as the time of induction.

By manipulation, one may bring about the aggregation of already induced cells. This must be distinguished from induction of cells to competence for aggregation. Failure to do so has brought much confusion in the literature. The mixing of already competent mating types is simply induction by manipulation, and so is reflocculation of mechanically dispersed flocs or of heat-killed cells.

How may the process of aggregation in a culture be separated from the development of competence? The two processes may be separated from each other by their kinetics; whichever is slower is rate limiting. When aggregation follows immediately upon mixing of mating types (see Figure 2, for example), it is safe to assume that the development of competence is not rate limiting during the demonstration of aggregation, that indeed the cells have been competent prior to the demonstration. When after mixing of mating types there is a long lag before the appearance of aggregation, it may be assumed that the rate-limiting event is the induction to competence. Such is the case with sexual agglutination in *Saccharomyces cerevisiae*. In the homothallic fission yeast system, the process of reflocculation of flocs dispersed by sonication is virtually over in a few minutes, but the induction to competence in a culture, as indicated by the time-course of flocculation, covers a period of a few hours (see Figure 7 or 8). The aggregation process itself is mechanical in many systems, induced cells killed by heat or formalin being quite capable of forming aggregates among themselves.[98,99,161,162]

The induction discussed thus far refers simply to the act of bringing on or about, of causing, of initiating, of effecting, without any attempt to invoke the mechanism of induction in molecular terms. Sooner or later, however, the induction of cell aggregation must be described and explained in terms of the Jacob-and-Monod model of induction or some such paradigm. Presumably, the change on the cell surface is the result of information that resides in the genetic material, which is DNA. The processing of this bit of information into a message for change, a messenger RNA, must involve a turning on of structural genes by regulatory genes. The message is translated by the ribosomal complex into proteins. These final gene products may be enzymes as well as structural proteins. The enzymes catalyze the attachment of the structural proteins onto the cell surface, membrane, wall, or appendage. The modified surface confers on the cell the competence to aggregate.

The description of the development of competence for aggregation in these terms is the ultimate objective in the study of induction. One must show the need for energy, the need for RNA and protein synthesis, the final gene product in the form of a protein or a glycoprotein component of the cell surface, the specific messenger RNA species, the DNA segment that codes for aggregation. No microbial aggregation system to date has come close to this objective. All systems taken together, however, point to some progress. The requirement for energy has been shown in many systems, confirming that induction is a vital, rather than mechanical, process. The requirement for RNA and protein synthesis has been shown in some systems.[84,98,100,163-165,428] In *Hansenula wingei*, the final gene products are complementary glycoproteins.[98,166-170] Peptides, proteins, or glycoproteins have been implicated in *Saccharomyces cerevisiae, Schizosaccharomyces pombe, Chlamydomonas* spp., *Blepharisma intermedium,* and *Paramecium* spp.[86,99,165,171-180] The genes for mating types are known and have been mapped in yeasts and chlamydomonads.[181-187] The fertility factor in bacterial conjugation has been well characterized.[188-190] Genes for flocculation of brewers' yeast have been identified[87,127,191] and so have the genes for aggregation in the cellular slime molds.[476-479]

VI. NUTRITIONAL CONDITIONS

As would be expected of a developmental system, microbial aggregation is profoundly influenced by nutritional conditions. Because of the variety of the systems and the variety of the mechanisms by which the multicellular condition is achieved, the nutritional factors are many and varied. Unfortunately, the more a system gets studied, the more contradictory are the observations.

A list of nutritional factors of microbial aggregation is shown as Table 2. It is essentially a rearranged version of an earlier one.[27] It is also somewhat shorter and more compact, some entries having been combined, others left out. Only a few items have been added to it.

The list has been simplified by grouping nutritional factors into two columns of opposite effects. One column is positive, i.e., the condition is reported to allow, enhance, hasten, increase, induce, intensify, magnify, stabilize, stimulate, support, or be required. In fine, it favors aggregation. The other is negative, i.e., the condition is reported to antagonize,

Table 2
NUTRITIONAL FACTORS OF MICROBIAL AGGREGATION

System	Positive	Negative
Bacteria		
Mating-aggregate formation during conjugation in *Escherichia coli*		Cell extract[192]
Donor cell	Anaerobiosis, rich medium, glucose as C source, Casamino acids, growth factors[194,195]	Starvation[194,195] Zn[28,193]
Recipient cell	Aeration, Zn[194-196]	
Star formation in rhizobia, pseudomonads, agrobacteria	Nutrient broth[197-199] Peptone plus yeast extract[200]	Minimal medium[197]
Pellicle formation in fimbriate bacteria	Aerobic plus static culture[201-207] Nutrient broth[201-205,209]	D-mannose[202,203] α-methyl mannoside[203,205]
Agglutination associated with transformation in bacteria		
Streptococci	Peptides, free amino acids[209]	
Pneumococci	Glucose or sucrose as C source[210,211] Amino acids[210-212] Uridine, choline, Mg[210]	No energy source[210] Maleate[213]
Bacilli	Aeration[214,215] Yeast extract[215,216,218] Casein hydrolysate[216,218] Glucose as C source[218] Sucrose as energy source[220] Ammonium salt as N source[218] N as nitrate or ammonium salt[220] Amino acids[216,218-223] Indole[216,218] β-mercaptoethanol[218,221,224] EDTA[218] Mg[214,216,221,222,224] Ba, Sr, Ca[214]	Complex, rich medium[216,217] Casein hydrolysate[217,219] Acetate, succinate and glycerol as energy sources[220] Amino acids[216,218,219,222,223] Phosphate[224] Cu[214,217,218,221] Fe, Mn[220] Ni, Zn[214]
Plaque formation and agglutination in *Streptococcus mutans*	Dextran[37,82,225,226] Sucrose[37,82,226-230] Ca[231]	Other polysaccharides and sugars[82,225,226]
Clumping in *Corynebacterium xerosis*	Inorganic salts[97]	Hydrogen, nitrogen gas[97]
Cell clumping in *Nocardia corallina*	Nutrient agar plus fructose[232]	
Aggregation in myxobacteria	Starvation[233] Limited concentration of required amino acids[241] Steroids[243] Divalent cations[244,245] K, Na[244]	Rich medium[234-238] Peptone[234,239] Hydrolyzed casein and amino acid[233,240-242]

Table 2 (continued)
NUTRITIONAL FACTORS OF MICROBIAL AGGREGATION

System	Positive	Negative
Yeasts		
Sexual agglutination in *Hansenula wingei*	Yeast extract, glucose as energy source[98] Peptone, gelatin, and protamine sulfate[94] Divalent cations[94,246]	Malt extract, glycerol as energy source, sorbose[98]
Diploid to 5	Aeration, exhaustion of medium[98,125,247] Growth limiting, culture filtrate[125] Yeast extract[98,125,247] Glucose as energy source[98,247] Vitamin limitation[125] Pyridoxine, biotin, and thiamine[125] Low Ca and Mg[125,247] Vanadate, molybdate, tungstate[125,247]	Excess aeration[125] Minimal medium[98,247] Adenine, guanine, cytosine, thymine, and uracil[125] High Ca and Mg[125] EDTA[125,247] Chromate[247]
Diploid to 21	Aeration, yeast extract, trace-element limitation[125] EDTA[125,247]	Amino acid, excess vitamins[125] Ca[247] Metal ions[125]
Aggregate formation in *Phialophora dermatitidis*	Nutrient limitation[248]	
Sex-directed flocculation in *Schizosaccharomyces pombe*	Aeration[65,66,84,131] Nutrient limitation[65-67,76,84,121,131] Peptone, vitamin-free Casamino acids[84] Salts[65,99]	Yeast extract[65,66,84] Yeast nitrogen base medium[66,84] Glucose[65,66,134] Aspartate[187] High cyclic AMP[134]
Sexual flocculation in *Schizosaccharomyces japonicus*	Glucose[68]	
Sexual agglutination in *Saccharomyces cerevisiae*	Aeration[69,249-251] Culture fluid[173] Peptidyl sex factor[173,174,252-255] Peptone[253] Glucose[100,250,251,256]	α factor[250] No fermentable sugars[254] Maltose, glycerol, or acetate as C source[254]
Flocculation of brewers' yeast	Aeration[259] Wort and wort components[85,258-261,263-266,268,269,271] Synthetic medium[77] Peptone yeast extract[127] Sugar mixtures[259] Glucose[273,274] Maltose, fructose, mannose, sucrose[274] Glucosamine[267] Amino acids[85,110,258,271] Ethanol[79,81,114,275,276] Methanol[81] Isopropanol, *n*-butanol, isobutanol, acetone[81,275] Salts[81,102,275] Mg[102,106,108,258,267,274,275,278] Ca[77,79,81,91,102,103,106,108,111,258,267,271,277,279-286] Other divalent cations[108,258,259,267,275,276,280,282]	Wort and wort components[77,91,114,259,262,263,267] Mimimal medium[258] Carbohydrates[77,91,103,267,272] Fermentable sugars[257,267] Glucose[81,127,267,272] Maltose[77,79,81,91,267,272,276] Fructose[81,260,275] Mannose[77,81,267,272,275-277] Sucrose[81,91,103,267,272,276] Galactose, lactose[81,272] Arabinose[275] Sorbose[272] Glucosamine[81] Arbutin, salicin, cellobiose, gluconic acid[272] Inositol[278]

Table 2 (continued)
NUTRITIONAL FACTORS OF MICROBIAL AGGREGATION

System	Positive	Negative
	Low Na + K[282]	Assimilable N[259,262,273] Amino acids[110,262] Urea or ammonium salts as N sources[110,278] High salt[279] Carbonate, phosphate, fluoride, oxalate[267] Diglycollate, citrate[279] EDTA[258,277,279] Na[81,275,279] Ba, Sr[258,279]
Cellular slime molds		
Aggregation in		
Dictyostelium	Oxygen[303,304] Nutrient depletion[117,287-293,295-300,302,305] Bacterial extract[306,307] Serum albumin[308] Glucose[309-311] Histidine[305-309,314] Other amino acids[309] Cyclic AMP and other cyclic nucleotides[117,294,298,301,316-329,338] ATP and other nucleoside triphosphates[331,332] Folic acid and other pteridines[313,333] Steroids[288,334,335] Ionophore A 23187[338,339] EDTA[337-339] Inorganic salts[302,304,308-310,315,340-343] Ca[298,304,308,315,319,320,340-343] Mg[310,338,340-343] Other divalent cations[308]	Nitrogen gas[304] Growth medium[298] Peptone[298] Glucosamine[312] Glycerol, citrate[305] Cysteine[305,315] High cyclic AMP[298,317,330] Adenine, guanine[305,310] Steroids[336] Diethylstilbestrol[312] Progesterone[336-338] EDTA[340,341]
Polysphondylium	Casein[333] Peptide acrasin[345] Folic acid[313,333,346] Pterin and other pteridines[313,346] Acetylcholine, eserine, 2,2-dichlorovinyldimethyl phosphate[347]	Excess nutrients[344] Dibutyryl cAMP, cGMP, and cAMP[330] Atropine[347]
Acytostelium	Steroids, alkaloids[348]	No air[348]
Filamentous fungi		
Strand formation in		
Merulius lacrymans	Nonnutrient medium with food-base[349,350]	Nonnutrient medium, food-base removed[349] Fresh food-base[349]
Helicobasidium purpureum	Malt extract[351] Nutrients in agar[351]	
Cultivated mushroom	Nonnutrient medium with food-base[352,353] Low-nutrient medium[353] Malt extract agar[352]	

Table 2 (continued)
NUTRITIONAL FACTORS OF MICROBIAL AGGREGATION

System	Positive	Negative
Rhizomorph formation in		
Armillaria mellea	Air[354]	Excess dextrose[355]
	Potato dextrose agar[354]	Excess peptone[355]
	Malt extract plus dextrose[355]	Excess asparagine[356]
	Peptone[355]	
Marasmius androsaceus	Oatmeal agar[357]	
Sclerotium formation in		
Sclerotium	Low-nutrient medium[360,361]	Low O_2 or high CO_2[24]
	Onion agar[363]	Carbon dioxide[358]
	Potato dextrose agar, oatmeal agar[364]	Fresh medium[359]
	Glucose-nitrate agar[359]	High-nutrient medium[360-362]
	Glucose as C source[362,364]	Plain agar[359]
	Sucrose as C source[364]	High glucose[360,362]
	High C:N ratio[78]	Cysteine[366,368–371]
	Lactose[358,365,366]	Cystine[367,368,371,372]
	Amino acids as N source[268]	Methionine[368,371]
	Threonine[358,366,367]	Other amino acids[367,368]
	Phenylalanine[367]	Glutathione[368]
	Organic N[373]	Urea as N source[78]
	Nitrate as N source[362]	High nitrate[362]
	Thiamine[362,375]	Ethanol[365,366,372,374]
	Iodoacetic acid[358,366,368-370,372]	Iodoacetic acid[371]
	Potassium iodate[369]	Potassium iodate[371]
	Low dimethylsulfoxide[376]	High K, high phosphate[362]
	EDTA[369,370]	High dimethylsulfoxide[376]
		EDTA[371]
		Cu[369]
Corticium rolfsii	Diamalt agar[377]	
	Mycelial extract[378-381]	
Sclerotinia	Carbohydrates[382,383]	As C source: D-xylose, cellobiose, L-sorbose, trehalose, melibiose, L-rhamnose, D-ribose, L-xylose, D-arabinose[384]
	High glucose[78]	Sugar alcohols as C sources[384]
	As C source: raffinose, sucrose, maltose, lactose, D-mannose, D-glucose, D-fructose, L-arabinose[384]	Mannose, glycerol, soluble starch as C sources[385]
	Maltose[383]	Lactose, galactose, mannitol[383]
	Amino acids as N source[386]	Certain amino acids[386]
	Asparagine as N source[382,386,387]	Cysteine, methionine[371,386]
	Ammonium N[382,383]	High peptone[78]
	Certain amino acids[386]	Urea as N source[78]
	Cystine[371]	Iodoacetic acid[371]
	Urea as N source[382]	High phosphate[382]
	Thiamine[78,388]	
	Pyridoxine, biotin[388]	
	Potassium iodate[371]	
	EDTA[371]	
	Zn[371,387]	

Table 2 (continued)
NUTRITIONAL FACTORS OF MICROBIAL AGGREGATION

System	Positive	Negative
Aspergillus	Oxygen[389] Peptone[390] Glucose as C source[390] Sucrose as C source[391] Nitrate as N source[391]	Glucose as C source[391] Urea as N source[391]
Rhizoctonia solani	High glucose or sucrose[78]	Urea as N source[78]
Botrytis cinerea	High glucose or peptone[78]	Urea as N source[78]
Pyronema domesticum	Potato dextrose agar, malt agar[392]	
Microsclerotium formation in *Verticillium*	Oxygen[393] Diffusible factor[395,396] Sucrose as C source[394,397] High C:N ratio[397] Ammonium N[397] Nitrate as N source[394,397] Biotin, desbiotin[398] Mn[395]	Low nutrient medium[394] High asparagine[394]
Synnema formation in *Hirsutella gigantea*	Liver fraction, yeast autolysate as N sources[399] Phosphoglyceric acid, gibberellic acid[399]	As N source: yeast extract, peptones, casein hydrolysate, glutamic acid[399] High N[400]
Isaria cretacea	Starvation[401] Sugars as C sources[400,402] Low C and N[400] Biotin, thiamine[400,401] Indoleacetic acid and related auxins[400,401] Phosphate, calcium carbonate[401]	Sucrose, cellobiose, mannitol, sodium acetate[400-402] High glucose[400] High ammonium N[400]
Graphium ulmi		High glucose[400] High ammonium N[400]
Coremium formation in *Penicillium*	Malt agar[403-405] Mg[406,407]	Low Mg[406,407]
Sphaerocybe concentrica	Thiamine[408] Pyrimidine plus thiazole[408]	
Algae		
Mating reaction or sexual agglutination in chlamydomonads	Oxygen[409,410] Cell-free culture fluid[411,412] Gamones[86,176,411-413] N depletion[133,137,185,414-427] Cyclic AMP[428] Low sodium acetate[429] Citrate, oxalate[430] Gibberellin A7[428] Ca, Mg, other divalent cations[431-434]	Carbon dioxide[410] Glutamine, glutamate[416] Ammonium salt[413,416,426] Nitrate[413,416] Urea[416] High sodium acetate[429] Citrate,oxalate[430] EDTA[432] Monovalent cations[412]

Table 2 (continued)
NUTRITIONAL FACTORS OF MICROBIAL AGGREGATION

System	Positive	Negative
Agglutination reaction in *Dunaliella salina*	Nutrient depletion[435] Saline[435,436]	
Isoagglutination of gametes in *Pandorina*	N limitation[437] S limitation[437] Cell-free culture fluid[436,438] Ca, Mg[437]	EDTA[363]
Agglutination of gametes in *Ulva mutabilis*	Fresh growth medium[439–441] Vitamins[439]	
Protozoa		
Mating reaction in *Blepharisma*	Starvation[119,165,442] Cell-free culture fluid[32,119,443-445] Gamones[119,165,443-449] L-tryptophan, D-tryptophan, L-hydroxytryptophan, D-hydroxytryptophan[119]	
Mating reaction in *Euplotes*	Starvation after feeding[450] Nonnutrient medium[119,450] Seawater[33]	
Mating reaction in paramecia	Dissolved gases[453] Starvation after feeding[119,451,453,458,461-463] EDTA plus $MgCl_2$[464] Ca[83,465]	Complete starvation[119,161,165,186,442-460] Overfeeding[138,460]
Mating-aggregate formation during chemically induced conjugation in paramecia	Nonnutrient salt solution[466] Ca-poor condition[467-469] Heparin, cysteine, acetamide, propanamide, malonamide, ethanol, propanol, butanol, glycerol, urea, methyl urea, biuret, acriflavine, proflavine, acridine red, acridine yellow[466-469] EDTA[464,469,470] Monovalent, divalent, and trivalent cations[466-474] KCl plus acriflavine plus Ca-poor condition[469,472,475]	Extreme starvation[467] Cell-free culture medium[468] Ca[466-469]
Clumping in *Mayorella palestinensis*	Growth medium, peptone[476]	Anoxia, agar, high salt[476]

attenuate, decrease, delay, depress, destabilize, diminish, inhibit, prevent, repress. In fine, it is unfavorable. Indifferent responses have been left out for the sake of brevity.

Well-known metabolic inhibitors such as cyanide or azide or dinitrophenol, the antibiotics, metabolic analogs, and other drugs have been excluded. They will be treated in the next section. To be listed, a condition that inhibits aggregation must show a certain degree of specificity. It must not inhibit to the same degree the processes of growth or other gross cellular functions. An inhibitor earns a place in the list either because it is not commonly recognized as such or because it is found in the medium as an ingredient or as an excreted metabolite. Or again, because it behaves as a stimulant. All stimulants, regardless of their repute, are included so long as they remain within physiological bounds.

Many difficulties there are in making this list. These difficulties arise from the definition of nutrition, from the study of cellular systems in vitro, from the nature of microbial aggregation, and from the literature on the subject.

The initial difficulty is defining a nutritional factor. The meaning of the term depends not only on the response of the organism but also on the objective of an investigation, which may remain nutritional or may conveniently become toxicologic, metabolic, pharmacological, or even molecular. What is convenient to the investigator is not necessarily so to the compiler. The objective of a nutritional investigation may be to find out how to get more aggregation, or how to get less of it or even eliminate it altogether. Or perhaps, it is to enable one to design a minimal medium specifically for aggregation. The objective may be to find out the mode of action of a drug or to unravel the mechanism of the phenomenon, to identify the specific inducer or the cultural conditions for the development of competence so that aggregation may be demonstrated at will. Or again, the objective may merely be the gratification of the curiosity instinct: let's see what happens when we throw in one more bit of the kitchen sink.

A nutritional factor eludes definition for another reason. In cellular systems studied in vitro, there is no clear-cut boundary between external application and internal consumption, what goes and does not go into metabolism, what items make up a diet and what items do not. It seems that the whole cell constitutes a digestive tract. A nutritional factor then may simply refer to a substance which when added to a culture medium produces an effect worth noting.

There is also the difficulty deriving from aggregation being mistaken for or confused with some more conspicuous component of the developmental program of an organism. Consequently, the literature is most difficult to evaluate. From this confusion, the retrieval of bits of nutritional information on aggregation demands from the retriever a bucketful of patience and a touch of masochism. A good portion of the entries were gleaned from reports not of aggregation per se but of associated events, such as chemotaxis, gametogenesis, and development of competence for transformation. For a nutritional study, the ideal is to monitor the appearance of competence for aggregation rather than scoring aggregation itself, which often is provoked by manipulation, but the compiler's job is not writing prescriptions. His is only to make sense out of what is already there, to bring about an ordered aggregation.

Only a few generalizations may be advanced, in spite of the long list. Starvation, complete or for some specific nutrient, is rather common as an inducing condition. For many systems, conditions that favor growth do not favor aggregation. Yet the development of competence for aggregation is vital and requires metabolic energy. If the requirement is met endogenously, total starvation may serve as a trigger for the processes of induction. Such a situation exists in well-fed paramecia in tap water. It may be surmised that the depletion of nutrients results in a generalized derepression which eventually switches on the program for development.

A discussion of aeration and inorganic ions, both already treated in the previous section, will not be repeated here, but it may be pointed out that neither aeration (which apparently fulfills the energy requirement of a vital function) nor inorganic ions (the reported effects of which are on the surface hopelessly contradictory) are likely to be specific inducers nor specific inhibitors. A discussion of pheromones, also found in the list, is deferred for a later section.

Macromolecules, proteins and polysaccharides, in particular, have been implicated in microbial aggregation. Except for dextran in dextran-mediated agglutination in *Streptococcus mutans*, they have been studied rather uncritically. It is not easy to appreciate their significance at the moment.

Sugars and sugar alcohols have been studied as carbon and energy sources, as precursors of repression, and as potential competitors for receptor sites on the cell surface. Save for sucrose in dextran-mediated agglutination in streptococci, none have been found to be specific inducers. Their effect, when inhibitory, is likely the consequence of catabolite repression.

Far more useful are the nucleotides. Pulses of cyclic AMP have been identified as the specific chemotactic signals in *Dictyostelium*. Although reported to be ineffectual in *Polysphondylium*, the nucleotide has been found to be active in other systems. The nucleoside triphosphates have also been found to enhance aggregation in the cellular slime molds.

Studies with amino acids as a class of nutrients yield contradictory reports. Many ecstatic claims have been made over the years for certain amino acids being specific inhibitors or inducers. Most of them have fallen by the wayside by now. More carefully defined cultural conditions may reveal that this or that amino acid, when inhibitory, merely serves as a mediator of a general metabolic repression. When stimulatory, it may serve to fulfill the requirements of other cellular functions, such as protein synthesis, which may be required for the synthesis of aggregation enzymes and aggregation molecules. The sex factors in *Saccharomyces cerevisiae* are small peptides.

Except perhaps for folic acid in the cellular slime molds, the vitamins as well as the steroids, purines and pyrimidines, and intermediary metabolites have been of very minor heuristic import. Certainly, reports abound as to their wondrous effects, but their significance is not readily comprehensible. There have been virtually no reports on the role of lipids.

There are indeed very few trustworthy and meaningful observations with regard to specific nutritional requirements for microbial aggregation. Other than cyclic AMP as a communication molecule in the cellular slime molds, there is hardly anything else worth a floc. If there is one general conclusion that can be made, it is that catabolite repression may after all play a much more substantial role than at present admitted. The entries in the table of nutritional factors then are mere ingredients to empirical recipes. They are many. In the many solutions, it is to be hoped, may eventually be found the ingredients for explanations.

VII. DISAGGREGATION AND INHIBITION OF DEVELOPMENT

A phenomenon may be studied by observing its existence or its unfolding. An insight may also be obtained by looking when it is not there or when the process is reversed. A study of the conditions required for the absence of a phenomenon may help define the conditions for its presence. Aggregation may be studied by looking at the process of disaggregation, the development of competence by inhibiting such a development.

As the cliché goes, only the uninhibited use inhibitors.[480,481] Judged by the glut of literature on inhibitors, experimental biologists, from Claude Bernard onward, must have been a very exuberant breed. Nonetheless, the metabolic map hanging on the wall would not be there but for the use of inhibitors. In a way, inhibitors bring out the "yin" to complement the "yang" in experimental research. They make a veritable stock-in-trade of the biological shopkeeper. The student of microbial aggregation is no exception. There is no doubt as to the usefulness of inhibitors, but their judicious usage and interpretation are something else again.

The inhibition of development to competence for aggregation may be viewed as the converse of induction. Such a study may be an extension of a nutritional study, except that the response looked for is negative, not the enhancement of aggregation but its inhibition, or the beginning of a genetic investigation.

The inhibition of microbial aggregation may be accomplished in many ways (Table 3). It may be accomplished genetically. The gene may be excised as in the loss of the fertility factor in bacteria or added as in recombination and in diploidization or altered as in mutation. Physical or chemical mutagenesis may be employed to enhance mutations to absence of flocculence or agglutinability. Mutations may affect directly the genes for aggregation or indirectly the genes for functions upon which aggregation depends. The aggregateless condition may be the result of mating-type switching, mutation to the "petite" condition in yeast, or even mutations interfering with such diverse functions as cell-wall synthesis, the cell cycle, conjugation, and sporulation.

Aside from tampering with the genes, one may bring about inhibition by an adjustment of the cultural milieu (Table 3). Physical and mechanical means may entail such conditions as low cell-population density, extremes of agitation (for liquid cultures and easily disruptible systems), anaerobiosis (for oxygen-dependent systems, such as the cellular slime molds), elevated temperature (for temperature-sensitive systems, such as fission yeast), darkness (for photosensitive systems, such as chlamydomonads), extremes of pH and of ionic strength (for many systems, including brewers' yeast).

As seen in Table 2, many kinds of substances have been reported to inhibit microbial aggregation in one obscure way or another. It only appears as though anything found in a bottle inhibits some system. In addition, many well-known metabolic inhibitors with fairly well-known modes of action have been found useful (Table 3).

In many systems, a transfer to fresh growth medium stops the processes of aggregation. The addition of excess sources of nitrogen to cultures of chlamydomonads interferes with gametogenesis.[415,416] Just a little more specifically, glucose has been used to inhibit sex-directed flocculation in fission yeast.[134] In general, substances which serve as carbon or nitrogen sources presumably act by way of catabolite repression.

Cell walls being contact sites in cell-cell interaction, a derangement of their synthesis resulting in defective walls may bring about an inhibition of aggregation. The fungicide 2-deoxyglucose at concentrations that allow population growth does inhibit sex-directed flocculation in fission yeast.[482,483] So do membrane-specific drugs, such as nystatin, polymyxin, and phenethyl alcohol.[84] Inhibition of energy-generating functions by cyanide, azide, and dinitrophenol prevents the development of competence for aggregation in many systems,[20,84,94,143] demonstrating the vital nature of competence development. In eukaryotes, the inhibition of mitochondrial development by chloramphenicol and the generation of petites by ethidium bromide may result in the loss of aggregation capacity.[84,85] The loss is attributable to a deficiency of respiration, but some mitochondrial function other than respiration might also be involved.[143] Mitochondrial protein synthesis may be separated from cytoplasmic protein synthesis by the use of chloramphenicol for the inhibition of the mitochondrial and cycloheximide for the cytoplasmic.[84,484] The inhibitory effect of actinomycin D, although not a proof for the existence of a specific messenger, nevertheless points to the requirement of competence development for RNA synthesis.[98,163] Inhibitors of the cell cycle and DNA synthesis might be helpful in defining the cell cycle stage at which the cells are most likely to be inducible.

These compounds and many more have been used in many laboratories to inhibit the development of microbial aggregation, too often, rather indiscriminately. By a judicious use of specific inhibitors, the induction of competence is separable from the appearance of aggregation. However, the objective of a study of inhibitors may simply be to find out what cultural conditions prevent aggregation so as to eliminate a nuisance, but having found out the negative conditions, one consequently finds out in the process what constitutes the cultural

Table 3
INHIBITION OF AGGREGATION DEVELOPMENT

Genetic
- Mutation to the aggregateless condition
- Change in ploidy
- Recombination
- Loss of fertility factor
- Induction to the "petite" condition
- Temperature-sensitive mutation
- Mating-type switch

Physical and mechanical
- Low population density
- Extremes of agitation
- Anaerobiosis
- Elevated temperature
- Darkness
- Extremes of ionic strength
- Extremes of pH

Metabolic
- Catabolite repression — fresh growth medium, C and N sources, sugars, amino acids
- Inhibition of cell wall synthesis — bacitracin, penicillin, 2-deoxyglucose, D-cycloserine
- Inhibition of membrane functions — nystatin, polymyxin, phenethyl alcohol
- Inhibition of energy generation — cyanide, azide, dinitrophenol, carbon monoxide, antimycin, oligomycin, rotenone
- Inhibition of DNA synthesis and the cell cycle — mitomycin, nalidixic acid, hydroxyurea, bleomycin, cytochalasin B, colchicine
- Inhibition of RNA synthesis — actinomycin, lomofungin, rifampicin, cordycepin, α-amanitin
- Inhibition of protein synthesis — cycloheximide, chloramphenicol, amino acid analogues, puromycin, streptomycin, erythromycin
- Inhibition of mitochondriogenesis — ethidium bromide, acriflavine, chloramphenicol, erythromycin

conditions for aggregation. The objective may be to find out what metabolic activities are specifically required for competence development or what other events are associated, even what genes or gene products are involved. Or the other way around, to show that aggregation is somehow connected with functions that are already well recognized and well characterized, thus bringing aggregation from the backwash into the mainstream of biological research.

One way to separate the mechanics of aggregation from the development of competence is to start with isolated aggregation structures, cause them to disperse into free cells, and then follow the kinetics of reaggregation. It is reasonable to presume that these cells have been competent prior to reaggregation — indeed, before disaggregation — and that they do not need endogenous induction the second time around. It is further presumed that the reaggregation process is comparable to the untampered situation.

Listed in Table 4 are agents that cause dispersion of microbial aggregates. The physical and mechanical methods have been found in a number of systems to be reversible.[81,99,296,485] Even if they may cause the death of cells, as heat does, they remain useful so long as the ability of the cells to reaggregate is not affected. Indeed, the reaggregation of dead cells confirms the nonvitality of the aggregation process itself. In fission yeast, flocs dispersed by heat into free (and quite dead) cells reflocculate when cooled to ambient temperature (Figure 14). The flocs derived from reflocculation may be deflocculated again and the deflocculation-reflocculation procedure may be repeated ad nauseum on the same material.

Most of the disaggregants are reversible in that dispersed cells can reaggregate among themselves, but enzymes, the proteinases in particular, have been found in several systems to cause irreversible disaggregation.[86,99,486-489] However, enzymic treatment does not kill the

Table 4
DISPERSION OF AGGREGATES INTO FREE CELLS

Physical and mechanical
- Mechanical shear
- Sonication
- Heat

Chemical and physiological
- Acids and alkalis
- Hydrogen-bond reagents — urea, guanidinium chloride
- Detergents — sodium dodecylsulfate
- Sugars
- Univalent lectins
- Univalent antibodies
- Univalent putative receptor substances
- Chelators (or deionized water) — EDTA
- Disulfide-cleaving reagents — dithiothreitol, β-mercaptoethanol, cyanide, bisulfite
- Periodate
- Enzymes — proteases, carbohydrases
- Spontaneous
- Culture fluid
- Addition of incompetent cells

cells. Consequently, in the cellular slime molds, the recapitulation of the developmental events may be observed. The finding that other kinds of enzymes are ineffective suggests strongly that the contact sites on the cell surface involve materials of a proteinaceous nature, these materials being capable of acting as substrates for proteinases, but sexual agglutination in chlamydomonads are sensitive also to α-mannosidase, suggesting a protein-carbohydrate interaction.[487] The irreversible action of sulfhydryl reagents, such as dithiothreitol and β-mercaptoethanol, suggests also the involvement of proteins.[98,99] Moreover, it points to the complicity of covalent sulfur-sulfur linkages. Hydrogen bonds are implicated by the use of 5 *M* urea or 3 *M* guanidinium chloride as reversible disaggregants.[99,486] The participation of hydrophobic interactions may be inferred from reversible disaggregation by sodium dodecylsulfate.[99] Disaggregation in deionized water or in the presence of chelators, such as EDTA, reveals the presence of repulsive forces, which are counteracted by ions.[87,98,282] Extremes of pH, effected by the addition of acids and alkalis, also cause reversible disaggregation. Periodate has been used to show the role of carbohydrates, which make up the bulk of the cell surface.[98] Monosaccharides have been reported to cause deflocculation in brewers' yeast,[272] but because the reported deflocculation is rather slow and the monosaccharides have been used on live cells, it can be politely asked if the observed effect is not due to catabolite repression or to processes associated with growth.

Lectins (sugar-binding proteins) such as concanavalin A have been used to block the agglutination of algal gametes.[86,490] Fragments of antibodies directed against the cell surfaces of aggregation-competent cells cause the disaggregation of cellular slime molds.[491] Putative agglutination substances recovered from the cell surface or the culture medium have been assayed by their capacity to prevent aggregation in yeast systems.[492]

Disaggregation occurs when aggregates are transferred to fresh growth medium. When cells in an aggregate begin to grow, they tend to break away from the confining social arrangement and to strike out on their own as free cells. Disaggregation may also occur spontaneously in a nongrowing aggregated culture. It occurs during disengagement of chlamydomonads and paramecia after the sexual reaction or of bacteria after conjugation. It occurs slowly but irreversibly as postdevelopmental deflocculation in fission yeast.[143] Culture fluids may cause disaggregation, irreversibly because they contain proteases or reversibly

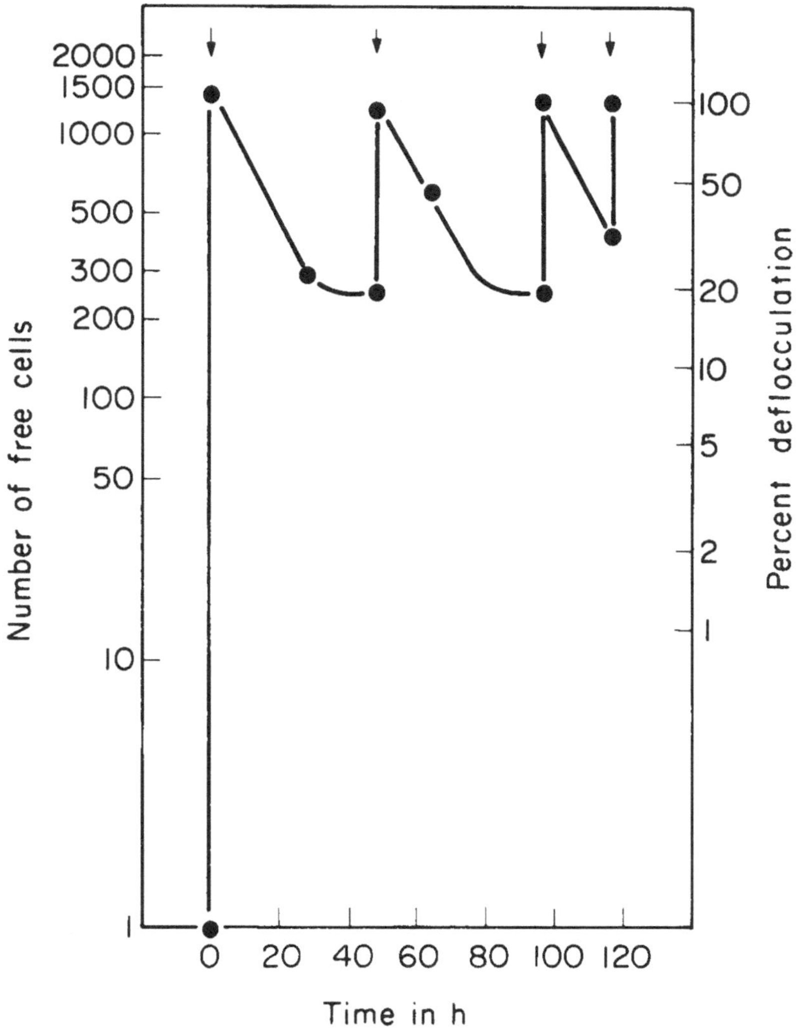

FIGURE 14. Reflocculation of fission yeast cells in deionized water after thermal deflocculation. A suspension of purified flocs was heated at 90°C for 2 min, then after it was sampled, allowed to stand undisturbed at ambient temperature until sampled again and reheated. Arrows indicate the time of heating. Cell concentration = 1.5×10^7 cells $m\ell^{-1}$. (From Calleja, G. B., *Can. J. Microbiol.*, 20, 797, 1974. With permission.)

because they contain monovalent factors which can competitively bind to contact sites on the cell surface. Addition of incompetent cells to isolated aggregates may also contribute to disaggregation, showing that whereas incompetent cells may not aggregate among themselves, they may weakly bind to induced cells, thus disrupting the aggregates by competing for binding sites which should have been occupied by competent cells.

In sum, the breaking apart of aggregates into their free components allows the cells to recapitulate the sequence of events and allows the investigator to follow the kinetics of reaggregation, to identify the putative receptor sites, to find out the molecular components of the interaction, the requirements for stability, and the participating forces. An assay for aggregation factors may entail the use of competent cells derived from aggregates and allowed to reaggregate in the presence and in the absence of the factors. Aggregates are literally dissected to reveal their anatomy and architecture, their composition with regard to mating types and conjugation figures, or for no other reason than to prepare them for photography, to enable one to plate the component cells, or to determine the number of cells within a structure.

VIII. PHEROMONES AND CHEMOTACTIC SIGNALS

Several microbial aggregation systems have been shown to be mediated by substances that may be collectively referred to as pheromones. Coined by Karlson and Lüscher[493] from the Greek words "pherein" (to carry) and "hormōn" (to excite), the all-inclusive term is more appropriate for the present purpose than such words as gamones, erogens, and ectohormones. Isolated from cell-free culture fluids, pheromones are diffusible metabolites that show aggregation-promoting activity. They include gamones, mating-type factors, and chemotactic substances.

If cells must gather together in a synchronous or coordinated manner or in a prespecified pattern, there must be a way of achieving this end other than by accident. There must be information as to what exactly is to be done, where, when, how to do it, and as to which cells in a population are concerned. The information is most likely communicated from one cell to another by chemical substances. If these substances have to act prior to the setting up of aggregation structures, they must traverse the distances between cells and therefore must be diffusible.

Many microbial sexual systems are known to involve diffusible metabolites that serve to bring about gametogenesis or chemotaxis.[493-495] Most of these sexual systems, however, are not aggregation systems by our definition and will not be discussed here. Indeed, among the aggregative microbes, only the yeast *Saccharomyces cerevisiae* and the protozoan *Blepharisma intermedium* are at present known to excrete identifiable pheromones that promote sexual aggregation.

Pheromones are bearers not just of sexual messages. In nonsexual systems, they may serve as signals for chemotaxis. In some systems, directed taxis constitutes the greater part of aggregation. It is most evident in those that are demonstrated on a solid substratum. Such is the circumstance of the cellular slime molds and the filamentous fungi. Where there is directed movement, molecular mediators may be assumed, but only in the cellular slime molds, among the aggregation systems, have the specific signals been identified.

In liquid cultures, where cell movement is stochastic and beyond endogenous control, chemotactic signals are of minor importance; even where movement is stochastic, signals for timing and recognition may be operative. The need for complementation must somehow be communicated: that there is present within the immediate environment of a given cell another cell that is complementary and ready for complementation.

The positive tropism of neighboring hyphae toward one another must be mediated by communication molecules. The species-specificity calls for distinct molecular species acting at a distance (approximately 10 μm) such as to prepare the walls concerned for mutual enzymic interactions.[25] However, no telemorphic substances, as they may be called, have been identified thus far, perhaps because of their volatile nature.

There is much confusion as to how many diffusible factors there are in the sexual agglutination of *Saccharomyces cerevisiae* and as to how many activities there are with respect to each of the factors isolated.[90] One thing, though, is most certain: the **α** factor excreted by **α** cells illicits a response from **a** cells, which in turn must produce an **a** factor, which affects the functions of **α** cells. The **α** factor has been isolated and identified as a peptide of 12 amino acids.[496-498] When added to a culture of **a** cells, it inhibits DNA synthesis and arrests the cell cycle at G1.[128,129] It causes **a** cells to swell (Figure 15). Aptly referred to in the literature as shmoos, these treated cells are now ready for conjugation. They certainly

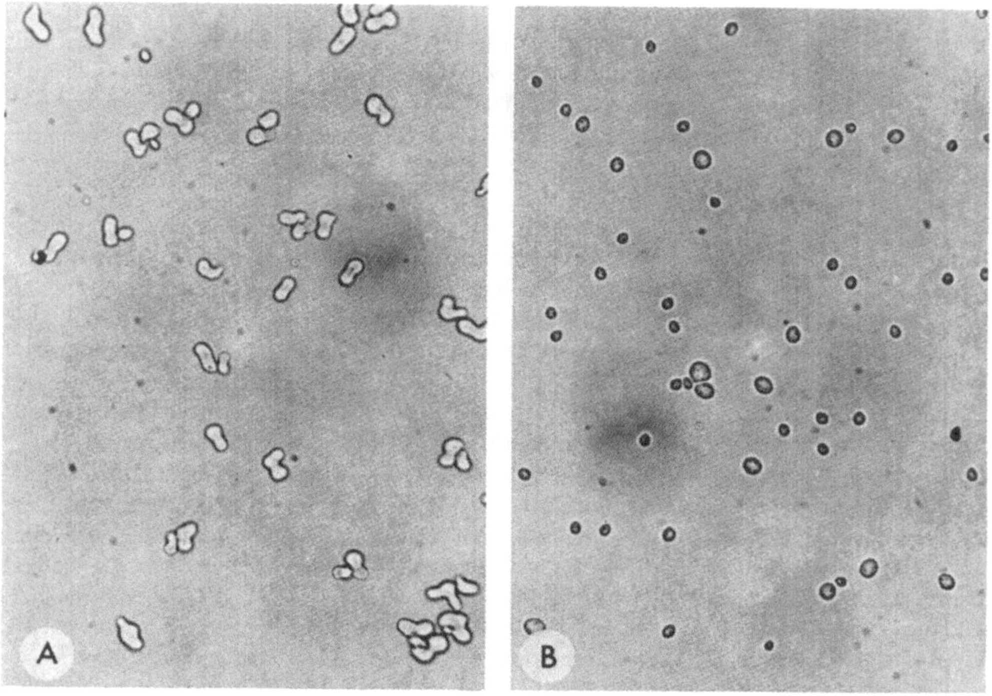

FIGURE 15. Formation of shmoos in *Saccharomyces cerevisiae*. (A) Mating–type **a** cells, 4 hr after exposure to α factor. (B) Untreated cells. Both cultures are in stationary phase. The α factor was prepared by the method of Duntze et al.[72] from a culture filtrate of mating-type **a** cells. (From Mackay, V. and Manney, T. R., *Genetics*, 76, 255, 1974. With permission.)

look like shmoos and are just as raring to breed as Al Capp's originals. The α factor also brings about an increase in the agglutinability of cells.[171-174] The peptidyl factor isolated as an agglutination inducer by Yanagishima's group is a modified molecular species of the α factor.[252,255] The putative **a** factor has not been isolated and may not be a pheromone at all but a cell-bound recognition substance.[90]

In *Blepharisma intermedium*, the current scheme calls for type I cells excreting gamone I molecules, which react with receptors on the surfaces of type II cells, which consequently are induced to excrete gamone II molecules, which react with receptors on the surfaces of type I cells.[119,500] When this amorous exchange of information is consummated, the cells become competent to form aggregates. Moreover, the gamones are chemotactic. Gamone I attracts type II cells, gamone II type I cells. Gamone I (blepharmone) is a glycoprotein of about 2×10^4 daltons,[443,444] a little too large for a pheromone. Gamone II, called blepharismone, is a much smaller molecule, the calcium salt of 3-(2′-formylamino-5′-hydroxybenzoyl) lactic acid.[446,501] Cell-free fluids of *Euplotes patella* have been reported to also contain conjugation-inducing activities.[450,502-504] Other gamones in ciliates are surface bound and will be discussed in the section on receptors and recognition.

The existence of a chemotactic attractant has been demonstrated in chlamydomonads.[419,433] When the tip of a capillary, filled with a suspension of *minus* gametes, is immersed in a suspension of *plus* gametes, the *plus* gametes swim into the capillary and agglutinate with the *minus* gametes, which remain inside. In the reverse experiment, the *plus* gametes swim out of the capillary and agglutinate with the *minus* gametes outside. Osmium-killed *minus* gametes agglutinate with and attract live *plus* gametes. Osmium-killed *plus* gametes also agglutinate with, but do not attract, live *minus* gametes. Finally, culture filtrates of *minus* gametes attract *plus* gametes, but not *minus* gametes (Figure 16). The reverse is not true: culture filtrates of *plus* gametes do not attract *minus* gametes. Thus, the chemotactic signal is unilateral. *Minus* gametes excrete an attractant to which *plus* gametes respond chemotactically. However, *plus* gametes do not particularly attract *minus* gametes at a distance. The *minus* attractant is not found in vegetative cells. It appears to be volatile and might be a small hydrocarbon.[434]

The chemical attractants found in cultures of the cellular slime molds are generically called acrasins.[290,505] In *Dictyostelium,* the acrasin is cyclic adenosine monophosphate (cyclic AMP).[117,294,301,316-329] At the end of the feeding period, some amoebae, located at what eventually become strategic points on a plate, start releasing rhythmic pulses of cyclic AMP. The frequency of the pulses, at first one in about 10 min, gradually increases to one every 2 min or so. The rest of the starving amoebae are induced to synthesize cyclic-AMP receptors on their surfaces and phosphodiesterase enzymes, which inactivate the signals. The pulsed signals travel a distance equivalent to no more than ten cell diameters before they are received and destroyed. Upon reception of a signal, an amoeba responds by moving about two cell diameters toward the signal and gradually stops. Before stopping, it shall have amplified and relayed outward the message to other amoebae by producing its own pulse of cyclic AMP. The amplified message diffuses further outward to be received by other amoebae which relay the message away from the center. A momentary pause ensues. Several minutes later, the process starts all over again. The overall result is a series of outward concentric waves of signals and an inexorable migration of amoebae toward the center, where it all begins and where the real mystery lurks. By dark-field photography, the rhythmic relay may be frozen in time and space as light and dark bands, alternating and concentric (Figure 17). A detailed description of this intricately orchestrated process entails a study of signal generation, reception, processing, and relay, the resultant taxis, and the final docking mechanisms.

The acrasins for other genera of the cellular slime molds are probably not related to cyclic AMP. In *Polysphondylium violaceum*, the evidence points to the participation of a peptide of less than 1500 daltons.[345] There is the possibility that the cyclic nucleotide, so widely used in nonmicrobial systems as to merit the title of second messenger, represents but one of many kinds of molecules used by aggregative organisms for purposes of communication. Yet there is also the possibility that other aggregation systems may not require communication by way of diffusible substances.

Diffusible substances may indeed serve in systems that require mutual coordination among components, as in heterothallic conjugation. They may also be functional where there is opportunity for preaggregation communication and where cell-cell interactions are weak and vulnerable. They do not seem to be necessary in systems that involve extraordinarily strong and stable interactions. Perhaps, it may be argued, their activities remain unobserved in the vigor of the interaction. On the other hand, these systems may just leave encounter to chance, instead of pre-encounter telecommunication. Cells may just rely on the strength and the specificity of the interaction for instant recognition and coordination. Diffusible substances might be the solutions to situations that involve directed movement, as in the cellular slime molds. In a Brownian milieu, such as a liquid culture, where movement is stochastic, where there is room for trial and error, diffusible substances are hardly an advantage. Instead, surface-bound recognition and coordination are bound to be more useful.

There is no common procedure for the isolation and the identification of pheromones and chemotactic signals involved in aggregation, precisely because very different kinds of substances appear to participate in different systems. The known mating-type factors, gamones, and acrasins differ considerably in structure. However, one thing is required: an assay procedure that will make detection possible and convenient. One can start with cell-free culture fluids, add them to a population of cells, and watch for aggregation-promoting

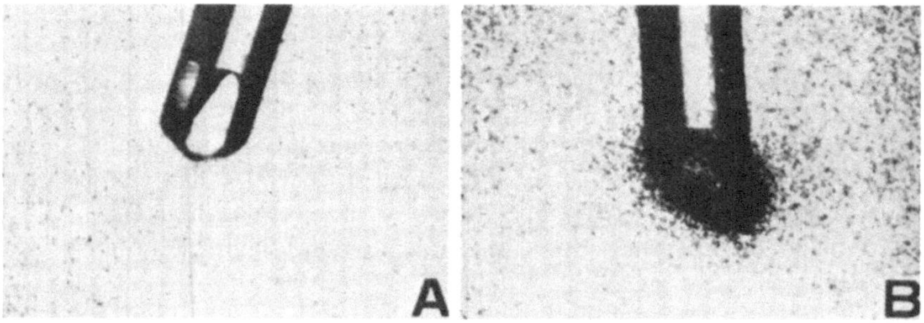

FIGURE 16. Chemotaxis in *Chlamydomonas* sp. A) *Minus* gametes show no attraction to culture filtrate of *plus* gametes in the capillary. B) *Plus* gametes attracted to the culture filtrate of *minus* gametes in the capillary. (From Tsubo, Y., *Bot. Mag.*, 70, 327, 1957. With permission.)

FIGURE 17. Light and dark bands of *Dictyostelium discoideum* on an agar surface.[117] Still amoebae are found in the dark bands, moving amoebae in the light bands. Two centers are shown. Line marker represents 1 cm. (Original photograph kindly furnished by P. C. Newell.)

activity. Or, in the case of heterothallic systems, one type of cells may be placed on an agar surface so as to allow a diffusible product to affect a nearby population of cells, the complementary type. The activity of the α factor of *Saccharomyces cerevisiae* was first demonstrated in this manner.[71] For the detection of acrasins in the cellular slime molds, one method consists of placing a drop of a cell suspension on an agar surface and a drop of test solution nearby.[506] If acrasin activity is absent in the test drop, the cells remain in their drop. If it is present, they break out. For quantitation, the concentration of the test solution and the distance between drops may be varied. The acrasin activity of cyclic AMP was first detected by this method.[316]

A suspected pheromone for aggregation must possess the following characteristics:

1. It must be diffusible. It is found in cell-free culture media and not an integral part of the surface of the cell.
2. It must be a metabolite. It is not found in the medium prior to inoculation. It is present in the medium only when cells have been present. Indeed, it is a cellular product.
3. It must have biological activity. Specifically, it promotes aggregation.
4. It must be isolatable. The isolated material promotes aggregation and is a distinct molecular species.
5. It must manifest some specificity with regard to source, target, and activity. In heterotypic systems, it may be specific to one mating type with regard to source and target. In both heterotypic and homotypic systems, the activity is not due to an enhancement of the general health and welfare of the cells nor due to such a gross effect as a drastic change in pH. Moreover, when it is added to a test culture, its effect is observed soon enough, in minutes, not many hours after addition.
6. It must be inducible. It is not synthesized or released or active at all times. In any case, its activity in a culture increases during the development of aggregation competence or during aggregation.

Many reports of aggregation factors have been heedless of the proposed criteria. The consequence has been a confusion in a literature that passeth nearly all understanding.

IX. FORCES AND MECHANISMS OF CONTACT

One cannot help but marvel at the efficiency and effectiveness of contact in microbial aggregation. Cell-cell interaction is characterized, in addition to its diversity and the multivalency of its components, by its selectivity, sensitivity, reversibility, rapidity, stability, and high affinity.

The interaction is quite selective. It is not just genus specific, but species specific as well. In heterothallic systems, it is specific between mating types. Moreover, it is specifically restricted to induced cells. In a population and a milieu that offer multiple choice, the cells are able to discriminate and recognize willing and able neighbors. This capacity for discrimination is related to its sensitivity. Where molecular structures have been isolated, they have been found to be active at very low concentrations. There is no need to swamp the system in order to get a measurable response. Furthermore, the sensitivity of the interaction is manifested by its facile perturbability. The interaction is responsive to many environmental effectors.

A most prominent feature of microbial aggregation is its reversibility. Of course, systems that have moved on to the next step in the developmental program — to copulation or conjugation, for instance, in sexual systems — are no longer reversible, but many an aggregation system, so long as it remains as such, may be caused to disaggregate and reaggregate again. Under appropriate experimental conditions that are rather mild by the standards of the chemist, the processes of disaggregation and reaggregation, although time dependent, are fairly rapid, nearly instantaneous in some systems.

Despite its sensitivity and its reversibility, a microbial aggregation system is reasonably stable to thermal agitation under ambient conditions. This stability, coupled with its selectivity and sensitivity, denotes a high affinity of the components for one another. Whatever molecular structures are involved, the selectivity and the strength must be due to high affinity of complementary surfaces.

These observations and considerations taken all together exclude some plausible mechanisms for cell-cell interaction. Gross electrostatic interactions, highly favored by many investigators in the past, can hardly offer much by way of specificity. Neither can hydrophobic interactions by themselves. Even just the single feature of reversibility excludes the participation of covalent linkages, except perhaps in the presence of elaborate enzymic arrangements. For sure, covalent linkages are forged in many microbial aggregation systems. After all, many are a prelude to sexuality and parasexuality. Cells in a sexual system eventually fuse, form dikaryons and zygotes, and lose their original identity; but these are events beyond aggregation. The cells prior to conjugation are bound together in a reversible way. They can be unbound and bound again without losing their cellular integrity. What aggregation has joined together, the investigator may still put asunder, and put the pieces back together again; but a zygote, or even a conjugation pair, once put asunder cannot be put back together again, not even with all the king's horses and all the king's men. Covalent bonds may confer stability, but it is stability without flexibility. They bring about rigidity and the loss of sensitivity. They are not easy to break and to reform.

The specificity, the sensitivity, the high affinity, and the reversibility of microbial aggregation, all these, are features shared by many biological phenomena. They are the reasons for the efficiency and the effectiveness of biological interactions. These features are due not to covalent bonding but rather to highly selective affinity of complementary surfaces. It is now a well-entrenched biological principle that the specificity of biological interactions derives from the stereochemical complementarity of molecular structures.[507-509] Complementarity is achieved by way of configurational fit, in short, by stereospecificity. It may not be the only mechanism involved in the interaction of surfaces, but it is simple and aesthetically pleasing. Besides, there are too many well-known examples to substantiate the general principle. To mention only a few, there are the two strands of the DNA helix, DNA replication, messenger and protein synthesis, induction and repression, the antigen-antibody reaction, the enzyme-substrate complex, protein structure, the pharmacological action of antibiotics and metabolic analogs, opiate receptors, hormone receptors, neurotransmission, histocompatability, fertilization, the sense of taste, and the sense of smell. Indeed, molecular stereospecific complementarity as the basis of specificity is one of the most unifying concepts in biology. It ranks as one of the more glorious generalizations in science, having been nurtured by a myriad of outstanding personalities, such as Pasteur, Ehrlich, Fischer, Landsteiner, Monod, Pauling, Burnett, Watson, and Crick. Is there any reason to believe that cell-cell interaction need be different, that it is not mediated by way of stereochemical complementarity of surfaces?

Stereochemical complementarity comprises three aspects: size and shape of the electron cloud that makes up the molecular structures, reactive groups that constitute the subsites within molecular structures, and configurational flexibility of ligand and receptor. The configuration of the electron cloud is not unchangeable. The ligand and the receptor must not be viewed as the rigid lock and key of Fischer nor the side chains of Ehrlich nor the male and female screws of Pasteur, but rather as flexible molecules subject to allosteric contortion and induced fit, particularly at their subsites.[510-512] The forces involved are van der Waals forces, coloumbic or ionic interactions, hydrogen bonds, and even hydrophobic interactions. Of the weak intermolecular forces, the hydrogen bond is the most significant.[507]

Having disposed of covalent bonds (including S-S bonds), which earlier were advanced in the literature as mediators of cell-cell interaction, one may proceed to find out the participating forces. The relative importance of the various forces may be worked out even before contact sites are isolated. It may be studied *in situ* by feeling, as it were, for contact sites still attached to and functional on the cell surface. Immobilized conveniently and naturally to a solid matrix that is the cell surface, the contact sites need not be spatially isolated. Instead, they may be isolated functionally, by heat-killing the cells, for instance. With no other cellular functions interfering, they are, for all practical purposes, isolated. The antigen-antibody reaction serving as a paradigm of noncovalent, high-affinity, high-specificity, stereochemical interactions, one may look in a microbial aggregation system for the intermolecular forces by using reagents for disaggregation.

The first obvious thing to do is to show the reversibility of contact by the use of reversible disaggregants. Cells in isolated and purified aggregates may be killed by heat or chemical treatment. They may then be reversibly disaggregated by physical means, such as heat, mechanical shear, or sonication. By chemical means, they may be reversibly disaggregated by acids and alkalis, detergents, chelators, sugars, and high concentrations of urea or guanidinium chloride (Table 4).

Many microbial aggregation systems are operative over a very wide range of hydrogen-ion concentrations, virtually excluding the participation of reactive groups ionizable within that range, if aggregation is simply the neutralization of charges. Aggregation may be viewed as the precipitation of cells due to cells having been rendered unstable as suspensions at their isoelectric point. Such is perhaps the case with agglutination associated with the development of transformation competence in bacteria. However, the concept of aggregation as simply the neutralization of repulsive forces between cells of identical sign hardly applies to any other system, because cells found in aggregates are not without a net charge. The contrary concept of ionic bonds being mediators of cell-cell interaction is likewise not applicable in systems that are functional in the presence of high concentrations of common salts. Invoked by workers who find a requirement for divalent cations in aggregation, the involvement of cation bridges is ruled out in systems that maintain their integrity even in deionized water or in the presence of chelators.

The efficient, reversible disaggregation of aggregates by heat and by high concentrations of urea or guanidinium chloride points to the involvement of hydrogen bonds. This has been concluded by workers on the yeast systems.[81,98,99] The slow, reversible disaggregation by sodium dodecylsulfate (SDS) suggests the contribution of hydrophobic interactions.[99] The detergent would disrupt the hydrophobic interactions, in effect reducing the binding forces primarily due to hydrogen bonds and rendering aggregates vulnerable to disruption by thermal agitation. If it may be assumed for the moment that there exist macromolecular structures covalently bonded to the cell surface, the forces that hold the cells together would be hydrogen bonds, sustained by hydrophobic interactions, between the complementary macromolecular structures.

If cell-cell interaction is mediated by protein ligands and carbohydrate receptors, or by protein receptors and carbohydrate ligands, then perhaps the addition of sugars, or of lectins (sugar-binding proteins), to a suspension of aggregates would result in a competition for ligands and receptors, causing disaggregation and inhibiting subsequent reaggregation. Although reported to be effective deflocculants in brewers' yeast,[81,272] sugars have been found to be ineffective in causing deflocculation of purified flocs of fission yeast.[99] Neither do they inhibit subsequent reflocculation. In general, molecules that exhibit high affinity for ligands or receptors on the cell surface should, provided they are univalent, help to disrupt an aggregate or to competitively inhibit its reaggregation.

The reaggregation of disaggregated cells is time dependent. Moreover, not all cells are able to subsequently reaggregate.[99] The hysteresis observed in reaggregation may be a mere indication of the partial denaturation of the molecular structures during the disaggregation process, but perhaps it suggests the specific molecular pairing of cell-surface components. A deeper understanding of the interaction in microbial aggregation demands the isolation and purification of the participant molecular structures.

X. RECEPTORS AND RECOGNITION

It is not enough that molecular structures in microbial aggregation possess high affinity for each other. High affinity alone will not confer specificity. There must be some element of uniqueness or near-uniqueness. The molecular structures must confer uniqueness to the cell surface by their complementarity, their number, their position, their distribution, or by their association with other structures. They must not just bind avidly, but must contain information to enable cells to recognize their complement.

A clear manifestation of biological specificity, microbial aggregation systems are selective systems, selective in that they are restricted to specific components that recognize one another. Recognition is accomplished by noncovalent, reversible, high-affinity binding of molecular structures that are stereochemically complementary over their surfaces of contact. It is not accomplished by mere groping for binding sites. Operationally, recognition is the binding of complementary sites. The complementary molecular structures, presumed to interact by hydrogen bonds, are generically referred to as receptors. The distinction between ligands and receptors does not seem fruitful from what is currently known about isolated molecular structures specific for aggregation. Besides, recall that the male and the female components may be neighbors on one side as well as on opposite sides of the cell-cell interaction (Figure 2, Chapter 2).

It must be pointed out that whereas cells in an aggregate are multivalent, the complementary molecular structures need not be. Because there are many of them, they individually need only be univalent, in the case of structures covalently bonded to the cell surface or with extremely high affinity that they are, for all practical purposes, integral parts of the cell surface. At most, they need only be bivalent, in the case of structures that interpose as ligands between two cell surfaces. In any case, the term receptors is made synonymous here to contact sites or binding sites or recognition sites. The contact and the binding and the recognition and the reception are one and the same.

An honest search for aggregation receptors must take into account the following considerations:

1. A receptor must be present on the cell surface, be it wall, membrane, or appendage, there is where the action is. A suspect must be at the scene of the crime. An alibi is not good enough.
2. The removal of the presumptive receptors from their site of action should constitute a loss of aggregation, or at least an observed decrease in the capacity of the bereft cells for aggregation.
3. A specificity as to source of the receptors must be established. The receptors are not found, or found only in low amounts, in other species or in other genera. They are not found in uninduced cells nor in aggregateless mutants. In a heterotypic system, they are not recoverable from the complementary cells.
4. A specificity as to target must be established. Target cells are limited to components of the system. In heterotypic systems, the target cells are the complementary cells.
5. A preparation of receptors must have biological activity. It must promote or inhibit aggregation. If it inhibits, it is univalent. If it promotes, it is at least bivalent.
6. The suspected material must be inducible. Its synthesis or its activity relates to the development of aggregation, but more important, the appearance of the material or of its activity comes before aggregation, not after.
7. Its sensitivity, *in situ* and in vitro, to inhibitors must parallel the sensitivity of aggregation.
8. The suspected material must bind to the cell surface of the source and target in a homotypic system, or the target in a heterotypic system.
9. It must be isolated and purified and the molecular complex with the complementary molecule demonstrated.

The above criteria are rather stringent. Hence, some modifying and accommodating statements are in order. The receptors need be present on the surface only at the time of aggregation. This means that it can be present elsewhere, in the culture filtrate or in the cytoplasm, for example. The agglutinin of *Hansenula wingei* is extractable from the cytoplasm and the gamones of the chlamydomonads are recoverable from the culture filtrate. A decrease in aggregation may not be observed upon the removal of the receptor material if it may be assumed that the recovered material constitutes a very small (and superfluous) percentage of the receptors on the cell surface and that it takes only a few receptors per cell for aggregation to occur. That excuse can be avoided by the recovery of more material. The specificity as to target may be less than strict. The isolated material may cross-react with related species. If it does, then that is informative and nothing to be ashamed of. The same accommodation may be made as regards the specificity as to source. Inducibility and its relation to the development of aggregation are not absolutely necessary. Because a cell-cell interaction is made up of at least two kinds of components (obvious in heterotypic systems but also conceivably appropriate in homotypic systems), only one component need be induced. The other may be constitutive during the developmental program. Such is apparently the case with sexual agglutination in *Saccharomyces cerevisiae*. The same consideration is applicable to inhibition sensitivity. One component may be sensitive, its complement indifferent. Such is apparently observed in certain species of *Chlamydomonas*. Moreover, a receptor detached from its site on the cell surface may become more susceptible to inactivation by inhibitors than a receptor *in situ*. An inhibition of aggregation by a univalent molecule need not be caused by a molecular fragment. The molecule may represent a whole receptor detached from a covalent link with the cell surface. A receptor need not be isolated. If it is an integral part of the cell surface, it should resist all but the most forceful wrenching.

Because of the diversity and the specificity of microbial aggregation systems and because of the molecular composition of the cell surface, the proteins or glycoproteins are the most likely candidates for receptors. Among the many models of stereochemical specificity, the nucleic acids and the carbohydrates are the only other molecular classes that deserve consideration. Nucleic acids are not typically located on the cell surface. Besides, they are not versatile enough in function. Complex polysaccharides, found in abundance on different cell surfaces, would make good receptor materials. There is more than sufficient variety in their chemical structures and in their three-dimensional conformations to accommodate stereospecific information: the antigenicity of carbohydrates is well known. There are more and better examples of the versatility of the proteins and much more is known about their three-dimensional structure.

The complementary molecular structures that are the receptors are no longer just hypothetical constructs in several systems. Despite the inherent difficulties, there has been progress in their isolation, purification, and characterization, however partial and tentative the characterization may be. They are listed in Table 5. It is not surprising to note the prominence of proteins and glycoproteins in the short list.

The best characterized of the putative receptors specific for aggregation are the complementary agglutinating factors of *Hansenula wingei:* the 5-factor from strain-5 cells and the 21-factor from strain-21 cells.[16,98,513-516] The 5-factor specifically agglutinates strain-21 cells. It does not agglutinate strain-5 cells nor the diploid cells of the hybrid 5 × 21. The 21-factor specifically inhibits the 5-factor-promoted agglutination of strain-21 cells. Each factor binds only to the cell surface of the complementary strain and not to cells from which the factor is derived nor to the diploid hybrid.

The 5-factor, originally derived from a cytoplasmic extract of strain-5 cells, is made of molecules of various sizes.[170,514] Recovered from subtilisin digests and purified by affinity chromatography, the 5-factor yields a homogeneous population of phosphomannan-protein molecules of close to 10^6 daltons.[169,517] The homogeneous material containing sulfur-sulfur linkages may be split by chemical reduction into subunits of about 1.25×10^4 daltons. The agglutinin activity of the 5-factor is destroyed by exo-α-mannanase and protease. The phosphate moiety is apparently not required for activity.[518]

Table 5
PUTATIVE RECEPTORS IN MICROBIAL AGGREGATION

Receptor	Source	Activity	Chemical composition	Mol wt
		Sexual agglutination in *Hansenula wingei*		
5-factor[16,98,513-517]	Subtilisin digest of strain-5 cells	Promotes agglutination of strain-21 cells	10% protein, 85% carbohydrate, 5% phosphate	9.6×10^5
5-agglutinin[169,517]	Chemical reduction of 5-factor	Same as 5-factor	28 amino acids, 60 mannose units	1.25×10^4
21-factor[98,170,515]	Tryptic digest of strain-21 cells	Inhibits agglutination promoted by 5-factor	Glycoprotein (carbohydrates ca. 25—35%)	4×10^4
		Sex-directed flocculation in *Schizosaccharomyces pombe*		
Wall-bound receptors[99,519]	Cell walls	—	Protein	—
Floc-factor fragments[516]	Culture filtrate	Inhibit reflocculation of heat-killed cells	—	$<3 \times 10^3$
		Sexual agglutination in *Saccharomyces cerevisiae*		
Agglutination substances[171,172]	Glusulase digest of cell walls	Inhibit agglutinability of complementary cells	Glycoproteins	10^6
α Agglutination substance[520-522]	Autoclaved **α** cells	Inhibits agglutinability of **a** cells	Glycoprotein (carbohydrates ca. 47%)	1.3×10^5
a Agglutination substance[520-522]	Autoclaved **a** cells	Inhibits agglutinability of **α** cells	Glycoprotein (carbohydrates ca. 61%)	2.3×10^4
		Aggregation in *Dictyostelium discoideum*		
Contact site A[529-532]	Cell surface	Binds Fab directed against competent cells	Glycoprotein	$1.2—1.3 \times 10^5$
Discoidin I[524-527]	Cytoplasm and cell surface	Agglutinates glutaraldehyde-fixed cells	Tetrameric protein	$4(2.6 \times 10^4)$
Discoidin receptor[527,528]	Cell surface	Binds discoidin	Protein (with carbohydrate?)	5.6×10^4
Glycoprotein 150[533]	Cell surface	—	Glycoprotein	1.5×10^5
		Aggregation in *Polysphondylium pallidum*		
Pallidin[534,535]	Cell surface	Agglutinates fixed differentiated cells	Protein	2.5×10^5
		Sexual agglutination in *Chlamydomonas*		
Plus isoagglutinin[86,175-178]	Culture filtrate of *plus* cells	Agglutinates *minus* cells	Flagellar membrane vesicles with glycoprotein	10^8
Minus isoagglutinin[86,175-178]	Culture filtrate of *minus* cells	Agglutinates *plus* cells	Flagellar membrane vesicles with glycoproteins	10^8

Table 5 (continued)
PUTATIVE RECEPTORS IN MICROBIAL AGGREGATION

Receptor	Source	Activity	Chemical composition	Mol wt
		Mating reaction in *Paramecium caudatum*		
Cilium-bound gamones[179,180]	Cilia from either complementary mating type	Agglutinate cells of complementary mating type	Cilial membrane with at least 4 kinds of glycoproteins	—

The 21-factor is a homogeneous extract derived from the cytoplasm or a tryptic digest of whole strain-21 cells.[98,170,515] The factor has a molecular weight of about 4 × 10^4 daltons. It is apparently univalent because it inhibits the agglutination induced by 5-factor. A glycoprotein, it contains carbohydrates (mostly mannose units) up to about 35%.

Molecular complexes due to the neutralization of 5-factor by 21-factor have been demonstrated in vitro.[170] The complementary factors being both glycoproteins, it is not known whether the formation of the complex is an example of a protein-protein or of a protein-carbohydrate interaction. There is evidence, however, that the protein portion of the 5-factor is required for specificity, whereas the sugar component is there to maintain the structural configuration.[169]

Sex-directed flocculation in the fission yeast *Schizosaccharomyces pombe* is presumed to be mediated by proteins because of its sensitivity to proteinases, its insensitivity to other kinds of enzymes, and its sensitivity to reduction by sulfhydryl reagents.[99,516,519] The complementary molecular structures are bonded covalently to the cell wall. That they are proteins with carbohydrates cannot be excluded — despite the ineffectiveness of carbohydrases as irreversible deflocculants — because the cell wall is mostly carbohydrate.

Putative fragments of the flocculation structures of fission yeast have been identified in the culture filtrate.[483,576] The fragments are univalent because they inhibit the reflocculation of heat-killed cells derived from purified flocs. They do not by themselves cause deflocculation, which suggests that they cannot compete for complementary sites when the cell-cell interaction is intact and operative. Their site of action is probably buried in the intact complex in vivo. The fragments are rather small as estimated by passage through Sephadex® G-25. They are not found in uninoculated media or in media that do not support flocculation.

Digested by Glusulase, cell walls of one mating type of *Saccharomyces cerevisiae* yield fractions that inhibit the agglutinability of the opposite mating type.[171,172] These fractions, further purified from either mating type, yield agglutination substances of about 10^6 daltons. The univalent activity is also detectable in extracts of autoclaved cells.[520-522] Purified by affinity chromatography on Con-A Sepharose followed by elution with α-methyl-D-mannopyranoside, these agglutination substances from autoclaved cells are smaller. The **a** agglutination substance, derived from **a** cells, has a molecular weight of 2.3 × 10^4 daltons, the **α** agglutination substance, derived from **α** cells, 1.3 × 10^5 daltons. Both substances are glycoproteins. From their sensitivity to proteolytic enzymes, it is assumed that the protein moieties are required for activity. The **a** agglutination substance is sensitive to reduction by β-mercaptoethanol or cysteine, but the **α** agglutination substance is not. The formation of a molecular complex between the complementary substances has been demonstrated in vitro.[521]

There are a number of surface proteins implicated in the aggregation of cellular slime molds.[491,523] Among these are the lectins discoidin and pallidin, a discoidin receptor, contact site A, and glycoprotein 150. Discoidin I is a carbohydrate-binding protein. First identified as a lectin that binds formalinized sheep erythrocytes, it preferentially binds *N*-acetyl-D-galactosamine.[524-527] Discoidin I agglutinates glutaraldehyde-fixed cells of *Dictyostelium discoideum*, to which it specifically binds with high affinity and from which it is derived. Developmentally regulated, it is most abundant during the aggregation stage. It is found on the cell surface as well as in the cytoplasm. A protein made up of four monomers of 2.6 $\times$ 10^4 daltons each, it may serve as a multivalent ligand between two carbohydrate surfaces. However, it is found in high amounts in aggregateless mutants.

The receptor specific for discoidin is identified by its binding of labeled discoidin.[527,528] The discoidin receptor co-precipitates with antibodies for the lectin. A cell-surface protein, the putative receptor is estimated to be 5.6 $\times$ 10^4 daltons. It is presumed to contain a carbohydrate moiety that is complementary to the carbohydrate-recognizing site of the lectin.

Contact site A is scarcely detectable during the growth phase and is most abundant during aggregation.[529-532] Found on the cell surface, it also binds the lectin concanavalin A. Contact site A appears to be a glycoprotein from its sensitivity to periodate and Pronase. Unlike discoidin, it is not detectable in the cytoplasm nor during the growth phase of certain strains (Ax-2 and v-12), when discoidin is found plentiful. Contact site A is identified by the neutralization of the aggregation-blocking activity of fragments of antibodies (Fab) directed against cell membranes of aggregation-competent cells. Blockage of other presumptive sites on the cell surface, contact sites B, by Fab or by EDTA inhibits aggregation but still allows end-to-end contacts in the form of rosettes or chains (Figure 18). (Contact sites B are not developmentally regulated.) When contact sites A are blocked by Fab, cells are still allowed side-to-side contacts. A preparation of polyspecific Fab blocks contact sites A and B. The consequence is a complete inhibition of cell aggregation. Extracted from the cell membrane by deoxycholate, contact site A is a particle of about 1.3 $\times$ 10^5 daltons.

Another glycoprotein that binds concanavalin A is that identified by Geltosky et al.[533] In low amounts during the growth phase, it is plentiful during aggregation. It is found on the cell surface, but not known if it is present in the cytoplasm. It has a molecular weight of 1.5 $\times$ 10^5 daltons. Aside from its feature of being developmentally regulated, glycoprotein 150 has no demonstrable activity that relates to aggregation.

The lectin found on the cell surface of *Polysphondylium pallidum* is called pallidin.[534,535] A protein of 2.5 $\times$ 10^5 daltons, it also agglutinates sheep erythrocytes, but unlike discoidin, it preferentially binds D-galactose. Pallidin agglutinates fixed differentiated cells of *P. pallidum*. Moreover, Fab directed against this lectin inhibits the spontaneous aggregation of the organism.

The gamones of chlamydomonads are isoagglutinins found in the culture filtrate.[86,175-178,413] From their sedimentation characteristic, their content of amino acids and sugars, and their susceptibility to trypsin, they appear to be large glycoprotein particles (10^8 daltons). The culture filtrate of one mating type promotes the isoagglutination of gametes of the opposite mating type. The isoagglutination caused by isolated gamones is similar to sexual agglutination of mating types in that it too involves flagellar interaction. Vegetative cells are unreactive.

Like aggregation in the cellular slime molds, sexual agglutination in chlamydomonads appears to be an example of protein-carbohydrate interaction. In *Chlamydomonas moewusii* and *C. eugametos*, the ability of *plus* gametes to react with *minus* gametes is destroyed by α-mannosidase, but that of *minus* gametes is resistant.[175] The *plus* gamone, derived from the culture filtrate of *plus* gametes, is destroyed by the enzyme, whereas the *minus* gamone remains active. The gametes of *C. reinhardtii, plus* and *minus*, are insensitive to α-mannosidase.

At a concentration that does not disturb locomotion, trypsin disrupts sexual agglutination.[175,486] The disruptive activity of trypsin is nullified by trypsin inhibitor. In *C. moewusii* and *C. eugametos*, only the *minus* gametes are sensitive, whereas in *C. reinhardtii*, both

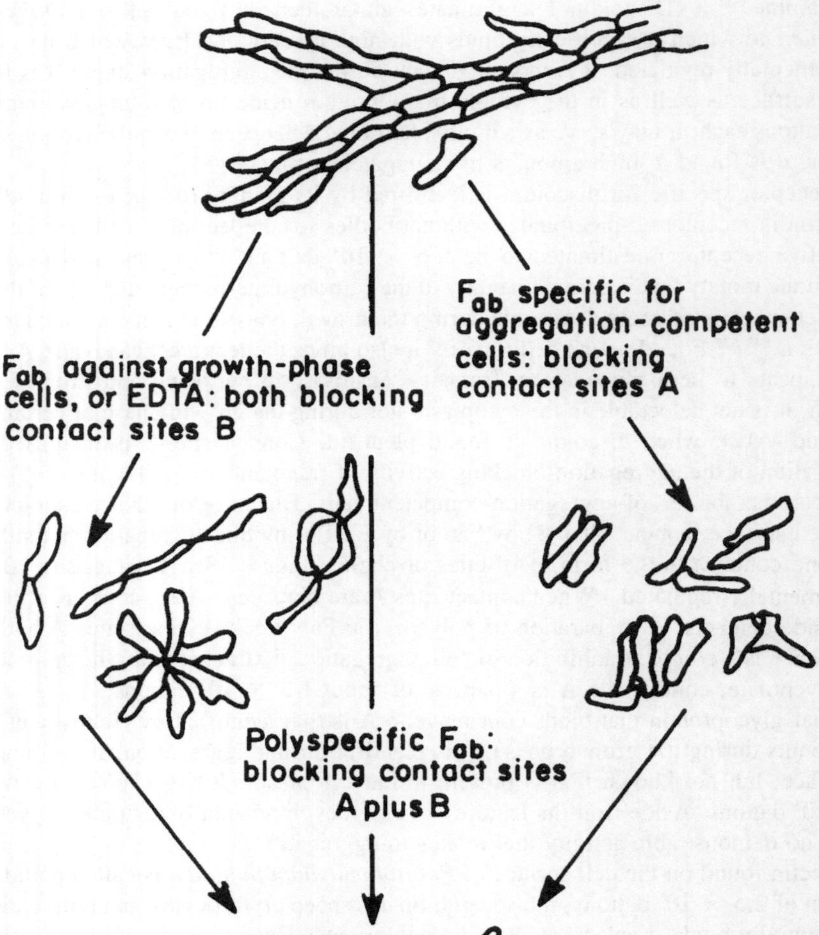

FIGURE 18. Inhibition of aggregation in the cellular slime molds by blocking contact sites A and B. EDTA or Fab (fragments of antibodies) directed against growth-phase cells block contact sites B and allow end-to-end contacts only. Fab specific for aggregation-competent cells block contact sites A and allow side-to-side contacts. Blocking contact sites A and B by polyspecific Fab causes disaggregation. (From Gerisch, G., et al., *Biology and Biochemistry of Eucaryotic Cell Surfaces*, Lee, E.Y.C. and Smith, E. E., Eds., Academic Press, New York, 1974, 49. With permission.)

minus and *plus* gametes are sensitive. The isolated gamones, *plus* and *minus*, are both inactivated by trypsin.

The gamone activity of chlamydomonads is associated with fragments or vesicles that bud off the flagellar membrane.[176] The membrane vesicles contain fuzzy material and mastigonemes. When isolated, mastigonemes prove to be inactive, but vesicles bereft of mastigonemes show strong isoagglutinating activity. The formation of membrane vesicles, however, is also observed in vegetative cells.

The mating reaction in the ciliates occurs by way of cilial contact between complementary mating types. Cilia detached from one mating type bring about the isoagglutination of cells of the complementary mating types.[161,536,537] Derived from complementary mating types, cilia themselves are observed to form an agglutinated mass. Fractionation of cilia reveals particles that turn out to be membrane vesicles, somewhat similar to the flagellar-membrane vesicles of chlamydomonads.[179,180] Preliminary work indicates that these vesicles contain a minimum of four kinds of glycoproteins, but polyacrylamide-gel electrophoresis does not reveal much difference between preparations from complementary mating types.

Other preparations reported in the literature may turn out to be receptors specific for aggregation. A notable example is blepharmone, an isolate from the culture filtrate of mating-type I cells of the ciliate *Blepharisma intermedium*.[443,444,448] Mentioned as gamone I in the section on pheromones and chemotactic signals, blepharmone appears more of a receptor than a pheromone. A glycoprotein of 4×10^4 daltons, it promotes the agglutination of mating-type II cells.

It is reasonable to presume that receptors specific for aggregation will be found where they are functional, on the cell surface. Because the receptors are developmentally regulated, they may be best isolated from the cell surface during aggregation.If an aggregation-active material is found in the cytoplasm, it may be a precursor for the aggregation receptor or an intact receptor before it gets attached to the cell surface. Either way, it must be found prior to aggregation. If the material is recovered from the culture filtrate, it may have fallen off its functional location and may just be a fragment of the intact receptor bound to the cell surface. Then, perhaps, it is most plentiful sometime past the period of aggregation.

Conceivably, receptors are integral parts of the cell surfaces of aggregation-competent cells. Hence, their isolation calls for procedures that ensure their release without destroying their activity. They may be solubilized from cell membranes by deoxycholate, as is contact site A from the cellular slime molds, or picked off the cell surface by proteolytic enzymes, as is the 21-factor from *Hansenula wingei*. It is assumed in any procedure chosen that their separation from the cell surface permits nonetheless the retention of a portion of their activity.

Recovered from the cytoplasm or from the culture filtrate or from the cell surface, a putative receptor must have biological activity. There must be a way of detecting its activity relating to aggregation. A substrate for a receptor preparation may be target cells that are whole (with cell surface intact and functional), already induced, and preferably fixed in some manner as to exclude the interfering participation of metabolic activity. Heat-killed cells dispersed from isolated aggregates may serve this purpose. If the preparation promotes aggregation, it is presumed to be at least bivalent. If it inhibits, it is presumed to be univalent. Either promotion or inhibition must be observed in minutes, not hours, after the exposure of the target cells to the preparation.

In a heterotypic system, the interaction is between two (or more) different cells with presumptively different surfaces or different molecular structures for aggregation. That an extract from one type brings about the isoagglutination of the other type is cause for concern. Such a demonstration is in effect a conversion of a heterotypic system to a homotypic system. For this reason, the promotion of isoagglutination in an obviously heterotypic system by an extract is somehow suspect, however specific it may be with respect to source and target. In a heterotypic system, a more reassuring assay for an isolated receptor is an inhibition.

When all attempts to isolate molecular structures for aggregation have failed, studies on contact and recognition need not be abandoned. Information about the molecular basis of recognition may be gained by the selective use of enzymes, in particular, proteinases and carbohydrases. The reactive groups may be probed by the use of sugars as competitive inhibitors of aggregation. Lectins, carbohydrate-binding proteins such as concanavalin A, may be used to locate the molecular structures and to determine their distribution on the cell surface. More specifically, fragments of antibodies directed against aggregation-competent cell surfaces may be used to block, and consequently identify, putative receptors *in situ*. If molecular structures cannot be isolated, they can nevertheless be monitored right where they are, provided there is a way of identifying, locating, and quantifying them.

XI. MORPHOLOGY OF THE INTERACTION

Sooner or later, a study of function becomes a study of form, and vice versa. Indeed, function is no more than form in action even as form is function frozen for the observer's visual perception. Microbial aggregation is no different from any other visible phenomena in that it can be visually observed. Visualization for the most part, of course, requires the use of the microscope, the individual components of a microbial aggregation system being too small for resolution by the unaided eye. Morphology, the study of form, is fundamentally a visualization.

Random taxis of microbial cells in liquid culture is not easily amenable to visual study, but directed taxis on a solid surface is fruitfully visualizable. Macroscopically observed, the rhythmic movement of the cellular slime molds toward aggregation centers (Figure 13) is followed by streaming and the eventual maneuvers for docking to form pseudoplasmodia. The myxobacteria show movement patterns no less intricate, although the physiology and the morphology of their taxis have not been as well documented. The elongation and the approach of one hypha toward another and the subsequent anastomosis may be visually monitored in filamentous fungi on an agar surface. Nondirected collision, however, as in a broth culture of bacteria, offers scarcely any opportunity for a visual approach.

The morphology of contact offers more opportunities. Even without the benefit of sophisticated gadgetry, one may ask some such questions as concern the size and shape of the aggregate, the number of cells that make up the aggregate, the valency of a cell, the arrangement of the cells with respect to one another. Do cells have preferred spatial orientations? Do they form chains or rosettes or palisades or sheets or flakes? Or do they tend to minimize their overall exposed surface and to seek the most thermodynamically stable shape? Are the contacts random, end-to-end, or preferably side-to-side?

Form and function go inseparably together, precisely because function derives from the material basis of form whereas form is determined and selected by function. The separation is achieved conceptually and operationally. In a microbial system, an approach that relies on the microscope more than on any other tool in the laboratory is essentially a morphological approach, even if the question at hand is essentially physiological. On the other hand, some morphological questions may be approached physiologically, or even in a theoretical manner.

Other than questions of gross phenomenology, which is to a large extent morphological, questions concerning form may be answered by looking into the microscope. Some questions are amenable to light microscopy. Others have to be resolved with the electron microscope or with the aid of cytochemical trickery. For instance, is cell-cell contact mandatorily between two differentiated cell types? Even in obviously heterotypic systems, such as the sexual systems, it still is worthwhile confirming whether an aggregate comprises roughly equal numbers of complementary cells and whether a cell of a given mating type preferentially interacts with cells of the complementary mating type. Where the mating types are morphologically alike, one type may be stained (or genetically marked) so as to render it morphologically distinguishable from the other. In presumptively homotypic systems, such as the cellular slime molds, there is the possibility of position effects with regard to the subsequent events. Do cells in one section of the aggregate end up eventually in a particular location in the fruiting structure? Are cells buried in the core of the aggregate less viable, less active, less prone to conjugate and sporulate?

What regions of the cell surface are involved? How close or how loose are the contacts? Are there visibly identifiable contact sites? Are appendages on the cell surface involved? Finally, are the contact sites the future sites for the next step in the developmental program, that is, of copulation, conjugation, or cell fusion?

It is obvious that microbial aggregation is a surface phenomenon. Thus, the focus of attention, as far as the student of microbial aggregation is concerned, is the cell surface. Although not as metabolically active as the cytoplasm, the cell surface is not an inert component of the cell. Aside from confining the cytoplasm and defining the limits of the unicellular organism, it maintains the shape and the integrity of the cell. It serves as an osmotic barrier, the selective gates for nutrient uptake and metabolite excretion, the Maxwellian mediator of ionic balance, as well as a sensor of the conditions in the culture medium. It is the cell's contact with the outside world. Involved in its own synthesis, it contains information for many other cellular functions, including cell division, locomotion, and tropism. More relevant to the subject at hand, it is the cell surface that is involved in cell-cell interaction.

The cell surface is not structureless, not a monotonic curved plane. Wall or membrane, it contains indentations and projections, ornamentations, and appendages. These appendages in various forms are described as pili, fimbriae, flagella, cilia, filaments, fuzz, or hairs (Table 6). In one way or another, they are implicated in microbial aggregation. It may be presumed that when two cells approach each other, it is these appendages which first come in contact. Their locomotor or their sexual functions disregarded, these filaments may be envisioned as threads that entangle cells together or as grappling hooks initiating as well as stabilizing the cell-cell interaction.

Morphology of contact is the visualization of these appendages as they appear to mediate aggregation. In the unicellular algae and the ciliates, their existence and their role in aggregation are undisputed. In bacteria, although their existence cannot be called into question, their role as aggregation structures remains inferred rather than proved. In yeast, both their existence and their role are moot and still subject to tentative appraisal.

Sexual agglutination in chlamydomonads is doubtless mediated by flagellar contacts (Figure 19). Complementary pairs of gametes interact by their flagellar tips. Vegetative cells also possess flagella but do not mate. The interaction is reversible in that flagellar tips change positions and cells change partners.[547] Indeed, it is this seeming hapchance that brings about the grossly visible clumping. Flagella are individually wrapped in a membrane which is continuous with the plasma membrane. The flagellum with its membrane is further encased in a sheath. Hair-like structures called mastigonemes extend from the flagellar membrane.[547,548] The initial contact at the flagellar tips progresses into a fusion of the flagellar sheaths. When the cells have fused, the flagella separate from each other again. The isoagglutinin activity of the membrane vesicles that bleb off the flagellar membrane has been demonstrated.[176]

The mating reaction in ciliates involves the cilia of complementary mating types.[455] This is concluded not only from a visual observation of the contact but also from the agglutinating activity of isolated cilia. The pellicle, however, cannot be excluded from having possible contact sites. It has been reported that any part of the cell surface is agglutinative.[549] There is better evidence now for believing that the ventral half of a cell is more agglutinative than the dorsal half.[536,537,550] Membrane vesicles isolated from cilia have been shown to possess agglutinative activity.[165]

Three kinds of appendages are found on the bacterial cell surface: the flagellum for locomotion, the pilus for sex, and the fimbria for some disputed function. The role of pili in conjugation of *Escherichia coli* is no longer contestable. The loss of F pili by mechanical, physiological, or genetic manipulation leads to a loss of conjugation activity.[551] Activity is restored by regrowth of pili, but whether pili are involved in mating-aggregate formation is

Table 6
MURAL APPENDAGES IMPLICATED IN MICROBIAL AGGREGATION

System	Appendages
Mating-aggregate formation in *Escherichia coli*	Pili[14,189,538-540]
Star formation in bacteria	Fimbriae[14,198,204]
Sexual agglutination in *Hansenula winei*	Surface filaments[125,516]
Sex-directed flocculation in *Schizosaccharomyces pombe*	Sex hairs[121,541,542]
Sexual agglutination in *Saccharomyces cerevisiae*	Surface fuzz[101]
Flocculation of brewers' yeast	Fimbriae[543,544]
Sexual agglutination in chlamydomonads	Flagella[86,411,414,545]
Mating reaction in paramecia	Cilia[455,546]
Mating reaction in *Blepharisma intermedium*	Cilia[119,500]

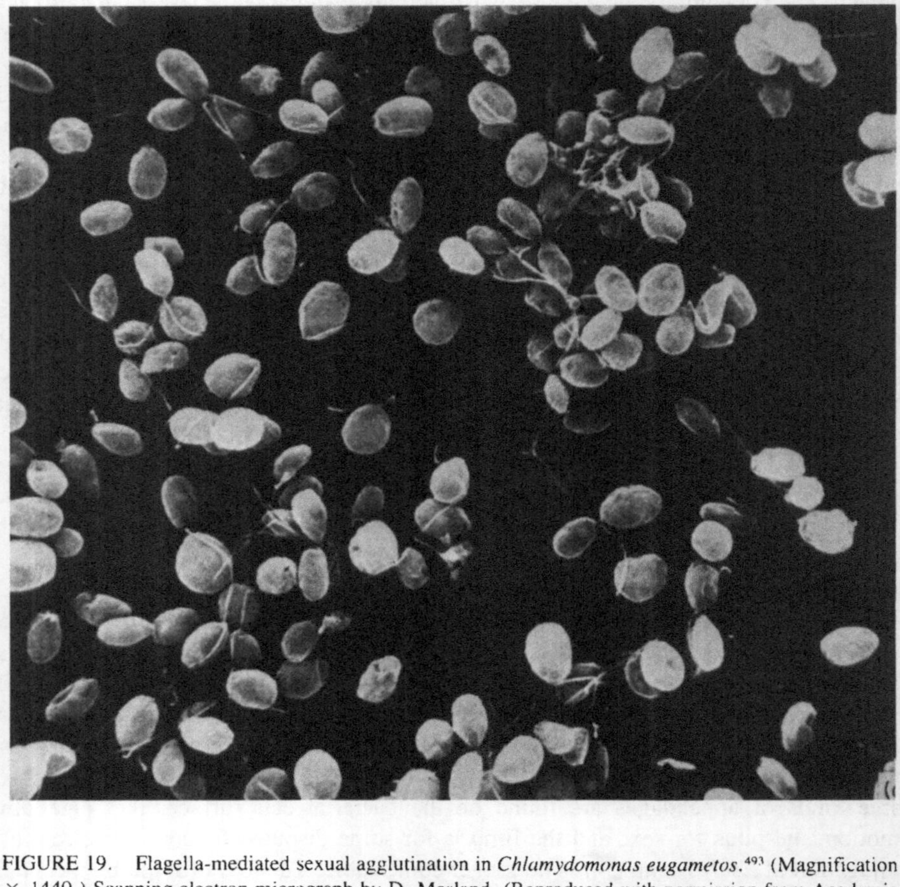

FIGURE 19. Flagella-mediated sexual agglutination in *Chlamydomonas eugametos*.[493] (Magnification × 1440.) Scanning electron micrograph by D. Mesland. (Reproduced with permission from Academic Press, London.)

another matter. In other bacteria, such as the starformers, fimbriae have been implicated in cell aggregation.[14,198,204]

Yeast cell walls have been shown in different laboratories to possess surface filaments. These filaments have been demonstrated in a variety of yeast species, including *Hansenula wingei, Saccharomyces cerevisiae*, and *Schizosaccharomyces pombe*.[101,121,541-544] Known by various names, they appear to be similar to one another at least in dimensions. A photograph of sex hairs implicated in sex-directed flocculation in fission yeast is shown in Figure 20. The proteinaceous nature of the sex hairs, which are found only on the surfaces of competent cells, is suggested by their appearance as an electron-dense material as well as by their susceptibility to removal by proteinases, which cause irreversible deflocculation.[121] Sex hairs are not seen on the walls of cells grown in cycloheximide at concentrations that inhibit the development of aggregation competence. Despite the agreement among yeast workers on the existence of fibrillar structures on the yeast surface, their nature and their role in aggregation are no better than assumed. The isolation of these structures with activities either promoting or inhibiting aggregation must be undertaken.

Much remains to be revealed by the electron microscope. Most of the published prints are focused on the events associated with aggregation rather than on the processes of aggregation. What is needed is not just higher resolution but also better fixing and staining procedures that maximize visualization and minimize disruption of what is hoped to be observed. Even more important, a morphological study must not be divorced from physiology, biochemistry, and genetics.

XII. GENETICS OF MICROBIAL AGGREGATION

A principal virtue of microorganisms as experimental tools is their genetic tractability. It more than compensates for one having to use the microscope in order to see them. For any microbial aggregation system to be amenable to genetic analysis, it must possess the following characteristics in common with other genetic systems.

1. The phenotypic character must be reasonably defined. The conditions for its appearance are identifiable such that its demonstration is reliable and quantifiable. Moreover, the phenotype is not confused with other phenomena nor are other phenomena mistaken for it.
2. The phenotypic character must be heritable. If it is not, then there is no cause for a genetic analysis.
3. It must be mutable. Heritable deviations from the phenotype are available. In the alternative state, or when the genotype is modified, the phenotype is not expressed or is identifiably modified.
4. It must be fairly stable. Although unstable traits can be monitored, as will be shown later, it helps quite a bit to start with a fairly stable one.

In some instances, the definition of the phenotype is in itself a task. The range of variability of the phenotype must be defined so that a mutation can be identified. The conditions under which the phenotype is observed must be arranged operationally such that the absence or a modification of the phenotype is conveniently ascribable to a genetic change and not due to physiological or environmental manipulation. Induction of aggregation by genetic manipulation has already been discussed in Section V of this chapter. We shall not repeat that discussion here but rather cite further examples of genetic findings relevant to aggregation.

All the currently known microbial aggregation systems may eventually be analyzed genetically, but some are more tractable than others. With regard to the definition of the trait of aggregation competence, some systems are more vague than others. Others are more

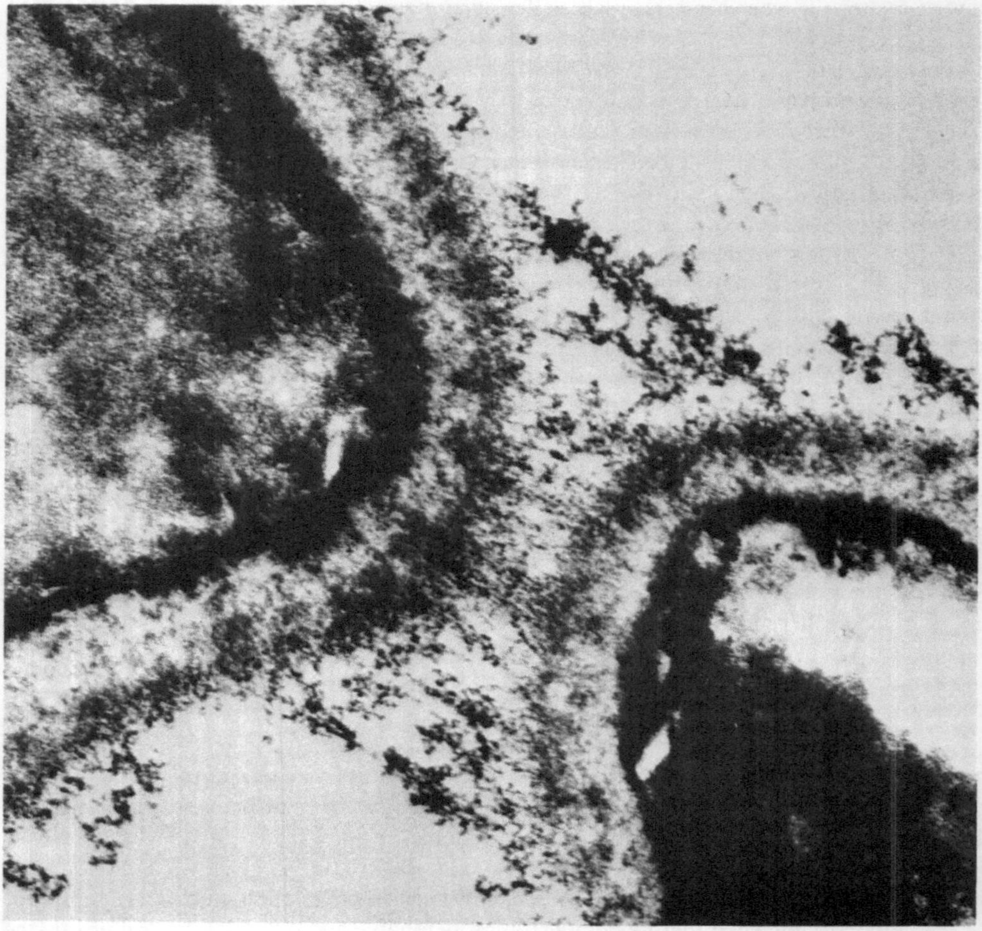

FIGURE 20. Sex hairs between two cell walls of *Schizosaccharomyces pombe*.[121] Cells were fixed in 2% aqueous $KMnO_4$ for 1 hr. The fixed cells, after dehydration with acetone, were embedded in a mixture of Araldite and Epon. Thin sections were stained with lead citrate. (Magnification × 75,500.) Transmission electron micrograph by B.Y. Yoo.

complicated in that they involve more associated events and functions and, consequently, more genes. Furthermore, only a few are found in organisms with well-defined genetic systems. The genetics of aggregation are most profitably studied in organisms which have been dissected for other genetic markers.

Genetics has become the study of the material gene. The objective of a genetic study of microbial aggregation is to assign genes specific for aggregation, locate them in the genome, identify the gene products, and ultimately to isolate the genes and characterize them in terms of their nucleotide sequence. The genetics of microbial aggregation is far from realizing that ultimate goal. In the best studied systems, it has not gone further than tentative assignment of genes and mapping them. Identification of gene products specifically derived from aggregation genes has not been achieved in any system.

Any attempt to assign a gene for aggregation will have to use aggregation itself as the phenotype, not an event or a phenomenon associated with it. However, the associated phenomenon, be it gametogenesis or synthesis of flagella or development of mitochondria, will likely contribute to the expression of the phenotype.

There is no doubt that aggregation in any of the systems we shall describe in detail is heritable. Be that as it may, the question of heritability of flocculation in brewers' yeast, for instance, preoccupied early workers on the system. That a particular phenotype is inherited, all that must be shown is its transmission from generation to generation. In a microbial culture, which contains many generations, the judgment of whether or not a phenotype is inherited is not too difficult to make, provided the phenotype is well defined. Doubt arises when there is too great a variability in phenotype and when there is genetic instability.

Given a definable and fairly stable phenotype that is aggregation, one may then proceed to select for mutants defective in aggregation or in aggregation competence. Being a complex series of events, aggregation is expected to be controlled by a number of genes. The genes for mating types in sexual systems are presumably involved; most likely they are not sufficient to specify aggregation competence. At present, the genetics of microbial aggregation is largely the genetics of mating types. Because of the nature of current genetic methods, the best studied systems are sexual systems, which involve mating types. However, parasexual phenomena may be used to advantage, as has been done in the cellular slime molds.

Mutants deficient in aggregation have been identified among the cellular slime molds.[552,553] Analyzed by parasexual techniques, they fall into ten complementation groups in *Dictyostelium discoideum*.[478,479,554] It is estimated from statistical considerations that approximately 50 genes code specifically for functions absolutely necessary for aggregation and another 100 genes for related functions. About the same number of aggregation-specific genes has been arrived at from complementation studies, followed by statistical analyses, of aggregation-deficient mutants of *Polysphondylium violaceum*.[477,555] Mapped by mitotic recombination techniques, genes *aggA*, *aggF*, and *aggI* fall into linkage group II of *D. discoideum*.[553] Gene *aggB* is found in linkage group I, gene *aggJ* in linkage group IV.

The sexual stage in the cellular slime molds occurs during macrocyst formation.[556] Aggregation precedes the sexual stage. Two mating types have been identified in *D. discoideum*: mating type A_1 (strain NC4) and mating type A_2 (strain V12).[557-559] In *P. violaceum* and *P. pallidum*, two syngens, each with two complementary mating types, are known. Four mating types are found in one syngen in *D. giganteum*. Homothallic strains include *D. mucoroides* and one strain (AC4) in *D. discoideum*.

Although there are 16 linkage groups in *Chlamydomonas reinhardtii*, only eight chromosomes have been cytologically identified.[185] The mating-type locus of this organism is mapped about 24 cM from the centromere on the left arm of linkage group VI (Figure 21). It controls the mating reaction between mating types *plus* and *minus*, which are assumed to be alternative states of the locus. Taxonomic relations among taxa may be derived by examining the compatibility relationships of the mating reaction. For example, the complementary mating types of *C. reinhardtii* mate only between themselves and not with the mating types of other species.[560] On the other hand, *C. moewusii eugametos* cross-reacts with *C. moewusii* syngen I, both of which are nonreactive with *C. moewusii* syngen II.

Mutant strains defective in gametic differentiation have been isolated from mating type *plus* of *C. reinhardtii*.[561] No flagellar agglutination reaction is observed in some of these mutants called *imp* (for impotent) when mixed with mating types *plus* or *minus*. The *imp* mutations appear to be unlinked to the mating-type locus. There are at least four loci involved in gametogenesis, which is a prerequisite to sexual agglutination.

Two complementary mating types are found in each of the 14 syngens of *Paramecium aurelia*.[562] They are referred to by Roman numerals or, in general, by the letters *O* and *E* (for odd and even). The mating type of one species may cross-react with that of another, but mating occurs only between mating type *O* and mating type *E*.

Three general patterns of mating-type inheritance are known in the ciliates.[186,187,454-456] In the simplest pattern, termed synclonal, the genotype directly determines the phenotype in that the clones from an exconjugant are of identical mating types. A recessive gene *mt*

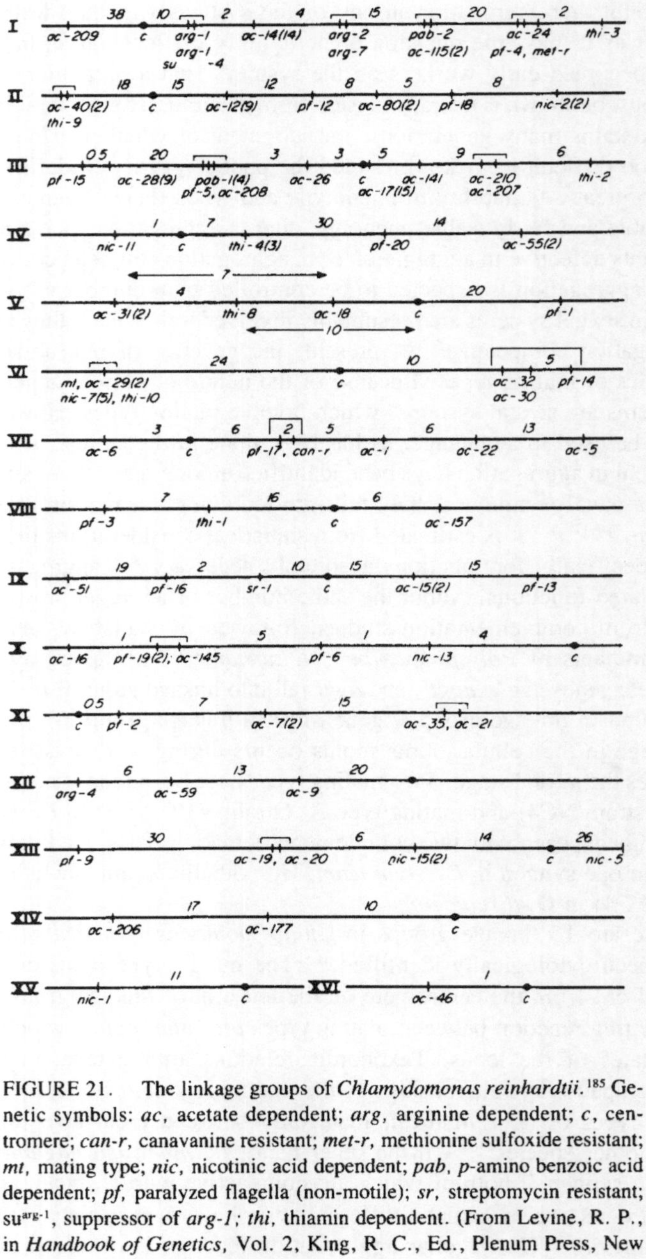

FIGURE 21. The linkage groups of *Chlamydomonas reinhardtii*.[185] Genetic symbols: *ac*, acetate dependent; *arg*, arginine dependent; *c*, centromere; *can-r*, canavanine resistant; *met-r*, methionine sulfoxide resistant; *mt*, mating type; *nic*, nicotinic acid dependent; *pab*, *p*-amino benzoic acid dependent; *pf*, paralyzed flagella (non-motile); *sr*, streptomycin resistant; su$^{arg\text{-}1}$, suppressor of *arg-1; thi*, thiamin dependent. (From Levine, R. P., in *Handbook of Genetics*, Vol. 2, King, R. C., Ed., Plenum Press, New York, 1974, Chap. 17. With permission.)

specifies the mating type *O* and *Mt*, its dominant allele, the mating type *E*. The heterozygote is mating type *E*, which at first may behave as mating type *O* for hours before finally becoming mating type *E*. The two other patterns of mating-type inheritance, termed caryonidal and clonal, are more complicated and therefore more interesting. In both these patterns, clones from exconjugants can be of two mating types. In the caryonidal pattern, members of a caryonide are of one mating type, but the original caryonides, derived from fission of an exconjugant, may differ in mating type. That an original caryonide is of one mating type or the other appears to be a matter of chance. Influenced environmentally, the probability of an original caryonide being mating type *O* decreases with a rise in temperature.

Nuclear differentiation is apparently the basis of both patterns. The clonal pattern, however, is a trifle different in that the seemingly random choice that is characteristic of the caryonidal pattern no longer appears to be random but appears to be controlled in the clonal pattern by a cytoplasmic agent. The clones from exconjugants have the phenotypes of the cytoplasmic parent. The mating type of a clone is determined by the cytoplasmic agent, which is presumed to be synthesized under the direction of the old somatic nucleus. This apparent instability of mating type is also found in yeasts.

The mating-type locus in the yeast *Saccharomyces cerevisiae* comprises two alleles, **α** and **a**, which specify the complementary mating types of haploid cells. In haploid cells, 19 chromosomes have been identified.[182] The mating-type locus is found about 20 cM from the centromere on the right arm of chromosome III.[562,563] It is complex, being probably made up of two cistrons corresponding to the mating types. The mating type **a** can mutate to the mating type **α**, and vice versa. Although no reciprocal recombination at the mating-type locus has been reported, gene conversion is known to occur with high frequency.[564]

In the homothallic strains of *Saccharomyces cerevisiae* and related species, haploid cells arisen from a single spore may mate to form a diploid whose genetic constitution is **a/α**. Homothallism genes are known to influence the homothallic character of a clone. Designated by various names, these nuclear genes are not linked to one another nor to the mating-type locus, except for *HM***α** and *HM***a**, both of which are also mapped on chromosome III, as is the **α/a** locus.[565-569] The interaction of the homothallism genes and the mating-type locus may involve the following relationship. *HO* (also known as D), a nonspecific dominant gene for homothallism, interacts with *HM***a**, a homothallism gene specific for mating type **a**, resulting somehow in the conversion of **a** to **α**. This conversion, as well as the reciprocal conversion mediated by *HM***α** from **α** to **a**, occurs frequently within a clone of a few cells in a few generations.[122,567,568] The consequence is a homothallic clone made up of heterothallic cells, which are either **a** or **α**. An analogous situation pertains in fission yeast.

Three basic mating types are known in *Schizosaccharomyces pombe*: the heterothallic h^+ and h^- and the homothallic h^{90}. Cells of a homothallic culture flocculate and conjugate among themselves. The complementary h^+ and h^-, both sterile in pure culture, mate with each other as well as with the homothallic h^{90}. There are several variations on this basic pattern.[64,183,515,571] For instance, h^{+N} and h^{+R} are subtypes of the h^+ mating type. The wild type is h^{+N}, whereas h^{+R} is a recombinant derived from a cross between h^{+N} and h^-, a cross which also yields the homothallic h^{90} as the other recombinant. The three basic mating types are also interconvertible by mutation. These interconversions are ascribable to the complexity of the mating-type locus, which consists of alternate genes very closely linked but nonetheless resolvable by recombination. The genes for mating types h^+ and h^- are located as adjacent or possibly overlapping sites on the right arm of chromosome II.[183,184] As in *Saccharomyces cerevisiae*, homothallism in *Schizosaccharomyces pombe* is the manifestation of the frequent interconversion of complementary mating types in a developing clone.[122,123] Thus is possible sex-directed flocculation, followed by conjugation, in homothallic cultures. Unlinked to the mating-type locus, genes called *mmo* (mating-type modifiers) have been identified to influence the expression of the mating type.[183,572]

The original complementary mating types of *Hansenula wingei* are strains 5 and 21. Other subsequent isolates hybridize with either of these strains. Tetrad analysis of crosses between 5 and 21 demonstrates the 2:2 segregation ratio expected of alleles.[515] The mating-type locus has not been mapped in this species, wherein sexual agglutination in yeast was first described.[62]

Mating in *Escherichia coli* K12 is also heterogamous.[538,573-575] It is consummated between interconvertible sex types. A conjugated pair consists of a donor cell called F^+ and a recipient cell F^-. The genes controlling the mating process are located on extrachromosomal elements called plasmids. The donor cell F^+ contains a plasmid, F factor, which is passed on to the recipient cell F^-, which heretofore contains none. In the process, F^- is converted to F^+.

F^- cells do not mate among themselves. $F^+ \times F^+$ matings do occur but are markedly less frequent and less efficient than matings between F^+ and F^-. The F factor may be lost spontaneously or by irradiation or chemical treatment, thus converting an F^+ to an F^-. Superfertile strains called Hfr (for high frequency recombination) are three orders of magnitude more efficient donors than F^+. However, the fertility factor of Hfr is normally not transferred to the recipient nor lost by chemical treatment. Hfr nevertheless contains the F factor because of its rare reversal to F^+ and the very rare conversion of the recipient to Hfr. The Hfr state is due to the integration of the F factor to a chromosomal site (Figure 22). The F factor is an example of an extrachromosomal element. It is the first and best known of the plasmids, which are DNA pieces that can replicate autonomously in the host cell. It is in fact an episome because, in its Hfr state, it replicates as an integral part of the host genome.

It is assumed that the mating-type locus in any of the genetic systems described above is somehow involved in the control of the aggregation process, which makes up an initial step of the mating reaction. Other genes, not specific for aggregation, and probably not part of the mating-type locus in sexual systems, are known to influence aggregation. For example, mutations in flagellar functions may influence sexual agglutination in the chlamydomonads. And again, mutations of nonnuclear genes that specify mitochondrial functions may likewise influence the aggregation of systems that require those functions.

Besides the genes for mating types in sexual systems and the aggregation genes identified in the cellular slime molds, there are the genes specific for flocculation of brewers' yeast. Several of these genes are known.[87,127,191] Referred to as *Flo,* they are numbered from 1 to 5. *Flo 3* is believed to be recessive, the rest are dominant. The best characterized of these is the dominant *FL04*, which is located 39 cM from the centromere on the right arm of chromosome I (Figure 23). Note that it is not linked with the mating-type locus, which is found on chromosome III. A recent report concludes that *FL01* and *FL02* are allelic to *FL04* and therefore proposes that all three be consolidated into a single locus henceforth to be known as *FL04*.[577] Earlier studies indicated that flocculation of brewers' yeast is determined by a multiple-gene system including modifier genes.[91,578,579] The complexity of the genetics of flocculation is illustrated by a modifier gene, *sufl*, which is a suppressor for *FL04*.[115] The suppressor gene is linked neither to *FL04* nor to the mating-type locus. It is recessive in some diploids and dominant in others.

In more ways than one, all cellular functions and structures have a genetic connection, even as all functions have a morphological basis and all structures have a physiological role. Genetics may be viewed as an organism's way of ensuring more physiology and physiology as an organism's way of ensuring more genetics. If genetics were narrowly viewed as the replication of the material gene, then it may be simply identified with mitosis and meiosis. In microorganisms, it seems that life is one continuous reproduction. To the physiologist, the mitotic or meiotic events are mere markers in a physiological continuum. Moreover, the genes are the center of cellular activity. The selfish genes are not there simply to make copies of themselves. They are there to direct traffic, too, and to bring order to a seething biochemical microcosm. If genetics is an attempt to decipher the mechanics of heredity, of passing on traits from one generation to the next, then it must involve a great deal of physiology and biochemistry.

The genetics of microbial aggregation has still a long way to go. The initial strategy is to identify mutations that specifically affect a function that is aggregation. Later, a correlation of these mutations with altered gene products must be established. The final objective of a genetic dissection of microbial aggregation is the explanation of the phenomenon in terms of the molecular biology of regulation and development.

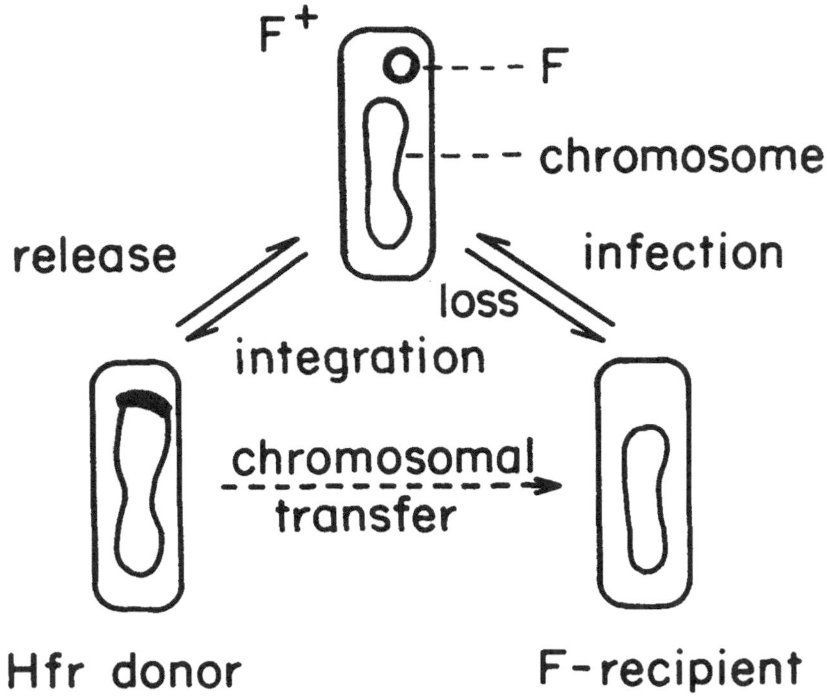

FIGURE 22. Interconversion of sex types in *Escherichia coli* K12. Adapted from Sermonti.[573]

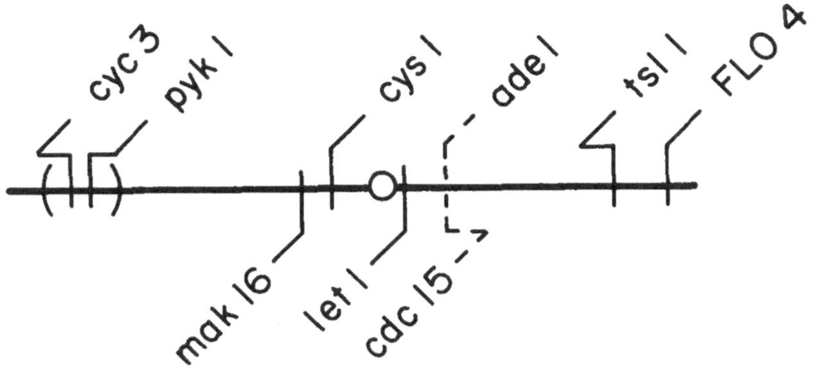

FIGURE 23. Chromosome I of *Saccharomyces cerevisiae*.[182] Genetic symbols: *cyc*, cytochrome c deficiency; *pyk*, pyruvate kinase deficiency; *mak*, maintenance of killer plasmid; *cys*, nutritional requirement for cysteine; *let*, recessive lethal; *ade*, nutritional requirement for adenine; *cdc*, cell cycle arrest at nuclear division; *tsl*, temperature-sensitive lethal; and *FLO*, dominant flocculence. Information concerning *cyc3* was obtained by personal communication from F. Sherman, *mak16* from R. B. Wickner, and *tsl1* from D. B. Kaback.

XIII. MOLECULAR BIOLOGY OF REGULATION AND DEVELOPMENT

Like many a trendy phrase, molecular biology is often defined so as to unequivocally include the interests and the aspirations of the one proposing the definition. It will not be defined here but will be used in such a way as to include the activities and the aspirations of workers on microbial aggregation. Undefined, the term implies the characterization of the gene and the gene product in terms of molecules and the conscious identification of their relationships in terms of molecular reactions.

The complete description of the phenomenon of microbial aggregation must involve morphology, physiology, genetics, biochemistry, and, ultimately, molecular biology. Perhaps, a bit more: because the phenomenon concerns whole cells and the association of these cells, the final approach will have to be synthetic and holistic. Each of the various approaches has been discussed thus far in a fragmentary manner. They must be put together now to constitute a synthesis of cellular and social biology.

Two levels of regulation are found in microbial aggregation: the level of the individual cell and the level of the community of cells. Both connote communication: information transfer within the cell and its immediate environment and exchange of information among cells. It is as if a cellular government that is represented by the genome comprises two super ministries: a ministry of interior and a ministry of external affairs. Or again, the two levels may be viewed as individual and social development.

All biological phenomena must be reducible to structures and functions, to materials and mechanisms, to molecules and reactions. All cellular structures and functions will eventually fall to the onslaught of molecular biology. Hence, there is no need to show that microbial aggregation has a molecular biology in its future. The question is whether a part, however small, of that future is now arrived.

Because it is inducible, microbial aggregation obviously is regulated. It is regulated in time and space. Genes are switched on and off at specific times and expressed as specific components of the cell surface by members of the community of cells. The material bases and the mechanisms of regulation are found in molecules that mediate the transfers of information. They include structural genes for aggregation, regulatory genes, suppressors, repressors, messengers, the processing of messages, protein synthesis, transport of gene products, assembly of aggregation structures, intercellular communication, timing, environmental control, etc. — in short, the developmental program.

The realization of the developmental program is determined by the genetic constitution of the organism, the physiological state of the cell, the environmental situation, and experimental manipulation. All these must come into play for the phenomenon to be observed. The information for aggregation competence is contained in the genetic program, the relevant sequence of nucleotides in the DNA supermolecule. This information needs to be sorted out from other bits of information. The individual genes, structural and regulatory, must be activated and deactivated at particular times in the life cycle of the organism and in the cell cycle of the individual cell. Genes not functional prior to induction will have to be turned on and genes not needed for the developmental program, for the sake of efficiency and order, will have to be turned off. The turning on and the turning off are mediated by repressors, which are products of regulatory genes, and by inducers, which are small-molecule metabolites or exogenous components of the medium. The information encoded in the newly activated genes needs to be processed. The DNA segments specific for the program are transcribed into messenger species of RNA which, in turn, are translated by the protein-synthesizing machinery, which comprises ribosomal complexes of protein and RNA, into gene products that are the proteins. The gene products that will become aggregation structures are transported to the cell surface where they will be installed by those gene products that are enzymes, which may or may not be specific for aggregation or development. Only then will a cell become competent for aggregation.

While all these processes are going on in the individual cell, there must be information transfer to and from the outside, information passed on from cell to cell concerning the situation outside, the readiness of the other cells, or the general condition of the community.

There may be messages for synchrony, as the α factor in *Saccharomyces cerevisiae*, or timing signals for chemotaxis, as the pulses of cyclic AMP in cellular slime molds. These communication substances (or telemolecules, if you may) are somehow excreted into the medium, where they are picked up by target cells. Aggregation is a social phenomenon and therefore cells must be informed members of the community. They must be able to send signals and receive them. The gathering together to make an aggregative structure is not just an individual effort but the concerted effort of the community of cells.

The molecular-biological approach will find ready access through the aggregation genes and the complementary molecular structures of aggregation. The interconversion of mating types is one particular area that appears ripe for dissection with the incisive tools of molecular biology. Provided the mating-type locus is central to sexual agglutination or sex-directed flocculation, it will likely lead to a better understanding of the aggregation phenomenon. The cassette model of Hicks et al.[570] and the flip-flop model of Egel[123] on the regulation of the mating-type locus are worthy of mention but will not be discussed here yet. Instead, they will be treated in a subsequent chapter on yeasts. For the meantime, what may be pointed out is the exciting prospect of finding some novel ways by which organisms turn on and off the signals for sex and development.

Notice that the molecular biology of bacterial conjugation mediated by the plasmid F came to the fore long before a general awareness of a unified subject called microbial aggregation, which now includes mating-aggregate formation during bacterial conjugation. Plasmids and extrachromosomal elements rank very high among the subjects that are central to molecular biology, and microbial aggregation, being an associated event to bacterial conjugation, easily makes the connection. Indeed, because of the connection of aggregation to sex and development in general, advances in these fields will have a most valuable impact on the study of microbial aggregation. Conversely, the study of microbial aggregation will undoubtedly influence the advances in these fields and others.

The complexity of the many microbial aggregation systems is admitted. Yet they are not exactly inscrutable. As varied approaches are brought to bear on them, the systems, though not made any simpler, appear to yield to experimental dissection and now appear decipherable. The advances of developmental biology will be made in many systems. As the recent flurry of activity attests, microbial aggregation systems will surely contribute to these advances. Already, the study of microbial aggregation is approaching a most exciting period. There are now new systems awaiting a concentrated attack toward elucidation. The older systems have yielded encouraging results concerning ultrastructure of contact, isolation of aggregation factors, mapping of aggregation genes, identification of chemotactic signals, and many more. The many disciplines that seek the elucidation of the aggregation phenomenon will eventually converge to a molecular biology of regulation and development, to the grand design of reducing biology to molecules and reactions, to the final reduction of development to structures and functions.

When all the genes for aggregation have been identified, mapped, and characterized, there will be surprises yet. Are the genes in clusters, do they form functional units? What ultimately turns on the developmental program? If structural genes are turned on by regulatory genes, what turns on the regulatory genes? There may still be surprises when the gene products have all been identified and chemically analyzed, when the intercellular communication system has been unraveled, when messages have been identified and deciphered. How are the receptors assembled? In the case of the cellular slime molds, what makes a center cell start a signal? Even more pressing, how does a center cell become one in the first place? When all the conditions for physiological competence and all the environmental conditions have been specified, there will be surprises yet. In yeasts and paramecia, what drives the interconversion of mating types? There are many intriguing mysteries that will have to be explained in molecular terms. When all the molecular components and their relationships have been identified, then we shall arrive at a fairly comprehensive understanding of the developmental program. Is there a better way of uniting intracellular regulation and intercellular control than through molecular biology?

The strategy of science is looking for opportunities. An area advances by improving on and expanding the findings made in others. In that way, it advances not only toward a particular direction but laterally as well, in the process enlarging and solidifying its foundations. The advances in the study of microbial aggregation are characterized that way. Thus, it is as much a matter of bringing molecular biology to microbial aggregation as a matter of bringing microbial aggregation to molecular biology. The entry of microbial aggregation into an area of the most sophisticated activity is not merely to update itself. It is hoped that microbial aggregation has something to contribute to the molecular biology of the internal affairs of the cell and — perhaps, more significantly — of cell-cell interaction and multicellular development — not just to the description of the cell surface and its activities, but to the understanding of social organization.

XIV. THEORETICAL CONSIDERATIONS AND COMPUTER SIMULATION

For the most part, approaches discussed thus far have been experimental. There is a body of theoretical work that now needs to be considered. Despite the limited application of colloidal theory to problems of microbial aggregation as a biological phenomenon, we shall borrow from colloid science some concepts relevant to the understanding of some microbial aggregation systems, in particular, systems that are in aqueous suspension.

The aggregation of particles may be viewed as the result of instability of a colloidal suspension.[580-584] Particles in an aqueous suspension bear a net electric charge, the sign of which is either positive or negative. For most cells under physiological conditions, it is negative. The sign and the magnitude of this charge are influenced by the ionic strength and the pH of the medium. The suspension itself is neutral. The net charge on the particle is counter-balanced by counter-ions. The result is an electric double layer at the interface between the particle and the aqueous phase. The double layer is made up of the charged particle and the counter-ions attracted electrostatically to it. The presence of the counter-ions brings about a concentration gradient stretching to some distance from the particle surface, where the ion concentration is highest (Figure 24a). A diffuse layer is established from the electrostatic attraction and the opposing process of counter-ion diffusion due to thermal motion. At a high ionic strength of the aqueous phase, the diffuse layer is compressed.

An electrostatic potential exists at the interface because of the net charge on the particle (Figure 24b). Greatest at the surface and decreasing with distance, it drops to zero at a shorter distance from the interface at a higher ionic strength. Of the same sign, two particles approaching each other are repelled by a repulsive force, V_r, which increases as the distance between the particles decreases (Figure 24c). V_r decreases with an increase in ion concentration (Figure 24d). It is counter-balanced by van der Waals forces, attractive forces between two particles of any kind. These attractive forces make up the attractive potential energy, V_a, which increases as the distance between two particles decreases (Figure 24c). In general, V_a is independent of ion concentration. Whether the interaction between two particles is in effect an attraction or a repulsion is determined by the net interaction energy, $V_r - V_a$, which is simply the difference between the two opposing forces. At a low ionic strength, there is a net repulsion, which may be viewed as an activation barrier that must be hurdled before aggregation can take place. At a high ionic strength, the interaction is a net attraction, as the repulsive force is diminished while the attractive force remains unaffected (Figure 24d).

The observed ion effect in various microbial aggregation systems can be partially explained by the compression of the double layer, resulting in the destabilization of the cells in a

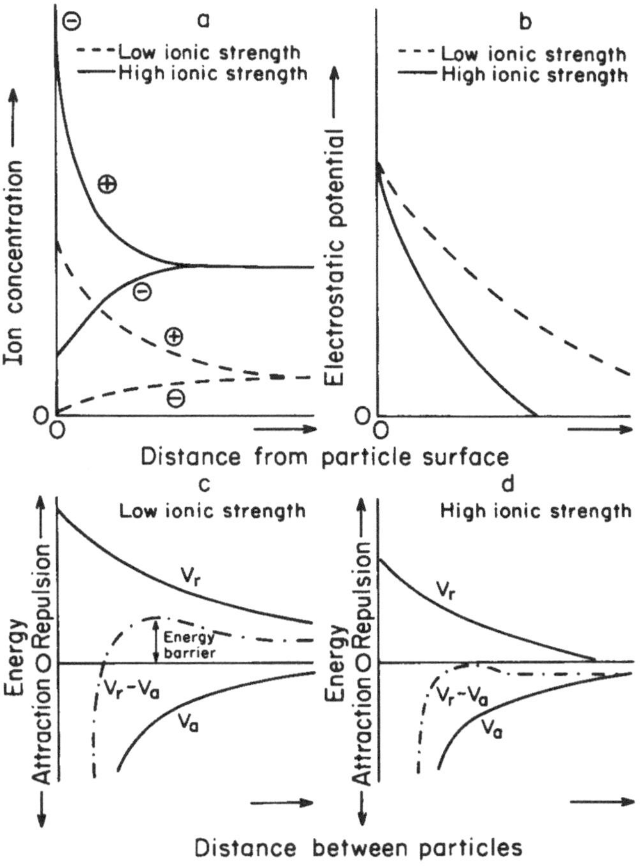

FIGURE 24. Diagrammatic representation of the electric double layer (a), the electrostatic potential (b), and the energy of interaction between two particles (c and d). Adapted from O'Melia.[580]

suspension, and allowing more fruitful collision. If the interaction in an aggregation system were purely electrostatic, it would be expected to follow the Schulze-Hardy rule: aggregation should be ion dependent and the effectivity of ions should increase with valency (Figure 25). Trivalent ions should be more effective than divalent ions, divalent ions more than monovalent ions. In fission yeast, however, this has not been found to be so.[99]

Aggregation may be explained by other mechanisms: adsorption followed by a neutralization of the charges, enmeshment in a precipitate of a coagulant, and interparticle bridging by polymers.[580,582] The last mechanism appears pertinent to microbial aggregation (Figure 26). One end of a polymer molecule adsorbs onto a particle. The rest of the polymer is free to adsorb onto another. The consequence of this two-step interaction is a particle-polymer-particle structure. The mechanism explains adequately many a polymer-provoked aggregation, but in its simplest version, it is hardly applicable to microbial aggregation systems. The theory when slightly modified appears applicable to interactions that are mediated by such filamentous structures as flagella, cilia, and fimbriae. The base of the filamentous structure is anchored covalently to the surface of a cell while the other end is free to adsorb specifically onto the surface of another cell or to interact with another filamentous structure. The mating reactions in unicellular algae and in the ciliates are obviously mediated by flagella and cilia, respectively.

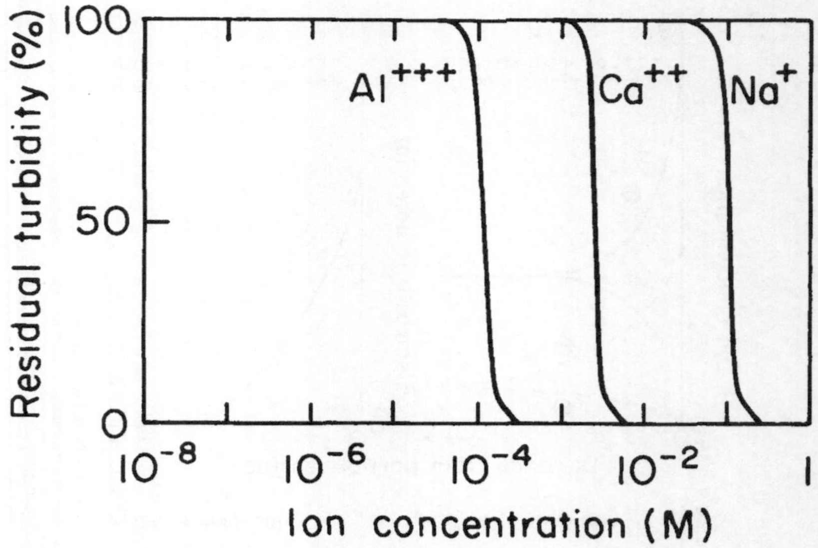

FIGURE 25. Schematic illustration of the effect of ion concentration and valency on flocculation.

The aggregation process, as already pointed out, is made up of two phases: movement and stable contact. The stochastic collision of two particles may be achieved in several ways. When collision is due to thermal (Brownian) motion, it may be called perikinesis (after perikinetic flocculation in the literature), and when collision is due to laminar and turbulent flow (as in mechanical agitation), it may be called orthokinesis (after orthokinetic flocculation).[582,584] Collision can be the consequence of other forms of motion, such as when particles of varying sizes overtake or intercept one another during gravity sedimentation. Other particle movements may come after aggregation. An example is syneresis, or the rearrangement of particles in an aggregate to a more stable interaction and to a more dense and compact aggregate.[582]

Originally formulated by von Smoluchowski for rapid coagulation, the kinetics of aggregation due to perikinesis may be described by the following equation:[580]

$$J_p = \frac{dN}{dt} = \frac{-4\alpha KTN^2}{3\mu} \tag{1}$$

where J_p is the rate of change with time in the total number of particles, N is the total number of particles at time t, α is the collision efficiency equivalent to the ratio of the number of fruitful collisions to the total number of collisions, K is Boltzmann's constant, T is absolute temperature, and μ is the fluid viscosity.

It is clear that aggregation depends on temperature, the initial particle concentration, and the collision efficiency. Collision is dependent upon thermal motion and the attractive interaction between particles. The repulsive component may be ignored when the system contains electrolytes in sufficient concentration. Taking the boundary condition that $N = N_o$, the particle count at $t = 0$, integration of equation 1 yields the equation:

$$N = \frac{N_o}{1 + (4\alpha KTN_o/3\mu)t} \tag{2}$$

Let the time in which the number of particles is halved,

$$t_{1/2} = 3\mu/4\alpha KTN_o \tag{3}$$

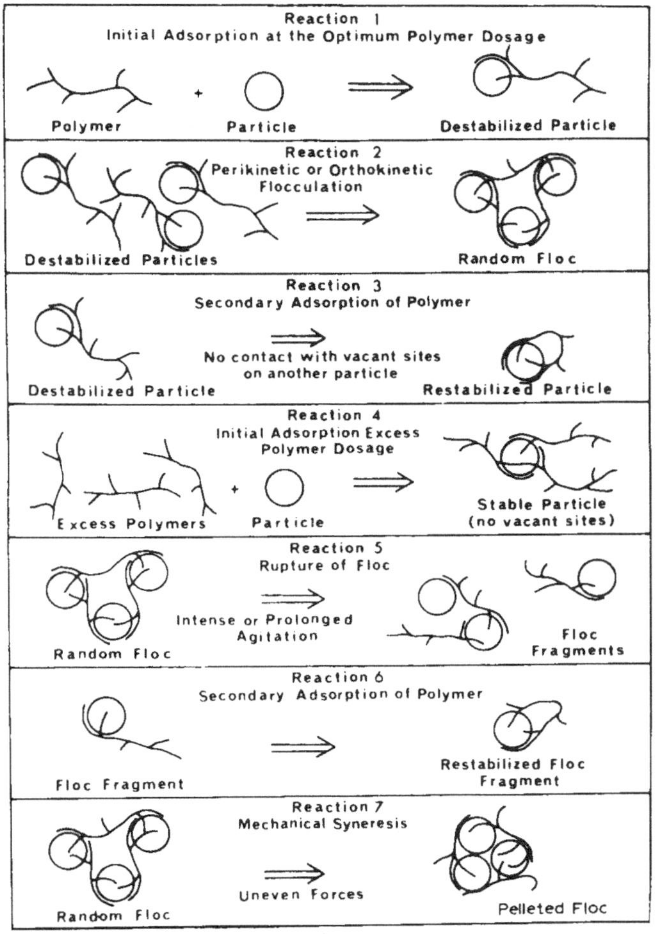

FIGURE 26. Interparticle bridging by polymers. Based on O'Melia's original illustration,[580] the schema is shown as modified by Yusa. (From Yusa, M., *Int. J. Miner. Process.*, 4, 293, 1977. With permission.)

Then

$$N = \frac{N_o}{1 + (t/t_{1/2})} \tag{4}$$

In water at room temperature,

$$t_{1/2} = \frac{1.6 \times 10^{11}}{\alpha N_o} \tag{5}$$

where $t_{1/2}$ is expressed in seconds and N_o in particles $m\ell^{-1}$. Being a ratio, α is without dimensions. It will take no longer than 3 min for half of a bacterial suspension of 10^9 cells $m\ell^{-1}$ to fall into the aggregated fraction, provided collision efficiency approaches unity. If the initial cell concentration is reduced to 10^6 cells $m\ell^{-1}$, $t_{1/2}$ increases to about 2 days. The time-course of aggregation due to perikinesis is graphically shown as the decay of N in Figure 27.

Also shown (Figure 27) are the particles of different sizes as they appear and disappear in the particle population. The number of these particles as a function of time may be expressed thus:

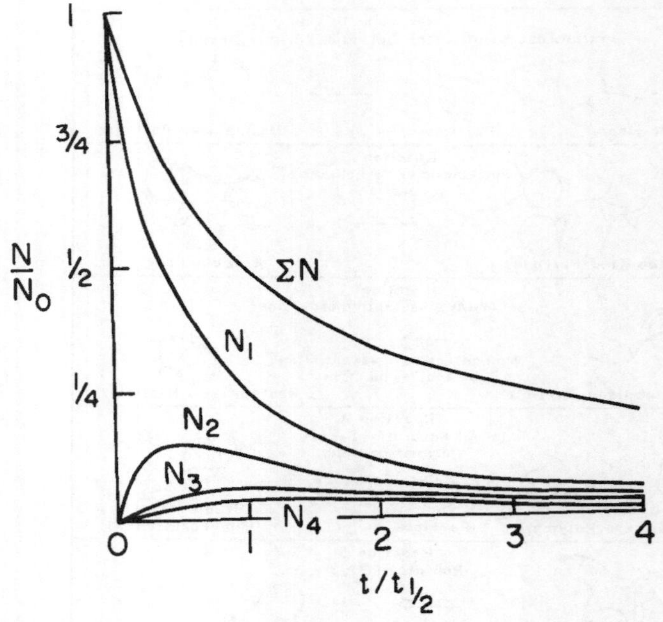

FIGURE 27. The kinetics of aggregation due to perikinesis. ΣN corresponds to all kinds of particles. The subscripts to N refer to aggregation numbers.

$$N_1 = \frac{N_o}{[1 + (t/t_{1/2})]^2} \tag{6}$$

and

$$N_2 = \frac{N_o(t/t_{1/2})}{[1 + (t/t_{1/2})]^3} \tag{7}$$

where N_1 represents the number of isolated primary particles and N_2 the number of secondary particles consisting of two primary particles in stable contact. In general, where k is the aggregation number,

$$N_k = \frac{N_o(t/t_{1/2})^{k-1}}{(1 + t/t_{1/2})^{k+1}} \tag{8}$$

The difference, as originally enunciated by von Smoluchowski, between slow and rapid flocculation is due to α, the collision efficiency factor. In rapid flocculation, α approaches 1. It is markedly less than 1 in slow flocculation. This means that only a fraction of the total number of collisions are fruitful collisions that result in fairly stable contact. In microbial aggregation systems, not all collisions are expected to be fruitful. Moreover, there are processes that bring about disruption of the growing aggregate. In other words, aggregation is complicated by processes of disaggregation.

Orthokinetic flocculation, caused by mechanical agitation, is described by the following equation:[580]

$$J_o = \frac{dN}{dt} = \frac{-2\alpha G D^3 N^2}{3} \tag{9}$$

where J_o is the rate of change with time in the total number of particles, G is the mean velocity gradient due to bulk fluid flow (agitation), and D is the particle diameter. Independent of temperature, aggregation due to orthokinesis is dependent on collision efficiency, agitation, particle size, and the initial concentration of particles.

Equation 9 divided by Equation 1 yields a ratio which is helpful in determining whether an aggregation system is enhanceable by agitation:

$$\frac{J_o}{J_p} = \frac{-2\alpha GD^3N^2/3}{-4\alpha KTN^2/3\mu} = \frac{\mu GD^3}{2KT} \qquad (10)$$

For agitation to help an aggregation system at all, the components must be larger than 1 μm. Components of microbial aggregation systems are of greater dimensions and hence the positive effect of agitation.

The discussion above on some aspects of colloidal theory does help in elucidating the effects of temperature, agitation, and ion concentration on microbial aggregation. However, colloidal theory is only of limited usefulness because microbial aggregation systems are characterized by exquisiteness and specificity, by a higher degree of order. They are more complicated than nonliving colloidal systems because the particles involved are larger, because of chemotaxis, because of special recognition factors on the cell surface, and because some of them are heterotypic. Aside from collision efficiencies being modified by recognition factors, cell movement may not be stochastic but directed. Furthermore, there are morphological constraints, notable of which are the shape of the cells, the distribution of effective contact areas, and the presence of surface appendages. There are other complicating factors, such as the modulation of the cell surface when stable contact is achieved. Then there is the occurrence of associated events, such as competence development, conjugation, or conjugation-induced lysis.

The movement of particles and their consequent structures have been studied by model-making and computer simulation. This may be called the paper-and-pencil-and-computer approach. It consists of defining a set of rules that are consistent with observations and presenting this set of rules in a form amenable to computer analysis. The output is a display of the morphological features of the resultant structures.

One such simulation of a simple model for floc formation in a colloidal suspension will be described presently.[585] A model floc is assumed to grow by the successive, completely random addition of primary particles, which are spheres of a unit radius, to a growing floc. The first primary particle is situated at the origin. The initial location of an incoming particle is specified by the spherical coordinates θ and ϕ, which are pseudo-random numbers between 0 and 2π and 0 and π. The radius vector R of this incoming particle is set 2 units larger than the maximum distance between the origin and the center of a particle in the floc. Somewhere within the sphere (of radius R) encompassing the growing floc, a point is specified by three additional pseudo-random numbers serving as Cartesian coordinates. From its initial location, the incoming particle is now moved along a straight line toward the point specified by the Cartesian coordinates until it comes in contact with a particle in the floc or until it is past the floc. A contact is assumed to be stable and subject to no rearrangement.

The interesting consequence of this set of rules is shown as a two-dimensional projection of the floc structure (Figure 28). A typical floc consists of a core, where the particle population density is greatest, and tentacle-like projections. Its resemblance to a minifloc of yeast is striking. Thus, although the original computer simulation was designed for application to silica sols in organic solvents, the model appears to approximately account for the initial dynamics and morphology of microbial aggregation in an aqueous suspension. Presumably, the pellet-like appearance of aggregates found in old flocculated cultures of *Schizosaccharomyces pombe*, for instance, is due to the smoothening of the initial flocs by syneresis, component rearrangement, and disaggregative processes caused by mechanical shear.

A growing corpus of computer simulation studies has been generated for biological systems.[586-590] Noteworthy is a random-walk model designed to explain the sorting out of cells

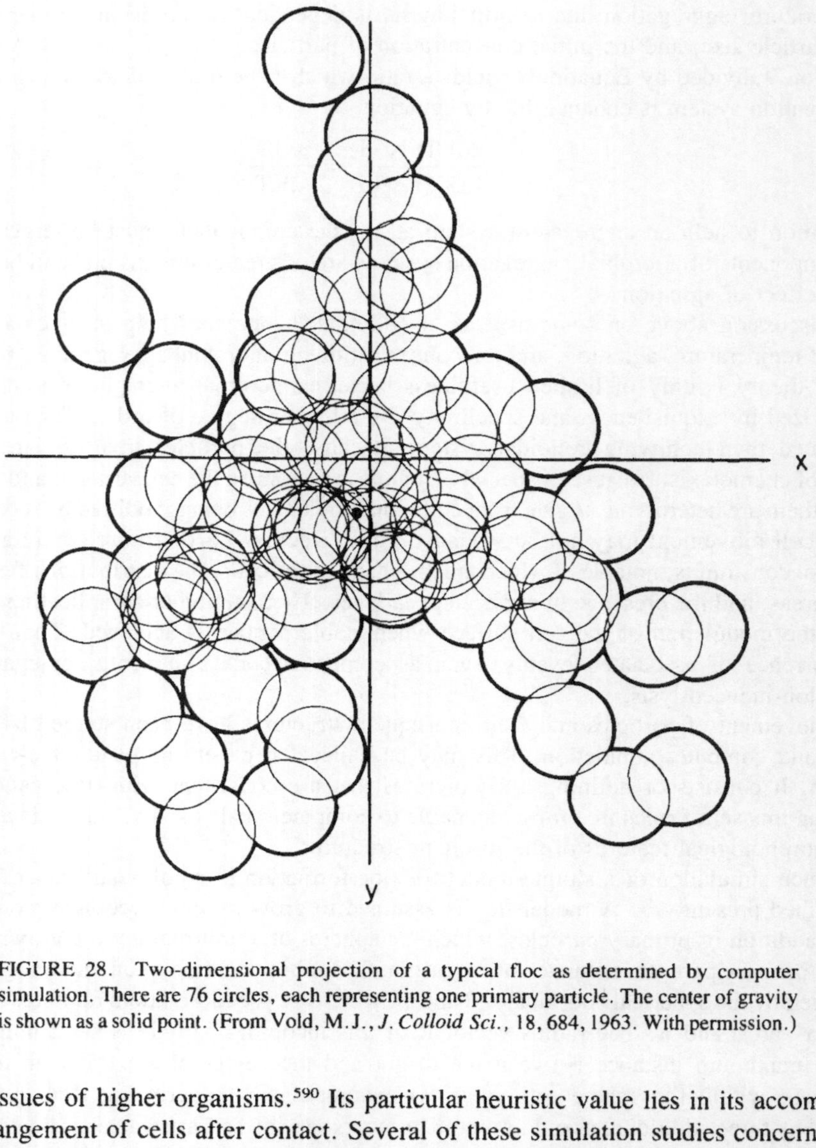

FIGURE 28. Two-dimensional projection of a typical floc as determined by computer simulation. There are 76 circles, each representing one primary particle. The center of gravity is shown as a solid point. (From Vold, M.J., *J. Colloid Sci.*, 18, 684, 1963. With permission.)

in the tissues of higher organisms.[590] Its particular heuristic value lies in its accomodation of rearrangement of cells after contact. Several of these simulation studies concern models for the aggregation of cellular slime molds.[591-599] One such model by MacKay[599] is briefly discussed below.

A population of amoebae (up to 1600) is randomly distributed in two dimensions within a circle contained in a drop (up to 1 mm in diameter). All amoebae are assumed to be capable of responding to signals of cyclic AMP. One or more amoebae are initially selected to pulse a signal. The others respond to the pulse by either chemotaxis or signal relay. After a response, an amoeba becomes refractory for awhile (3 to 10 min for relay, 100 sec for chemotaxis). All amoebae follow the same algorithm of behavior. The state of an amoeba at any given time is specified by its x,y position and time t, the time it received the latest signal. If t is greater than the refractory period, an amoeba is ready to respond to another signal. The response depends on the local concentration of cyclic AMP. If the concentration is above the threshold for chemotaxis (10^{-9} *M*), the amoeba moves a specified distance for a specified duration (20 μm in 100 sec) toward the direction of the maximum in the concentration gradient. If this concentration is above the relay threshold (8×10^{-9} *M*), the amoeba becomes refractory for awhile (15 sec) before it releases its own pulse and makes a chemotactic step. Aside from chemotaxis, a random motion (5 $\mu m\ min^{-1}$) is also taken into account. The x,y,t coordinates of the amoebae are arranged in a firing queue. As an amoeba fires, it takes its place in the queue and the queue grows as the simulation proceeds. Because of diffusion and of degradation of the signal by the enzyme phosphodiesterase, the effective strength of a signal in time becomes progressively negligible. After some time, amoebae in front of the queue are removed.

The model further assumes certain values for such parameters as the number of molecules released per amoeba (10^7), the diffusion constant of cyclic AMP ($9.7 \times 10^{-6}\ cm^2\ sec^{-1}$), and phosphodiesterase activity ($K_m/K_{max} = 1 - 10$ sec). These and the other parameters, such as the concentration thresholds for relay and for chemotaxis, the delay time for relay, the refractory periods for relay and for chemotaxis, duration of each chemotactic step, etc., are derived from the literature and are varied. The above values in parentheses are typical. The computer output is in the form of two-dimensional graphic animation comparable to time-lapse photography. A sample of the graphic output is shown in Figure 29. Based on the behavior of individual cells, the model has proved to be successful in simulating phenomena observed in cellular slime molds: pulsatile waves of migration, branching streams, entrainment of centers, and spiral patterns.

Computer simulation is not only useful in verifying the validity of the assumptions in a model, but also in guiding experimental approaches. In a way, it is a thought experiment. The experimenter sets the rules, the computer as an obedient technician performs the more tedious portions of the thinking process. The experiment lies in the manipulation of the parameters. One advantage of a computer experiment over an experiment in a test tube, aside from its rapidity, is that parameters can be varied at will, precisely, without the investigator having to contend with the imprecision of pipettes. In the making of a model, one simplifies, hoping to find what parameters count in explaining a phenomenon. Modeling and simulation are one more synthetic and analytical tool in the study of microbial aggregation systems.

XV. COMPARATIVE APPROACHES

Still another approach is comparative. The objective is to universalize findings that derive from rather circumscribed circumstances. The comparative approach may be made consciously or unconsciously. However it is made, in the end, there is a synthesis, a simplification, a generalization, perhaps, a new way of looking at things.

Of course, the comparative approach, when not at the forefront, is always at the back of the mind of the investigator. Comparisons are made routinely inside and outside the laboratory. The dialectical process is in essence the comparison of thesis and antithesis. The inductive method calls for a jump from the particular to the universal. A requisite to the jump is the opportunity to compare particulars. A comparison implies the presence of more than one item. The concepts of a standard, of an average, of a control, of an exception, all derive from the comparison of at least two items. Experiments are repeated with the end in view not only of confirming an observation but also of comparing results from experiment to experiment and of arriving at what is normal or the average. After all, one swallow does not a summer or a meal make. Comparisons are made by varying conditions and situations, a comparable situation serving as a control. Quantitative procedures are devised with the hope of comparing results with greater precision and objectivity. Tables and graphs are devices for rapid comparison of items. The detection of a well-defined mutant is an act of instant comparison. A mutation is a deviation from the norm, with which it must be compared.

Aside from all these and more, there are approaches that are consciously comparative. They include the comparison of systems and the comparison of isolates and strains within

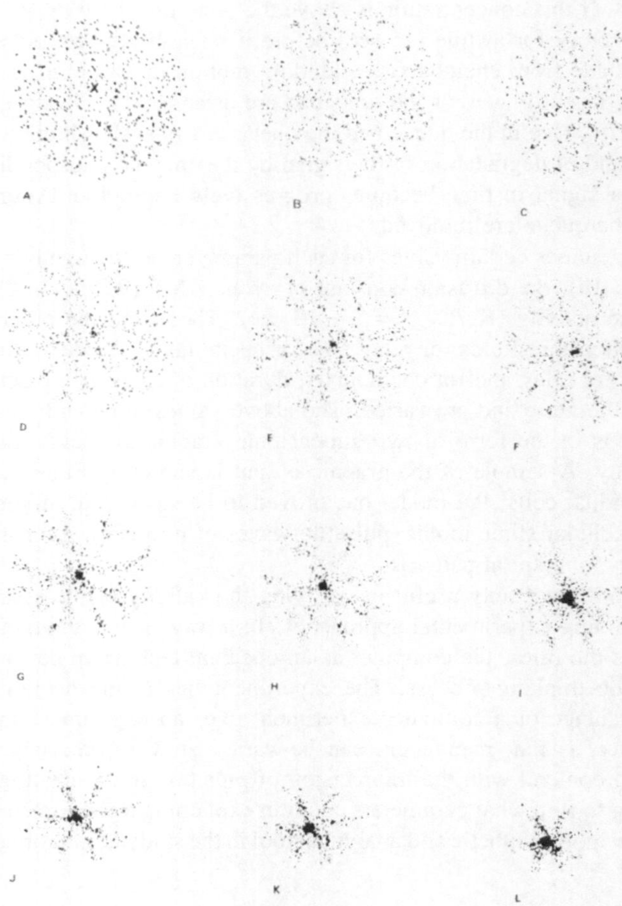

FIGURE 29. Computer simulation of aggregation in the cellular slime molds.[599] Graphic output from a simulation with one center, marked with an X in frame A. Each dot represents an amoeba. Each frame corresponds to 20 sec. Frames A to L are 0, 20, 40, 60, 80, 100, 150, 200, 250, 300, 350, and 400 sec, respectively. (From Mackay, S. A., *J. Cell Sci.*, 33, 1, 1978. With permission.)

a system. Specifically now, there is the approach to determine the cross-reactions of related species with regard to their aggregative capacities. The cross-reaction may be observed at the level of organisms, living or dead: the complementary reaction between mating types, for example. Comparisons may be at the level of aggregation structures: the morphology and the biochemistry of contact. The patterns of cross-reactions among the various species of *Chlamydomonas* and among the syngens of *Paramecium aurelia* have already been mentioned in a previous section.

There is some advantage to working with only one system, one strain: it is less complicated and more convenient. Narrow it may be, it is a concentrated effort, free from the distractions of having to compare strains. However, a comparison of this one strain with others is hard to avoid. First, an obviously heterotypic system is made up of two strains, not one. Second, a genetic analysis is hardly possible without the use of mutants. Besides, one would like to know whether what is true of one strain is also true of others. One would like to know if he is working with an exception or the rule.

There are other reasons for a comparative study. Isolates may be compared as to their aggregative capacities in order to determine the range of the phenotype, to find out which among them are inducible, which among them are nonaggregative. Strains in the laboratory and strains freshly isolated from nature are compared with the object in mind of ascertaining the validity of the experiments as demonstrations of what actually occurs in nature and, further, of assessing the distribution and the ecological significance of the aggregation phenomenon. Then there is the utility of the aggregation character as a taxonomic marker. Some of the developmental structures in filamentous fungi are the consequence of aggregation. They serve as morphological indicators useful in the identification and the systematics of fungi. Classifiers of yeasts use the aggregation character as a supplemental criterion for determining whether an isolate belongs to a particular taxon.[600] The mating-type character shows the way not only to the genetics of an organism but also, in a limited way, to its systematics. The various species of the yeast genus *Hansenula* may be grouped into those that are homothallic and those that are heterothallic and thus may be derived a phylogeny of the genus (Figure 30). The mating type is used as a taxonomic character in defining syngens and species of ciliates. Also, it must be recalled, the species concept is based on the ability of members of a species to hybridize exclusively among themselves. It just happens that hybridization in a number of systems is preceded and facilitated by aggregation.

Beyond strains, species, and genera, comparisons of microbial aggregation systems may be made at even higher levels of the taxonomic hierarchy. Cross-reactions, perhaps, are not expected at these levels. Instead, one may expect in a comparative inventory to discern some common morphological and physiological features, some common biochemical and genetic patterns. At a still higher level of comparative focus may be seen the common thread that runs through all the diverse forms of microbial aggregation. Such an eclectic overview may be attempted theoretically as well as experimentally, but the most profitable approach, it seems at the moment, is through scholarship, the schoolmen's trade. A comprehensive synthesis is the function of a review, of a monograph, of this book. The objective of a comparative approach is to see unity despite seemingly incongruous diversity, to discern the universal from among the particulars, a veritable process of conceptual aggregation. The success and failure of the present enterprise will be measured by the extent to which the field of microbial aggregation has become unified, how many common items have been identified, how many particular observations have been correctly reduced to general concepts, and how well, or how badly, the synthesis has been made. Finally, there is the attempt to see in microbial aggregation a model for aggregation in higher organisms.

The search continues. Having found the average and having made the generalization, one starts looking for the exception, the deviation, the variation, all the while hoping to discover a new phenomenon. Once the exceptions begin to accumulate, it is time for another round of generalization.

XVI. TELEOLOGICAL CONSIDERATIONS

When all the facts are in (an end hopelessly unattainable), there is yet one other aspect of microbial aggregation that will remain to reckon with. One may ask the final question: of what use is microbial aggregation? It is a most difficult question. As is, it is not quite a scientific question, for scientific questions lead to answers that are verifiable or falsifiable by current methodology. Rather, it is speculative, even philosophical in that it is important enough but does not yield definitive answers. Some parts of the question, when dissected, are answerable scientifically, other parts are not. Before we attempt to dissect the question, let us first look at the general nature of teleological questions.

Teleology has had a bad reputation, thanks to its unnecessary connection with mysticism and with theistic apologetics. Thanks, too, to the success of reductionism. Most materialists

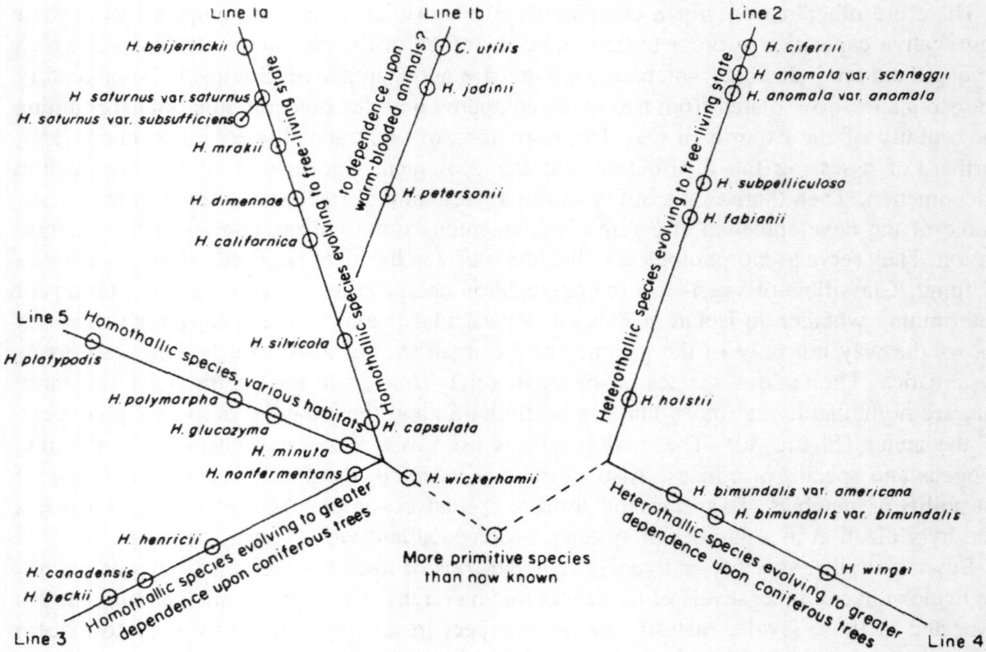

FIGURE 30. A phylogenetic tree proposed by Wickerham for the yeast genus *Hansenula* H. et P. Sydow. (From Wickerham, L. J., in *The Yeasts: A Taxonomic Study*, 2nd ed., Lodder, J., Ed., North-Holland, Amsterdam, 1970, 226. With permission.)

confuse teleology with theology and superstition. However, it need not be doomed to such a mean fate. Teleology looks for the final cause. Reductionism when reduced to its most extreme position boils down to looking for the first cause. Neither the first nor the final may be knowable, and looking for one is no less quixotic than looking for the other.

Reductionism interprets higher levels of organization in terms of lower levels.[602] Teleology reverses the interpretation: from lower to higher levels. Reductionism attempts to unify things to a point, to an origin. Teleology connects them to a grand design, to an end. Reductionism looks for a direction from which all things originate, teleology in which all things tend to become. By reduction, things are simplified so they are comprehensible. Teleology enlarges things so they they become comprehensive. Reductionism looks inward for the more simple, the more primitive, the irreducibly small: small is beautiful. Teleology looks outward for the more developed, the higher level, the incomprehensively large. To the teleologist, the infinitely large is beautiful.

These are the diametric (to some, ridiculous) extremes. Somewhere between is a well-traveled middle ground that is more reasonable (to some, trivial). Just as reductionism need not worry with the irreducibly primary, teleology need not concern itself with the very far but with the nearer ends, not with ultimate causes, but with the not-so-final. The explanation need not be complete or absolute nor brought to its logical (or illogical) extreme. Failures in complete reduction have been truly useful, so have been failures in total teleology. Either approach is valid. When one approach falters, the other may be of help. Cosmology and the search for elementary particles do profit from each other, as do the question of the origins of life and the study of evolution.

Teleology should not be an embarrassment. Contrary to some viewpoint,[603] it is not an approach unique to biology among the sciences. The Copernican, Galilean, and Newtonian corpus was inspired by teleological considerations, a grand reduction teleologically inspired;

so was the Darwinian assumption, teleology in nonteleological terms, materialistic, mechanistic, even statistical. Indeed, mechanism and materialism are components of a limited teleology.

The teleological approach seeks the meaning of events in a series without having to resort to a divine plan or a cosmic design. Teleology can be limited to explanations short of the cosmic. However, it cannot be so limited as to merely offer an immediate explanation. A question or an explanation is teleological only if it dangerously approaches the bounds of science.

All developmental systems — the development of an organism, of a star, of a galactic system, of rivers and mountains and continents, of society and of the universe, of life, of everything that unfolds — all these can be approached teleologically as a developmental program. The objective is to look for a direction, an organization, a pattern, a design, a program.

In an open series, every event, provided it is not the initial, is a consequence of a previous event. Every event, provided it is not the last, is also the cause of the next. When the series is closed, every event is both cause and consequence. The relationship between cause and consequence need not be absolute. It may only be, as observed, inevitable. Having observed its program unfold, we assume that an organism in the course of time seeks, or is somehow compelled to seek, the realization of that program. Common sense tells us that organisms strive to survive, that waters in the rivers inexorably get to the sea, that a seed eventually becomes a tree. Events associated with the realization must be shown to be dispensable or indispensable, accidental or essential.

There is a hierarchy of programs. One program is contained in another program which, in turn, is contained in yet another, and so on. The survival of an individual may be viewed as a part of the survival of the community, of the species or of a higher taxon, of an ecological system, of life, all of which may be viewed as parts of a cosmic inevitability.

Teleology is used here in consonance with its philosophical roots and differs markedly from the antireductionist vitalism of Driesch and his school. The limited teleology excludes entelechy, *élan vital,* and such concepts, not because they are mystical, but because they have proved to be unproductive. Entropy is no less mystical and dogmatic than entelechy, but entropy has proved its usefulness. The limited teleology refuses the burdensome task of looking for final causes. It is only interested in things amenable to current methodology and understanding. It excludes divine purpose or the guiding hand of a creator. That, incidentally, is an extreme case of reduction as well, for the guiding hand of a creator is not just the final but also the primary cause. Neither does the limited teleology include man's interests, his economics, aesthetics, politics, ethics, and aspirations. The purpose of an organism is not the glorification of man nor of his assumed creator, but to be, to exist, to become, to unfold. The dynamic of a cell is internal and external. It is internal in that the material of the cell, its molecules, determines its future. It is external only in the sense that the environment allows the cell to become. The limited teleology just falls short of being metaphysical and anthropocentric. In the end, it has to be materalistic and mechanistic.

The success of molecular biology lies in its being immediate and local. Teleology, just short of being metaphysical, is larger, less immediate. Parts of what used to be teleology have been reduced to molecular biology, the metabolism of the cell, for instance. The more far-out parts remain in metaphysics. Teleological questions are mere extensions of immediate questions of function. The farther an explanation goes out on a limb, the weaker its explanatory power, the more tenuous its hold on facts. When it is too far out, it steps out of the bounds of science and into metaphysics.

To what end, for what purpose, of what use, why? These are questions that constitute teleology. The end, the purpose, and the use, of course, all derive from man's own search for purpose, for meaning, from man's search for order, for explanation, from man's instinct for games, for solving jigsaw puzzles.

Indeed, there is a valid place for teleology so long as it remains with the questions. Because only the verifiable and the falsifiable are admissible in the courts of science, the answers must be trivialized in terms of a materialistic, mechanistic, and statistical universe. Probably the greatest teleological question that has succumbed to materialism and mechanism is evolution: the unfolding, the emergence of life not by a divine plan but by differential survival and reproduction, by natural selection.

After this long digression, we are now ready to dissect the question: of what use is aggregation? The question does not infer a consciousness or a conscious effort on the part of the organism to reach a predicted goal or a preestablished harmony.[604] One need only invoke inevitability. The question is clearly teleological, but the answers must nevertheless be materialistic and mechanistic.

"Of what use?" must be rephrased in more tractable ways. Is aggregation a vital function? In ordinary physiological situations, it is not. There is no microbial aggregation system that is known to be indispensable. Nonaggregative cells are healthy cells. Furthermore, the existence of nonaggregative species argues against the essentiality of aggregation. Does aggregation help in the survival of the organism? This is testable. Aggregation mutants might be at a disadvantage when forced to compete with the wild type for the same space and the same nutrients in a mixed culture. Does it help the organism to evolve? This might be more difficult to test, but must be observable in a continuous culture. If aggregation enables the individuals or the species to compete better, then differential survival and reproduction will enable them to withstand natural selection. Mere tenacity of an organism improves its chances of evolving. What is the role of aggregation in the general welfare of the cell? Does it help in any other function in the cell, even as that very function is teleologically being questioned?

What is the role of aggregation in development? Clearly, aggregation is an integral part of a series of events in the life cycle of an aggregative organism. Because it helps complete the life cycle, it is both cause and consequence of events in the life cycle. Moreover, an aggregate is obviously an intermediate structure in the continuity that extends from the unicellular organism to the multicellular condition. As a consequence of the aggregation process, aggregation structures are formed which may be more resistant to adverse conditions. That is quite evident in filamentous fungi. Furthermore, resistant structures farther along the developmental cycle — spores, for example — may be indirect consequences of the aggregation process.

The more immediate function of microbial aggregation is its contribution to sex. Although not indispensable to sex, aggregation serves to enhance initial cell-cell contacts and thus to facilitate the slow process of conjugation. In a turbulent environment specially, cells are brought together in a more efficient and more organized manner, rather than through random opportunities. Of what use is sex?[605,606] In lower forms, sex is not primarily for reproduction, but for genetic recombination. That aggregation contributes to sex and therefore to genetic recombination may be taken for granted. In so contributing, aggregation does not necessarily help the individual. It does help, however, the cause of the species for differential reproduction, natural selection, and evolution.

In a most primitive arrangement, recognition factors on the cell surface promote a match between cells that are genetically the same but for the genes that specify the complementary structures. Hence, a heterotypic match does not offer much by way of opportunities for meaningful recombination of genes any more than a match between two cells of the same mating type. Indeed, heterotypic sex is not any more efficient than homotypic sex, unless the change from homotypic to heterotypic sex is just the beginning of more mutational events, which would make the mating types more different than ever.

Of what use is aggregation? Of use to what entity? To the components of the cells? Such as the cell wall, the macromolecule, the smaller molecule, the carbon atom, the hydrogen atom, the elementary particle in the cell? Of what use is aggregation to itself? The answers to these absurd questions can only be absurd. One might as well ask the question: of what use is the cell to aggregation? No less absurd is the question: of what use is aggregation to life at large or to the unfolding of the universe?

Of what use is aggregation to the cell, then? Must it be assumed that a cell does not exist for other cells, but for itself alone? That if it happens to help the community of cells, that is merely an accidental consequence of the cell helping itself? The assumption is tenable only when a cell is isolated and when there can be no cell-cell interaction. Microbial aggregation, however, is a cell-cell interaction. Cells during aggregation must be viewed as components of a bigger entity, the community. The question must then be posited in such a way as to yield answers that connect not only to what is already known but also to appropriate organizational levels in the hierarchy of entities.

Of what use is aggregation to the community or to the species? A most convenient explanation, not quite specific, is that it contributes to the unfolding of a program, to development. Development is an unfolding of the potentialities inherent in the organized materials of the cells. Does aggregation indeed help a microbe to unfold, to attain a prespecified goal in the developmental program? The prespecification may reside only in the observer's mind. From single cells to an aggregation of cells, however, a higher organization and a higher order are achieved, higher in that the new order acquires more complicated structures and functions, thereby opening up more possibilities. That aggregation is part of development may then be taken for granted. Whether development is useful rather than merely just inevitable is another teleological matter.

Because of its global distribution, and in some cases, because of its intensity, aggregation to the species must be assumed to be of some advantage, however minimal it may be. Unicellular organisms can benefit from multicellular interaction and association. What cannot be denied is that the process has not yet perished in the competitive jungle. That aggregation exists at all is proof enough of its own worth. Worthy or unworthy, a human judgment, however, is merely a reflection of human values.

Finally, of what use is teleology? Of what use is trying to solve a jigsaw puzzle?

Chapter 4

BACTERIA

I. INTRODUCTION

The global collection of cellular organisms falls into two large groups: eukaryotes and prokaryotes. The presence of a membrane-delimited nucleus identifies a eukaryote. In contrast, a prokaryote does not have a true nucleus. The genophore of a prokaryote is a "nucleoplasm ... not bounded from cytoplasm by a nuclear membrane."[607] The protein-synthesizing machinery of a prokaryote consists of ribosomes (70 S) that are smaller than those of eukaryotes (80 S). There are many other points of differences, but the others are minor and not as clear cut. The structural unit for either group is nonetheless the cell. The prokaryotic cell is generally simpler and more primitive, lacking the many cytologically identifiable organelles of the eukaryote. Included among the eukaryotic microbes are the yeasts, the cellular slime molds, the filamentous fungi, the algae, and the protozoa. The prokaryotes comprise the blue-green algae and the bacteria.

As a group, the bacteria have not been well studied with regard to their aggregative properties. Because they serve, however, as models for a great many biological phenomena, it is expected that there will be many advances in the study of bacterial aggregation before long. Already, new systems are being reported while old systems are receiving renewed attention. An example of a fairly old system being reexamined is mating-aggregate formation during conjugation in *Escherichia coli,* undoubtedly the best studied cell. Other systems include star formation in *Pseudomonas* and other bacteria, pellicle formation in fimbriate bacteria, agglutination associated with transformation competence in *Streptococcus* and *Bacillus,* plaque formation and agglutination in *Streptococcus mutans,* aggregation in myxobacteria, and aggregate formation in *Streptococcus faecalis*. In this chapter, we shall presently examine each of these bacterial aggregation systems in some detail.

II. MATING-AGGREGATE FORMATION DURING CONJUGATION IN *E. COLI*

Bacterial conjugation was first demonstrated in *E. coli* by Lederberg and Tatum.[73,608] Since the initial report, the literature on the subject has known no letup. A number of references may be consulted for an overview of the subject.[189,538-540,573-575] Bacterial conjugation is the plasmid-mediated transfer of DNA from one cell to another by cell-cell contact. The transfer is unidirectional. It involves a donor and a recipient. It is not an exchange or a recombination of chromosomal DNA but the transfer of plasmid DNA from one cell to another. In the process, however, a small segment of the donor chromosome may sometimes be transferred with the plasmid.

"A plasmid is a replicon that is stably inherited (*i.e.*, readily maintained without specific selection) in an extrachromosomal state."[576] It is synonymous to an extrachromosomal element. A conjugative plasmid is one that brings about the transfer of DNA by conjugation. The best known conjugative plasmid is F, which is also an episome (a genetic element that replicates in either of two states: autonomous or integrated into the host chromosome[609]). An *E. coli* cell that carries the plasmid F is called an F^+ cell and can serve as a donor (Figure 22, Chapter 3). An F^- cell carries none and serves as a recipient. An Hfr cell is a superfertile donor in which the episome is integrated into the host genome and is a better donor than F^+. Upon release of the plasmid, Hfr becomes F^+. When F^+ loses the plasmid, it becomes F^-. In turn, F^- can be infected to become F^+. An F-derivative plasmid incorporating a segment of the host chromosome is called an F' plasmid.[576] A cell that carries an F' plasmid is termed an F'–containing cell.

The mating phenomenon consists of the donor, F^+ or Hfr, pairing with a prospective recipient, F^-, and transferring plasmid DNA to the recipient. The mechanism of transfer is not clear, but cell-cell contact appears to be necessary. The genetic event consists of the interaction between the complete genome of the recipient and a small fraction of the genetic endowment of the donor. The transfer of DNA is outside the scope of our present intent and will not be discussed. Instead, we shall focus on the aggregation aspect of the phenomenon.

From the very beginning, it was assumed that conjugation merely involved mating-pair formation. Although mating aggregates have been observed before,[610-612] only fairly recently has it been shown quantitatively that mating aggregates are formed during the conjugation process.[28] Indeed, it appears not only that most mating cells are found in aggregates, but also that aggregate formation is a phenomenon intimately associated with conjugation.

According to Achtman and Skurray,[575] the following stages are observable during the conjugation process: (1) binding of sex pili, (2) formation of wall-wall contacts, (3) stabilization of mating aggregates, (4) DNA transfer, and (5) active disaggregation. The F-mediated mating cycle of *E. coli* is shown in Figure 1. The complete cycle is divided between unstable and stable aggregates.

The first stage apparently involves the binding of sex pili of the donor to the receptors found on the cell surface of the recipient. F pili, specified by the F plasmid, are the best known of the various sex pili found in *E. coli.*[14,189,538-540,613] During logarithmic phase, an F^+ cell has 1 to 3 F pili, whereas an F^- cell does not have any. F pili are filamentous appendages of up to 20 μm in length (Figure 2). Recent physical studies show that they are hollow cylinders with an outer diameter of 8.0 nm and an inner diameter of 2.0 nm.[669] They are identified from other filamentous appendages on the cell surface of male donors by their exclusive property of binding male-specific phages. RNA phages bind to the pilus anywhere along its entire length, DNA phages only at the tip.[614,615] The F pilus is made up of protein subunits called F pilin. Each subunit of about 1.1×10^4 daltons contains about 124 amino acid residues, two phosphate residues, and one glucose residue.[539,616,670,671] About 69% of the highly hydrophobic subunit is in an α-helical configuration.[670] F pilin subunits are found 0.37 nm apart along the axis of the pilus, which is made up of four co–axial helices with a pitch of 12.8 nm.[669] Purified pilin preparations adsorb specifically to F^- cells of *E. coli.*[575]

The involvement of F pili in conjugation is demonstrated by several lines of evidence. Known pili mutants and tra^- mutants, which are resistant to male-specific phages and thus presumed to be without pili, are unable to transfer DNA.[617] Brinton[551] has shown that blender treatment of donor cells results in the loss of pili and of mating ability and that the kinetics of the reappearance of piliation parallels that of the reappearance of mating ability. Other treatments resulting concomitantly in reduced piliation and other donor activity include incubation at 25 or 50°C, SDS, Zn^{2+}, and cyanide.[28,193,540,618] Incubation at these temperatures depolymerizes F pili, and Zn^{2+} appears to cause shedding. In all cases, mating is inhibited.

A number of proteins are found on the cell surface. One of these could be an aggregation protein required for the formation of mating aggregates. Known by various names (II*, G, d, 3A, and pOmpA),[575,619-622] it will be spelled out here as outer membrane protein A, for lack of a common term. Isolated from cell membranes, outer membrane protein A is insoluble in water but soluble in SDS.[622,623] The native protein has a molecular weight of 2.8×10^4 daltons.[624,625] There are about 10^5 molecules of the protein per cell.[625]

The structural gene for outer membrane protein A is known as *ompA*, which maps at 21.5 min on the *E. coli* chromosome.[626-628] Also called con^-, *ompA* mutants are 50- to 5000-fold less efficient as recipients and contain reduced amounts of outer membrane protein A.[628,629] Some mutations mapping at *ompA* show a structural alteration of the protein.[627] Other con^- mutants have been shown to contain defective lipopolysaccharide that lacks

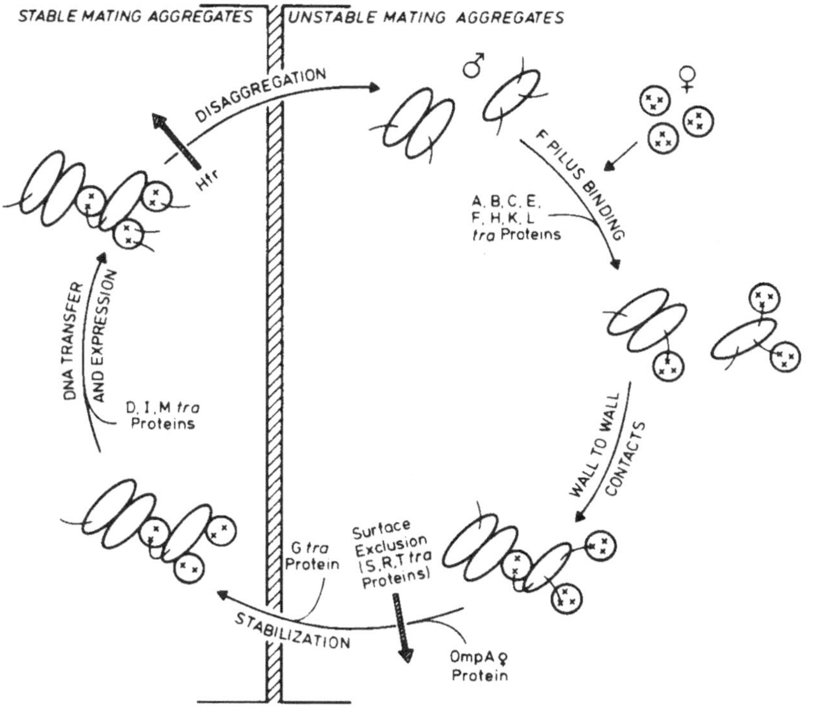

FIGURE 1. The mating cycle in *E. coli*. The donor (male) cells are shown as sausages with one to three hairs corresponding to F pili. The recipient (female) cells are shown as spheres with x's corresponding to F-pilus receptors. Two sites of inhibition are marked by hatched arrows: inhibition of stabilization by surface exclusion and inhibition of disaggregation when Hfr serves as donor. (From Achtman, M. and Skurray, R., in *Microbial Interactions*, Reissig, J. L., Ed., Chapman and Hall, London, 1977, chap. 6. With permission.)

heptose.[630] One of these heptoseless mutants also exhibits a partial deficiency of outer membrane protein A. The mutant is also defective in forming stable mating aggregates.[630,631] Still another group of con$^-$ mutants has been shown to contain galactose- and phosphate-deficient lipopolysaccharide.[632] These mutants are good recipients for F and F-like plasmids, but do not allow the transfer of R144, a plasmid belonging to the I compatibility group. One such mutant has been reported to be unable to form aggregates with R144-containing cells.[575] Matings involving I plasmid, but not F, are inhibited by lipopolysaccharide derived from F$^-$ cells. Thus, lipopolysaccharides seem to be implicated as receptors in matings coded for by I plasmid and mediated by I pili.[632] On the other hand, stable-aggregate formation determined by F involves protein receptors, one of which could possibly be outer membrane protein A. The binding of sex pili, the first step in the mating cycle and an aggregation step, may then be envisioned as the interaction between pili of the donor and their receptors on the cell surface of the recipient.

Whereas the information for cell surface receptors for pili is encoded in the recipient genome, the genes for the synthesis of pili are contained in the plasmid. The genetic map of the F plasmid is shown in Figure 3. F is a covalently closed circular DNA molecule with a molecular weight of about 6.2×10^7 daltons.[633,634] Mapping between 65 and 82 kb are nine *tra* cistrons deduced to be required for the synthesis of F pili.[617,633] These are *traA*, *traL*, *traE*, *traK*, *traB*, *traC*, *traF*, *traH*, and possibly a portion of *traG*. The synthesis of F pilin, the protein subunit of the F pilus, is specified by *traA*.[636] The other cistrons, whose definitive functions are yet unknown, may contain the information for the processing enzymes

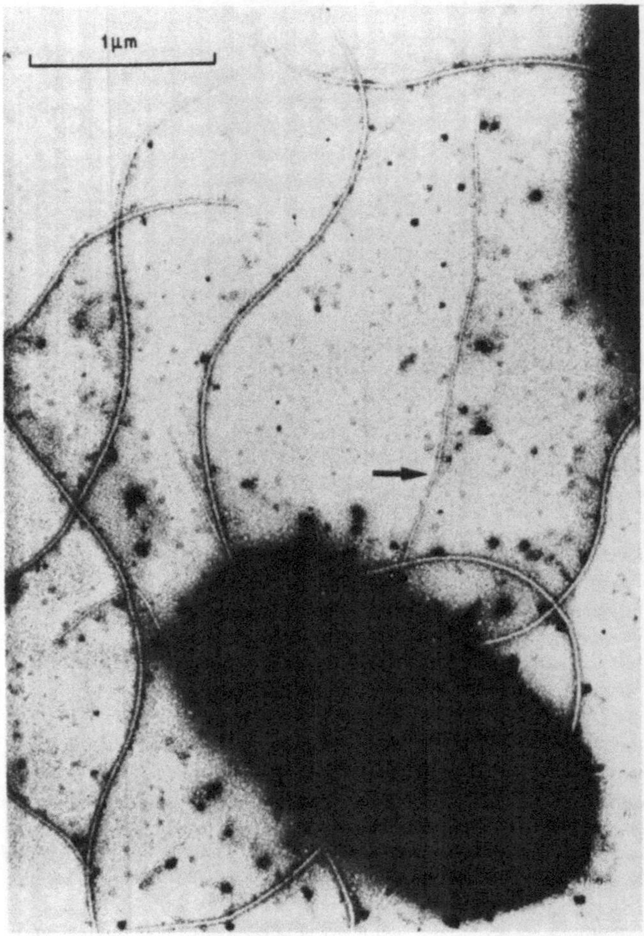

FIGURE 2. F pilus on the surface of *E. coli* K–12 (F*lac*)$^+$ cell. The arrow points to an F pilus. The thicker filaments are flagella, the thinner ones are the common I pili. Original photograph from R. Helmuth. (Reproduced with permission from Chapman and Hall, London.)

or for other components of the intact pilus, including a hypothetical basal structure. Other tra$^-$ mutants point to additional cistrons related to DNA transfer.[617,635,637] Although all are tra$^-$, some *traG* mutants do synthesize F pili. Whereas *traD* mutants synthesize F pili, they are resistant to infection by pilus-specific RNA phages. Pilus structure is unaffected by mutations in *traI* and *traM*. The cistron *traM* is found between 60 and 68 kb,[575] whereas cistrons *traG*, *traD*, and *traI* are bunched up between 80 and 92 kb.[633]

Aside from the functions of donor ability, also located in the *tra* region is the function related to the lack of recipient ability of F$^+$ cells. By a vague mechanism called surface exclusion, host cells bearing conjugative plasmids are somehow rendered poor recipients, thus preventing matings between F$^+$ cells. Surface exclusion is determined by cistrons *traS* and *traT*, both of which are located between *traG* and *traD*.[575,638] Mutations in these cistrons convert F$^+$ cells into more effective recipients, without affecting DNA transfer nor pilus synthesis.[575,639] A presumptive product of the *traT* cistron is an outer membrane protein with a molecular weight of 2.4 × 10^4 daltons.[575,640] Close to *traT* and *traS* is a site called *ilzA*, which codes for immunity to lethal zygosis, which is a phenomenon wherein high ratios of donor Hfr to recipient F^- cells in a mating mixture result in the loss of viability of recipient cells. Surface exclusion and lethal zygosis will be further discussed a little later.

The *tra* cistrons form an operon that extends from *traA* to *traI* (Figure 3). The operon is more than 15 × 10^6 daltons.[639,641] Located just prior to *traA*, in a clockwise direction, is a positive regulator gene, *traJ*, which controls the transcription of the *tra* operon.[637,642,643] Mutations in *traJ* affect synthesis of F pili, surface exclusion, and transfer of DNA.[635] Two cistrons, *finO* and *finP*, constitute a repression system.[644,645] They map, in a clockwise direction, just before *traJ*. The *finO* product interacts with the *finP* product to cause repression of the *tra* operon in some F-like conjugative plasmids isolated from nature.[646,647] The F plasmid itself is constitutive or derepressed and is assumed to make no functional *finO* product. The *finO-finP* system appears to act on *traO* to block the synthesis of the *traJ* regulator protein and consequently the transcription of the *tra* operon.[642, 648] DNA is transferred in a counterclockwise direction starting from the site *oriT*. Thus, the sites *inc* (for incompatibility), *frp* (for F replication), *oriV* (for origin of vegetative replication), and *pif* (for phage infection inhibition) are transferred ahead of the *tra* operon.

To participate in conjugation, a donor cell must synthesize sex pili. The *tra* operon is presumably transcribed and its transcript RNA translated into proteins that will make up the sex pili. The peak of pilus synthesis occurs during logarithmic growth or just prior to stationary phase.[613] For its part, the recipient cell must synthesize its receptors. When the two kinds of cells are allowed to come together by random collision, the initial event is presumed to be the binding of the pili of the donor to the receptors on the surface of the recipient (Figure 1). Binding is followed by wall-wall contact. Pili appear to serve as mere grappling hooks to facilitate the next event by converting a random collision to a fruitful one. In the transition from wall-wall contact to a stable aggregate, outer membrane protein A seems necessary. Also apparently needed is a protein product of the *traG* cistron. Surface exclusion inhibits the transition to stable mating aggregates. Upon stabilization of the mating aggregates, DNA transfer occurs, facilitated by protein products from *traD*, *traI*, and *traM* cistrons. After DNA transfer, active disaggregation occurs. Disaggregation is not observed when Hfr is used as donor.

The quantitative aspect of mating-aggregate formation has been examined by Achtman.[28,667] He showed by light microscopy and electronic particle-counter analysis that matings between donors with F or F-like conjugative plasmids and recipients result in aggregates, rather than merely pairs, as had been assumed by investigators before him (Figure 14, Chapter 3). Under optimal conditions for mating, most of the mating pairs are found in aggregates. Indeed, multiple matings as well as multiple genetic transfers per cell are commonly observed.

Mating aggregates in an Hfr × F^- mixture are quite stable.[28,667,668,672] The stable period lasts for at least an hour or more, so that in an asynchronous population, the greater portion may have been in aggregates for at least 10 min. Shear forces, as those that occur during pipetting or dilution, do not cause disaggregation. Stable aggregates become resistant to SDS, which has been shown to dissolve F pili, and to cyanide, which is believed to cause retraction. Thus, even when all pili are gone, stable contacts are maintained, indicating that at this point in the mating cycle, they are no longer mediated by F pili. Indeed, most of the cells in aggregates are found to be in wall–wall contact. The kinetics of mating-aggregate formation is shown in Figure 4. Within 20 min after mixing of donors and recipients at 37°C, up to 80% of the cell population are found in aggregates.[667] The figure also shows the inhibitory effect of SDS when added at the start of mixing. Cyanide and Zn^{2+} likewise prevent the occurrence of aggregation.

Aggregates vary in size. Small aggregates consist of two to five cells, large aggregates 6 to 20 cells (Figure 11, Chapter 3).[28] It is the latter size-range that qualifies mating-aggregate formation during conjugation in *E. coli* as an aggregation system by our definition. A recent

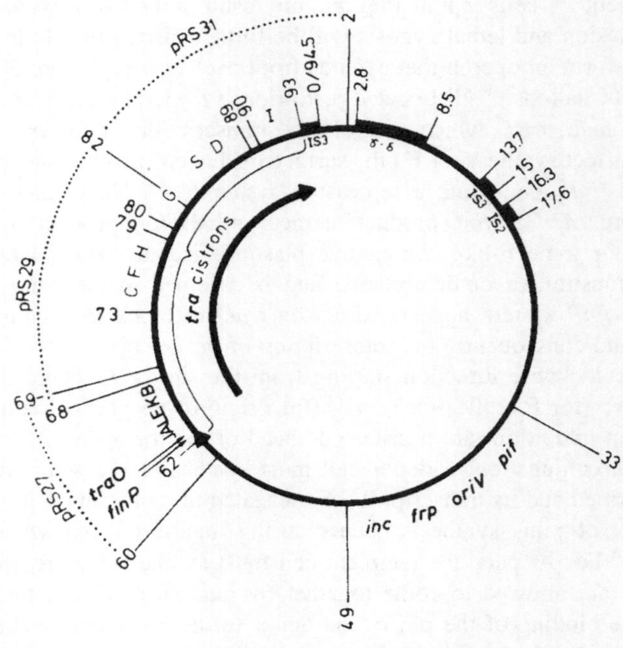

FIGURE 3. Genetic map of the F plasmid. Symbols (starting at 6 o'clock, counterclockwise): *inc* for incompatibility, *frp* for F replication, *oriV* for origin of vegetative replication, *pif* for phage infection inhibition; IS2, IS3, and γ — δ are insertion sequences; *tra* for DNA transfer; *traO* is the site defined by *cis*-dominant mutants resistant to repression by the *finO-finP* repression system. Three plasmid chimeras are also shown: pRS27, pRS29, and pRS31. All are pSC101 derivatives bearing *Eco*RI fragments of F DNA. Numerals around the circle refer to kilobase units. The clockwise order of the *tra* operon on the map is as follows: OJALEKBCFHGSDI. Other *tra* cistrons are omitted: *traM* (between 60 and 68 kb)and *traT* (near *traS*). The triangle at 62 kb indicates the site of *oriT*, the origin of DNA transfer. The heavy arrow inside the circle points to the direction of transcription of the *tra* operon. The map supersedes earlier maps.[664-666] (From Achtman, M. and Skurray, R., in *Microbial Interactions*, Reissig, J. L., Ed., Chapman and Hall, London, 1977, chap. 6. With permission.)

report claims aggregation numbers of up to 50 cells.[672] The aggregates are obviously heterotypic, at least in the genetic sense, being made up of donor and recipient cells. In practice, mixtures are deliberately made heterotypic by the use of morphologically identifiable strains. Thus, the donor may be long and slender, the recipient short and plump. The composition of an aggregate as regards donors and recipients may then be scored visually. Depending on concentration, input ratio, and physiological state of the cells, the ratio of donor to recipient in the aggregate can vary by as much as a factor of four.[28] When the input ratio is 1:1, about equal numbers of donors and recipients are found in aggregates. The majority of the cells belong to aggregates of seven or more cells each.

In mating mixtures of F^+ or R^+ donors and F^- recipients, the mating aggregates formed are not as stable as those in mating mixtures wherein Hfr is the donor. Soon after the autonomous plasmid DNA is transferred, active disaggregation is observed. The half-life of mating aggregates of F′-containing cells and F^- is 5 to 15 min.[672]

The requirements for mating-aggregate formation as well as its location in the mating cycle have been analyzed by the use of mutants. Nonpiliated mutants of F^+ are unable to

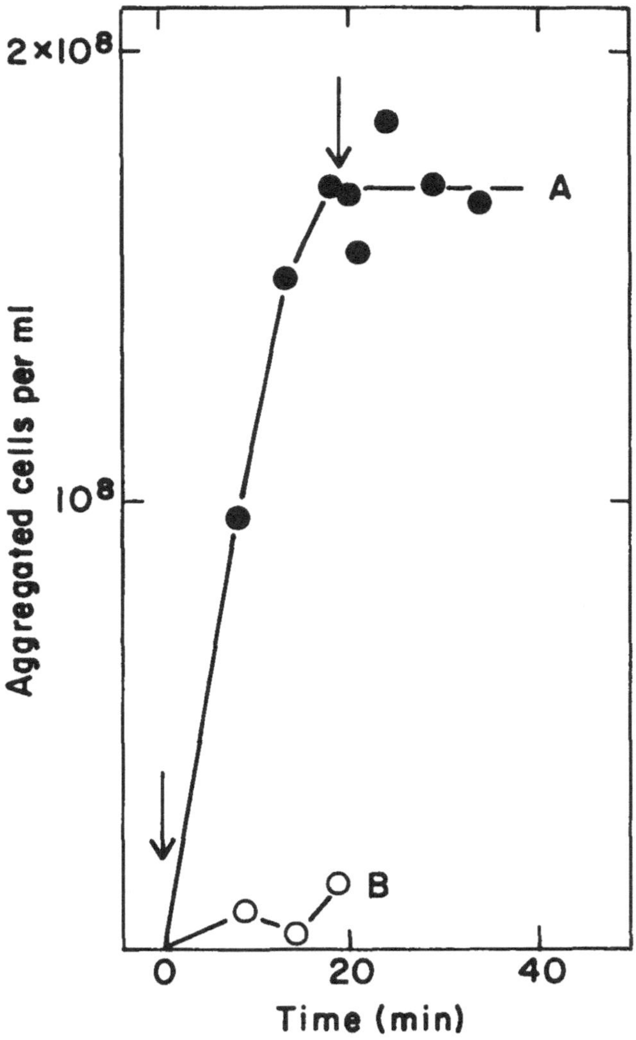

FIGURE 4. The kinetics of mating-aggregate formation and the effect of SDS.[667,672] The mating mixture consists of an Hfr Cavalli strain and an F⁻ strain. SDS (final concentration: 0.01%) was added at the times indicated by the arrows. (From Achtman, M., Morelli, G., and Schwuchow, S., *J. Bacteriol.*, 135, 1053, 1978. With permission.)

form stable mating aggregates with F^- cells,[575] but mutations in *traD, traI,* or *traM,* none of which affect pilus synthesis, allow the formation of stable mating aggregates, thus eliminating the transfer function from being a requirement for mating-aggregate formation. Most con^- mutants fail to form mating aggregates with F^+ cells, suggesting the need for surface receptors on the recipient, such as outer membrane protein A. The con^- mutant that shows a deficiency in outer membrane protein A does not form stable mating aggregates in liquid medium but can act as recipient on an agar surface.[630,631] The number of mating aggregates is much reduced in *traG* × F^- matings, implicating a *traG* product. Nevertheless, the cells in the few aggregates that are formed are observed to be in intimate contact with one another.[575] Taken all together, these observations point to the stabilization stage as occurring before wall-wall contact and prior to DNA transfer (Figure 1). It appears then that mating-aggregate formation is a prerequisite to conjugation, at least, in liquid culture.

With regard to the conjugation process itself, including DNA transfer, a role of the pilus as a conjugation tube[551] seems unlikely. DNA penetration is a problem not only in bacterial conjugation but in phage infection and in bacterial transformation as well. However, that aspect of conjugation is beyond the province of our present interest. A number of models call for channels or pores in the cell envelope through which DNA is transported, but we will go only so far as pointing to references on the subject.[649,650] It can be surmised that in the case of bacterial conjugation, some form of intimate wall-wall contact but obviously short of cell fusion, rather than mere contact by way of sex pili, is probably necessary. At any rate, right after DNA transfer, in matings wherein the donors carry the plasmid in its autonomous state, there is an active process of disaggregation, not merely a passive crumbling of the stable but transient aggregation structure.[575,672] It may well be that the process is enzymic in nature. The process is analogous to postconjugation disengagement in unicellular algae.

Because of surface exclusion,[651] cells bearing conjugative plasmids are poor maters among themselves. Thus, mating aggregates are not observed in pure F^+ cultures. Moreover, in matings in which the recipient is F^- but a positive mutant for surface exclusion, formation of stable aggregates is less than normal, whereas in matings in which the recipient is F^- but a deficient mutant for surface exclusion, aggregate formation appears to be normal.[575] The defect resembles that of *traG* × F^- matings. It appears that the protein coded by *traG* is necessary for stable aggregate formation and that surface exclusion is an inhibition of the function of *traG*.

One other interesting phenomenon associated with bacterial conjugation is lethal zygosis.[652-657] In Hfr × F^- matings, wherein the donor is in overwhelming excess, most of the recipients (up to 99%) are killed within 2 hr. The phenomenon is not observed when F^+ is donor or recipient.[658] Death, operationally defined as nonviability, may be due to a dysfunction in the cell membrane.[657] Surface exclusion accounts for much of the resistance of F^+ as recipient to lethal zygosis, but because F^+ cells that are deficient in surface exclusion are also resistant, an *ilz* function (for immunity to lethal zygosis) has been invoked.[659] Two *ilz* cistrons have been located on the F genetic map: *ilzA*, between *traO* and *traG*, possibly a cistron for surface exclusion, and *ilzB*, within the 90-0-33 kb region, possibly a cistron for active disaggregation. The exaggerated reaction that is manifested as lethal zygosis could well be an aberrant enzymic rearrangement of the surface structure, a pathological reaction due to prolonged contact.[575] Such a reaction may be prevented by surface exclusion and reversed by active disaggregation. The phenomenon is somewhat analogous to conjugation-induced lysis, which has been proposed to be due to overactivity or malfunction of lytic enzymes during conjugation in fission yeast.[121,660,661]

The two purified structural components of the aggregation system, F pilus and outer membrane protein A, are assumed to mediate the cell-cell interaction. The filamentous appendage is somehow involved in the formation of unstable aggregates, a preliminary event to the formation of stable mating aggregates. The membrane protein is somehow involved in the stabilization. Nevertheless, they do not seem to interact with one another. Purified F pili specifically bind to F^- cells, to F^+ cells that are positive mutants for surface exclusion, and to con^- mutants that are deficient in outer membrane protein A or heptose-deficient in their lipopolysaccharide.[575] Hence, the binding of the F pilus to the recipient surface does not require surface exclusion, outer membrane protein A, nor the putative *traG* protein. Purified outer membrane protein A at a certain concentration in the presence of lipopolysaccharide does inhibit conjugation.[662,663] The inhibited event appears to be the stabilization of mating aggregates.[575] Much higher concentrations of the protein prevent the formation of unstable aggregates, without inhibiting the binding of pili to the recipient cells. Purified lipopolysaccharide at low concentrations does not inhibit matings between F^+ and F^- cells.[632] Neither does it interfere with mating-aggregate formation, nor with the binding of F pili, nor does it reverse the inhibition of mating-aggregate formation by outer membrane protein A.[575] Although these observations do not point exactly to the site of action of conjugation-inhibition by outer membrane protein A with respect to the donor or the recipient, they suggest protein-protein interactions between the F pilus and the recipient surface, and on the other, between outer membrane protein A and the donor surface.

It must be noted that cell-cell contacts do occur in pure, nonmating cultures.[28,667,672] About 5 to 25% of the cells in pure F^- cultures and 15 to 45% in pure Hfr or F′ cultures have been reported to form clumps. The contacts may be mediated also by pili or possibly by other cell appendages. However, these spontaneous, unstable clumps of cells have been shown to be nonmating aggregates. Moreover, the clumps are quite small, consisting of merely a few cells. They do not qualify by our definition as an aggregation system.

Many events are involved in the plasmid-mediated conjugation in *E. coli*. Mating aggregates are formed during this process. The main features of this aggregation are now known and will form the basis for further dissection in the future. Cells bearing the F plasmid synthesize constitutively a minimum of 16 proteins specified by the *tra* operon of the plasmid. These proteins are needed for the synthesis of the pilus, for surface exclusion, for DNA transfer, and other functions. In the chromosome of the recipient cell, the other component of the heterotypic system, lies the information for the synthesis of surface receptors. Surface exclusion prevents fruitful interaction between donors. When a population of piliated donors, however, encounters a population of recipient cells, donors and recipients interact, initially by binding of the donor pili onto the receptors on the surface of the recipients. More intimate intercellular contacts are formed by wall-wall interaction later. These unstable aggregates are then made stable by the action of other putative receptors, such as outer membrane protein A. DNA transfer occurs among complementary cells in stable aggregates. After the transfer, the aggregates actively break apart.

Much remains to be unravelled, but with such a very well-studied cell as *E. coli*, there must be much excitement to look forward to. Note, for example, that the greater part of the discussion above concerns with the F plasmid only. There are other conjugative plasmids known. Their roles in conjugation-associated aggregation remain to be examined. Mating-aggregate formation during conjugation in *E. coli* is an example of an aggregation system that has been discovered by a careful reexamination of a supposedly well-known phenomenon.

III. STAR FORMATION IN RHIZOBIA, PSEUDOMONADS, AND OTHER BACTERIA

While it took quite awhile for investigators to concede the occurrence of mating-aggregate formation during conjugation in *E. coli*, there are other bacterial conjugation systems in which aggregation was recognized long before the elucidation of its sexual significance. One such system is star formation in a number of bacterial genera and species.

Star formation is an aggregation system wherein component cells form a cluster that resembles a star (Figure 5). A unit star consists of bacilli arranged radially from a center. The phenomenon is observed in many bacterial species found in the soil or water. It has been reported in species of *Rhizobium, Pseudomonas, Agrobacterium, Chromobacterium,* and *Xanthomonas*, among the better known organisms, as well as in other soil isolates that have not been taxonomically identified. It was known to microbiologists, among them Beijerink and Löhnis, as early as the first decade of this century.[674,675] The early literature, largely in German, has been summarized by Knösel.[674] In the German literature, the phenomenon is known as "Sternbildung."

Long before sexuality was known in bacteria, it was suspected that star formation precedes conjugation in a sexual cycle. The sexual nature of the phenomenon was confirmed a few years after the discovery of bacterial recombination in *E. coli*. Conjugation between cells

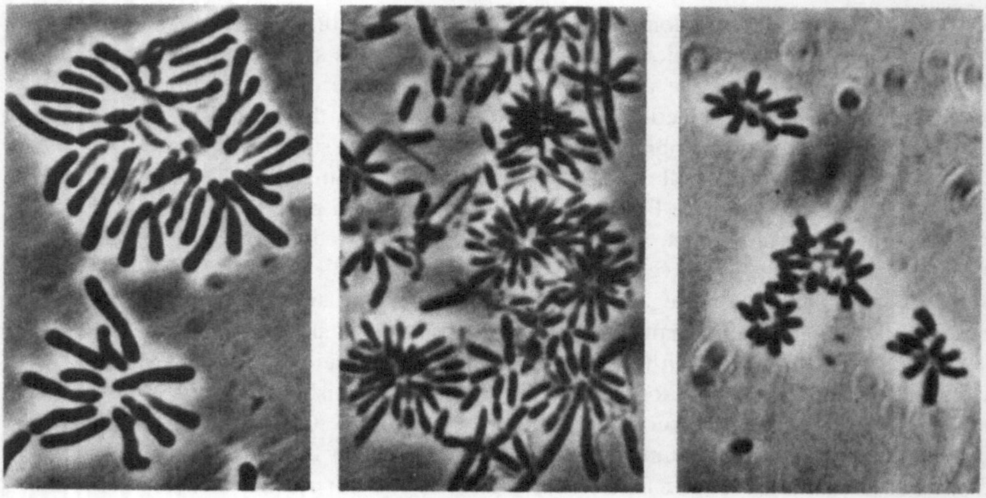

FIGURE 5. Star formation in three strains of bacteria. Left, strain B_6 of Rhizobiaceae; middle, strain B_{13} of Spirillaceae; and right, *Agrobacterium stellulatum*. (From Heumann, W., *Die Naturwissenschaften*, 14, 330, 1960. With permission.)

found in stars was first demonstrated genetically by Heumann.[197,673,675-679] From his studies came the first chromosome maps of *Pseudomonas echinoides*,[679] but the more interesting conjugation system is that of *Rhizobium lupini*, whose genetics of fertility genes is the best studied among the starformers.[680-682] Its usefulness as an experimental system partly derives from its having been purposely selected for its high activity in star formation and conjugation.

A sexual cycle for starformers has been described by Heumann.[675] Proposed more than 2 decades ago, the schema badly needs some modification by now. However, no improvement on it is found in the literature. In his schema, ''Sternbildung'' is followed by four other morphologically distinct events: ''Kopfbildung'' (head formation), 'Kopulation'' (copulation), ''Zygotenbildung'' (zygote formation), and ''Zygotenkeimung'' (germination). The entire cycle at 27°C takes 8 to 9 weeks, the events up to cell fusion lasting for a week or two. In his schema, star formation is the consequence of slime secretion (''Schleimabsonderung'') at the pole of a component cell and the chemotactic migration of cells toward sticky star centers. Although there might indeed be intimate cell-cell contact later for DNA transfer, it is most probably not as shown in the schema. Moreover, later investigators have found by electron microscopy no indications of cytoplasmic fusion.[683] It is then unlikely that zygote formation, analogous to what occurs in eukaryotes, ever takes place. The resting stage of the presumed zygote and its emergence as if from a spore remain to be verified by others.

Only motile, young, and healthy cells are involved in star formation.[675] There is no doubt that the cells do come together. Star formation is observed about 4 to 6 hr after inoculation of a rich medium. Chemotaxis, precisely because the cells are flagellate, is likely involved, and so is development of competence to aggregate. Within the star, however, cells continue to divide. There is also no doubt as to the temporal position of star formation in a developmental program. The conversion of cells from a vegetative state to a state of sexual development depends on the substrate. Addition of blood to the culture medium enhances this conversion. In minimal medium, the addition of nutrient broth enhances conjugation.[197]

Star formation has been shown to be mediated, at least, in its initial stage, by unipolar fimbriae.[198,684-687] In *Pseudomonas echinoides*, the fimbriae are found as a tuft at the pole opposite to that of the flagellum.[684,685] Fimbriae of one cell are observed to stick to fimbriae of adjacent cells. Thus, the cells in a star are initially held together by their fimbriae. After the initial event, contraction of the fimbriae follows, uniting a number of cells by their poles to a stable and apparently irreversible cluster that is more or less star shaped. A star comprises a few cells up to hundreds of cells. The unipolar location of fimbriae on each component cell is clearly responsible for the radial arrangement. The fimbriae of starformers can attach not only to fimbriae of kindred cells but to other cells as well. They bind to red blood cells, causing hemadsorption that is inhibited by D-mannose. Fimbriae of nonstarforming mutants appear to have lost their cohesive and adhesive properties as well as their ability to contract.[685]

Essentially the same events occur during star formation in *Agrobacterium luteum*.[686] The unipolar fimbriae of this peritrichously flagellated organism can be as long as 10.5 μm. Swarmer cells presumably cohere by their fimbriae and are pulled together, upon contraction of the fimbriae, into a star.

It is assumed that fimbriae found as a tuft at one pole of the bacillus serve as recognition sites and as points of initial contact prior to the irreversible union of competent cells.[687] The stable union thereby helps secure the transfer of DNA from one cell to another. It may be further assumed that the competence of cells to form stars is brought about by the presence of active fimbriae. Only fimbriae involved in star formation may be described as contractile.[685-687] The contraction that follows the initial contact appears to start at the base of the fimbriae.[687] In *Pseudomonas echinoides*, the fimbriae dissociate into tubelike fibers which are 2.0 to 2.5 nm in diameter. At the start of the contraction, three to six of these fibers are observed, but the number increases as the distance between cells decreases. While the single fibers shorten, their diameter remains constant. When the contraction is completed, the fibers lose their tubelike property.

Further elucidation of the nature of active fimbriae has been derived from a study of a nonstarforming mutant of *P. echinoides*. The fimbriae of this mutant are noncontractile. A single fimbria is a hollow cylinder with an outer diameter of about 6.5 nm and an inner diameter of about 2.5 nm.[198] The cylinder is made up of three helices, each containing six identical subunits, which are globular and proteinaceous. The subunit has a molecular weight of about 8×10^3 daltons as detected by SDS gel electrophoresis. Based upon a comparison of the contractile and the noncontractile structures, a model for the contraction of fimbriae has been proposed.[198] Contraction is thought to be a continuous rearrangement of the putative subunits with respect to one another.

Conjugation in the actively starforming *Rhizobium lupini* is characterized by its isosexuality.[680] A cell can serve as donor or recipient interchangeably. In spite of the interchangeability of roles, however, the transfer of genetic material is always unidirectional in a conjugation pair. Whether a cell serves as recipient or as donor depends on the physiological state of the cell. Recipient ability is observed only during the logarithmic phase of a culture, whereas donor ability is achieved only during stationary phase. A one-way transfer from a defined donor strain (male) to a defined recipient strain (female) has been achieved by selective elimination of fertility genes, which determine the role of a given strain as a donor or as a recipient.[680-682]

Unlike in *E. coli*, conjugation in *R. lupini* is not directed by a conjugative plasmid. Rather, it is a direct transfer of a chromosomal segment from a donor cell to a recipient.[681,682] It is specified by two fertility genes, called *Fc* and *Fd*, located diametrically opposite each other on the bacterial chromosome (Figure 6). Respectively, the fertility genes are found *c*lose or *d*istal to a pigmentation marker called *car* (for carotenoid pigmentation). They split the circular chromosome into two partially overlapping regions, *c* and *d*. A donor cell, in one conjugation event, transfers either one of these two DNA pieces. *Fc* leads the transfer of the *c* region, *Fd* that of the *d* region. Both are transferred in the same direction. The fertility genes are inactivated by mutagenesis, by SDS treatment, or spontaneously upon prolonged storage. By these treatments, four fertility types have been recovered: Fc^+Fd^+ (wild-type

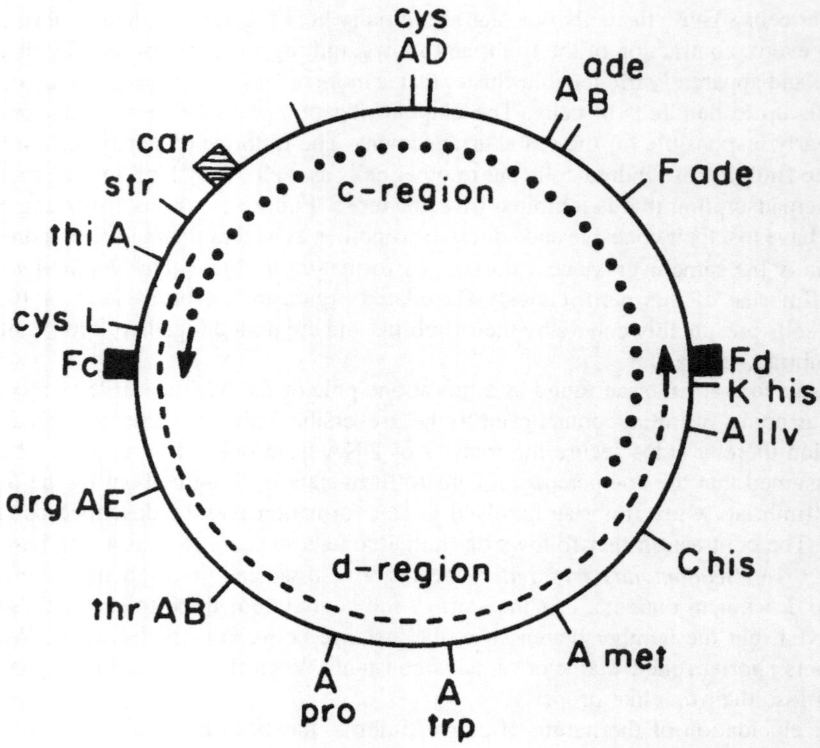

FIGURE 6. Chromosome map of *Rhizobium lupini*. The circular chromosome is divided into two overlapping transfer regions, c and d. Arrows point to the direction of transfer. The fertility genes, Fc and Fd, are transferred as leading elements of their respective regions. The c region is closely linked to car (for carotenoid pigmentation), d is distally linked to it. The symbol str stands for streptomycin resistance. The others are auxotrophic markers. (From Pühler, A. and Burkardt, H.-J., *Mol. Gen. Genet.*, 162, 163, 1978. With permission.)

donor for both regions), Fc^+Fd^- (donor for *c* region only), Fc^-Fd^+ (donor for *d* region only), and Fc^-Fd^- (recipient only, no donor ability). Thus, a donor needs only one of the fertility genes. The wild type has both, even as the recipient need not have any. These chromosomal genes have been shown to be inhibited by the resistance plasmid RP4.[688] An RP4-containing donor strain loses the capability to transfer chromosomal genes. It recovers its transfer capability only after the ejection of the plasmid.

The carotenoid marker has been used in the genetic analysis of conjugation in *R. lupini* to demonstrate the occurrence of merodiplonts.[682] From mutant haploid parents that are differently colored colonies on an agar plate, the merodiplont is identified by its additive pigmentation. Thus, a "yellow × pink" cross and a "yellow × white" cross both yield a brown merodiploid clone, and a "white × white" cross yields a yellow merodiploid clone. The parental classes are recovered upon segregation. As many as 5% of the transconjugants formed are merodiplonts. After chromosomal transfer, the donor DNA is somehow integrated into the chromosome of the erstwhile haploid recipient, now a zygotic cell. Genetic analysis of merodiploidy suggests that the two homologous regions are inserted in tandem sequence.[689]

For all their supposed differences, aggregation in starforming bacteria and mating-aggregate formation in *E. coli* do not differ much. Star formation is mediated not by plasmid-specified sex pili, but by a unipolar tuft of fimbriae. The interaction of fimbriae is an initial event in a sexual development that includes some form of copulation, conjugation, and zygote formation. The attachment sites are the fimbriae at one of two poles of a bacillus, enabling many cells to cluster by their poles around a center to form a structure that resembles a many-pointed star. Contraction of the fimbriae allows more intimate contact between cells. Thereafter occurs the transfer, by an unknown mechanism, of chromosomal DNA of the donor to the recipient. As more and more is learned about this system, it appears to be no more different from conjugation systems among members of the Enterobacteriaceae after all. Star formation, an obvious prerequisite to sex, is one aggregation system that points to a rule of thumb that cells in aggregates need only be examined more closely for one to uncover a sexual function. The phenomenon itself may be more widespread than at present known.

IV. PELLICLE FORMATION IN FIMBRIATE BACTERIA

A more general phenomenon than star formation is pellicle formation among fimbriate strains of bacteria. The pellicle consists of cells densely packed at the air-liquid interface of tube cultures. It is grossly visible in nutrient broth as a layer or a film or, in some instances, as a rim or a ring of cells. The phenomenon is observed soon after the end of logarithmic growth in aerobic, static cultures. It is not observed in static cultures incubated anaerobically or in cultures aerated by continuous agitation. Secondary population growth usually occurs following pellicle formation, but because pellicles are formed at the end of logarithmic growth, it may be assumed that the phenomenon is not merely due to the failure of cells to separate after cell division. Nor is it merely a case of colony formation in liquid medium under practically nondispersive conditions. Thus, pellicle formation is truly within the bounds of what constitutes microbial aggregation. It will become clear as the discussion develops that star formation is merely a special case of pellicle formation. Both phenomena are apparently mediated by fimbriae.

A digression on the current usage of the terms pili and fimbriae can no longer be avoided. Indeed, the digression at this point is just appropriate, for the experimental work on pellicle formation has been to a large extent a study of the fimbriate state. Between pili and fimbriae, which are often used interchangeably, the term fimbriae is the more general. It is used to refer to nonflagellar filiform appendages of bacteria.[14,204,690,691,698,701] Unlike flagella, fimbriae are not demonstrable with the light microscope. Even after treatment with a flagellar stain, they remain far below the resolving power of the finest optical system. It takes the electron microscope to define their shape (Figure 7). Generally thinner, straighter, and much greater in number than flagella, they appear to be unrelated to locomotor function. Nonfimbriate cells can be motile. Fimbriate cells are motile only if they possess flagella. Moreover, although both are made up of protein subunits, flagella and fimbriae are antigenically different from each other.[692-694]

The existence of appendages finer than flagella on the surfaces of bacterial cells was first reported independently by Anderson and by Houwink in 1949.[695,696] They were called filaments by Houwink and van Iterson,[697] who first clearly distinguished them from extracellular slime and flagella. They were called fimbriae (singular, "fimbria", Latin for thread or fiber) by Duguid,[690] pili (singular, "pilus", Latin for hair) by Brinton,[698] and are known by various other names. Avoiding the controversy as to the legitimacy, linguistic propriety, or priority of the terminology,[551,699,700] we shall adopt fimbriae as the more encompassing term and use pili to refer to a special type of fimbriae, implicated in sex, determined by conjugative plasmids, and further characterized by receptor activity for pilus-specific bacteriophages.[14]

Fimbriae have been found in many species of bacteria.[14,204,551,701] To date, however, most of the reported fimbriate species are Gram-negative bacteria, mainly among members of the

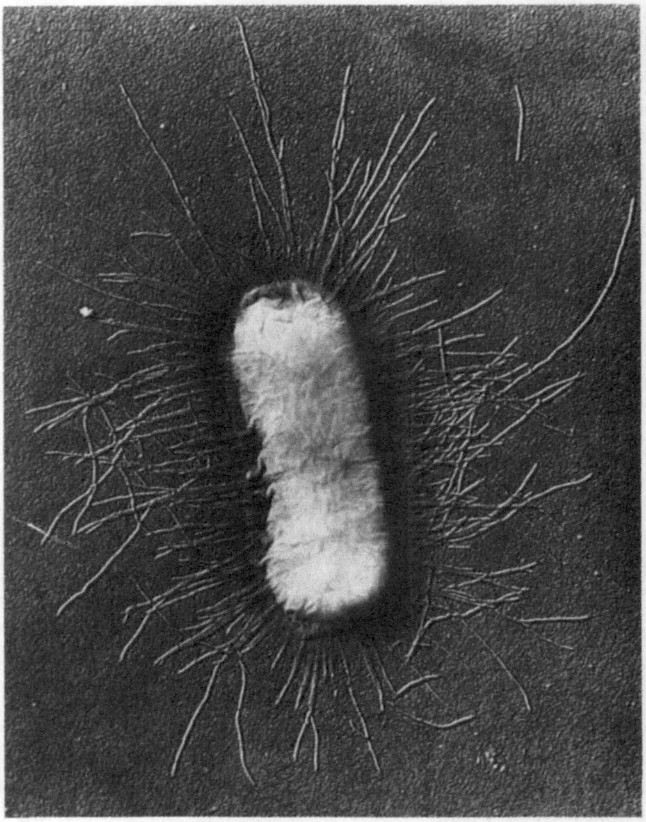

FIGURE 7. The fimbriae of *Klebsiella* sp. Electron micrograph of a cell from a 48-hr broth culture at 37°C. The specimen was shadow-cast with gold-palladium at 15°C. The capsule, barely visible and much shrunken by drying, obscures the cell wall and the basal portions of the fimbriae. (Magnification × 30,000.) (From Duguid, J. P., *J. Gen. Microbiol.*, 21, 271, 1959. With permission.)

families Enterobacteriaceae (*Escherichia, Salmonella, Shigella, Klebsiella, Enterobacter, Serratia,* and *Proteus*), Pseudomonadaceae (*Pseudomonas*), Vibrionaceae (*Vibrio, Aeromonas,* and *Photobacterium*), and Rhizobiaceae (*Rhizobium* and *Agrobacterium*).[197-204,208,551,684-686,697,701-713] They have been found also in species of *Caulobacter,*[204,714,715] *Chromobacterium,*[703] *Fusobacterium,*[716] *Moraxella,*[207,717-719] *Pasteurella,*[720] and *Neisseria.*[721-724] Thus, they are found in diverse species, among aerobic, facultatively anaerobic, or anaerobic rods or cocci. In addition, they have been reported in the Gram-positive *Corynebacterium renale*[725-727] and *Streptococcus sanguis.*[728,729] A more extensive survey will undoubtedly enlarge the number of known fimbriate bacterial species and may settle the conjecture as to whether the presence of fimbriae is principally restricted to Gram-negative species. It is often observed that strains freshly isolated from nature are usually fimbriate.[208,691,698,704,730,731]

The number of fimbriae per cell can range from 1 to 1000.[14,204] Fimbriae are arranged peritrichously (distributed all over the cell surface) or as polar tufts (monopolar, as in the starformers, or bipolar).[701] Sticking out from the cell surface, they are perhaps rooted in the cytoplasm (Figure 7). Their lengths, in general, can vary between 0.2 to 20 μm, their widths between 3 to 14 nm.[14] They may be tapered or uniformly wide along their entire length.

Fimbriae are made up of protein subunits.[551,705,708,747] Diffraction patterns of purified type-1 fimbriae from *E. coli* indicate that the subunits, each with a molecular weight of approximately 16 × 10^3 daltons, are arranged in a helix, with 25 subunits per 8 turns of the helix.[551]

Six fimbrial types (designated I to V and F), including F pili as a separate type, have been identified in *E. coli* by Brinton.[551] Seven types (types 1 to 6 and F) are included in the classification scheme proposed by Duguid and associates.[704] This slightly more comprehensive schema will be adopted here, despite the existence of a third and more recent list provided by Ottow.[14]

Type 1 fimbriae, distributed peritrichously on the cell surface, are hollow cylinders varying in length between 0.2 and 1.5 μm.[204] Their external diameter in *E. coli* is about 7 nm, the diameter of the axial hole 2.0 to 2.5 nm.[551] Between 50 and 400 are found per bacterium.[204] They are widely distributed among the enterobacteria. They are found in *E. coli, Shigella flexneri, Klebsiella aerogenes, K. pneumoniae, Enterobacter cloacae, Serratia marcescens,* and in most serotypes of *Salmonella*. Cells with type 1 fimbriae cause hemagglutination of red blood cells from various sources as well as the agglutination of fungal, plant, and other animal cells. Hemagglutination is sensitive to disruption by D-mannose and α-methyl-D-mannoside.[201,732] This fimbrial type is particularly interesting because it, perhaps alone among the various types, is implicated in pellicle formation.

Type 2 fimbriae, found in a few strains of *Salmonella* (*S. pullorum, S. gallinarum,* and *S. paratyphi* B), are indistinguishable from type 1 fimbriae in number, arrangement, and dimensions.[703,704] They are different in that they are nonadhesive and nonhemagglutinating. They are believed to be *crm* (for cross-reacting material) mutants of type 1 fimbriae.[204,692]

Type 3 fimbriae, found in *Klebsiella aerogenes* and *Serratia marcescens,* are thinner (approximately 5 nm in diameter) and more numerous (50 to 1000 per cell) than type 1 fimbriae.[202,204,705] They resemble type 1 fimbriae in length and in being peritrichously arranged. Cells with type 3 fimbriae cause hemagglutination of red blood cells treated with tannic acid or heated at 70°C. This hemagglutination is resistant to D-mannose.

Type 4 fimbriae are even more slender (approximately 4 nm in diameter).[711] Found in species of *Proteus (P. mirabilis, P. vulgaris, P. rettgeri,* and *P. morganii),* they differ from type 3 fimbriae in being implicated in the hemagglutination of fresh red blood cells (not treated with tannic acid).[703,711-713] Also peritrichously arranged, they are similar to type 3 fimbriae in length and in number.

Type 5 fimbriae comprise the unipolar appendages of the starformers (see Section III. of this chapter). They bring about, aside from star formation, hemadsorption onto red blood cells but not hemagglutination because of their unipolar distribution, which renders the bacterial cells functionally univalent.[198,684-687]

Type 6 fimbriae, found in a few strains of *Klebsiella ozaenae,* are thick (approximately 10 nm in diameter), very long (up to 10 μm), and scanty (4 to 40 per cell).[704] In some cultures, they exhibit hemagglutinating activity.

F and F-like pili, or sex fimbriae, are grouped together to form the seventh type (see Section II of this chapter). They fall into group 2 in Ottow's list, which includes six groups.[14] The first four fimbrial types of Duguid et al. are bunched together in group 1. Group 5 corresponds to type 5, but three new groups are not represented in the older classification scheme.

None of the three schemes of classification is comprehensive enough to accomodate all the kinds of fimbriae that have been described in the literature. All that is needed for now is to be able to make the general statement that mating-aggregate formation in *E. coli* is mediated by sex pili, star formation by the unipolar type 5 fimbriae, and pellicle formation by type 1 fimbriae, but even that minimal need is hardly served. Pellicle formation is observed chiefly among escherichiae, shigellae, klebsiellae, and salmonellae, organisms that have been shown to bear type 1 fimbriae. However, it has been observed also in vibrios[208] and *Proteus hauseri*,[712] neither of which is accomodated in the classification schemes.

A culture of a fimbriate strain usually contains cells that are fimbriate and cells that are not.[201-206] Depending upon the cultural conditions, the composition of a population can change from almost wholly fimbriate to almost wholly nonfimbriate. Strains of *Shigella flexneri*, for example, can be converted to the fimbriate phase by serial culture in tubes incubated aerobically but without agitation.[201] They are converted to their nonfimbriate phase by continued aerobic culture on agar plates, or in agitated broth, or in static broth incubated anaerobically. The fimbriate-phase change of a culture requires many generations (20 to 100). Fairly complete conversion from one phase to the other requires several (three or more) serial transfers (Table 1). At the level of the individual cell, it has been estimated from the ratio of the fimbriate to the nonfimbriate in a culture derived as a clone that the conversion from one phase to the other occurs at the rate of about 1×10^{-3} cell division. The rates are the same in either direction.

It is presumed that fimbrial phase variation is the consequence of a mutational event.[691] Conversion of a culture from one phase to the other is a matter of selection of mutants and not a direct influence of the environment on the synthesis of fimbriae. Aerobic and static conditions merely favor the growth of fimbriate cells and allow them to outgrow cells without fimbriae. Conversion to the fimbriate phase of a culture is indicated by pellicle formation and a three- to fourfold increase in the maximum population density. This increase can be attributed to the greater availability of molecular oxygen to the cells in the pellicle, thereby allowing them further growth, even as the nonfimbriate cells in the depths of the broth are inhibited by the anoxic conditions. Indeed, anaerobically grown fimbriate cells yield a biomass no greater than that of nonfimbriate cells and no pellicle. In agitated broth, where pellicle formation does not take place either (obviously due to the disruptive condition), maximum population density of a fimbriate culture is about the same as that of a nonfimbriate, because both cultures have equal access to molecular oxygen.

In the same table (Table 1) is shown the rise in hemagglutinating activity of cultures as they are converted to the fimbriate phase. Thus, the change to the fimbriate phase of a culture is accompanied by a rise in hemagglutinating activity and by pellicle formation, both of which are ascribable to the adhesive properties of fimbriae.

The growth curves of fimbriate and nonfimbriate strains of *Salmonella choleraesuis* under various conditions are graphically compared in Figure 8. The nonfimbriate strain shows no pellicle formation in any of the three cultures. The facultatively anaerobic trait of the organism and the enhancing influence of aeration on growth are shown by the difference in the maximum population density between aerobic (shaken) and anaerobic cultures. With regard to this parameter, the strains are indistinguishable. However, the population maxima are significantly different when the two strains are grown under aerobic, static conditions: the fimbriate strain attains a greater maximum; more important, it forms a pellicle, which becomes grossly visible at about 10 hr, at the start of secondary growth. Clearly, the additional growth may be ascribed to pellicle formation.

Fimbrial phase variation confers on an organism the advantage of added versatility.[204,691] With the ability to convert from one phase to the other, bacterial cells in the fimbriate phase can adhere to surfaces of animal, plant, or fungal cells, while their nonadhesive, nonfimbriate forms can disperse to get to fresh substrates. The potential for two forms improves the chances for survival of an organism as a commensal, a parasite, or a pathogen. Under anoxic conditions, saprophytic enterobacteria gain access to molecular oxygen by growing and interacting as a pellicle. It has been noted that nonfimbriate cells of *E. coli* K-12 are more efficient fermenters than their fimbriate forms.[773] The nonfimbriate, it appears, are better adapted for growth under anaerobic conditions.

Table 1
GROWTH, PELLICLE FORMATION, HEMAGGLUTINATING ACTIVITY, AND FIMBRIATION IN SERIAL, AEROBIC, STATIC BROTH CULTURES OF *SHIGELLA FLEXNERI* AT 37°C

	From colony 1				From colony 2			
N	G	P	H	F	G	P	H	F
1	0.29	–	0	0	0.28	–	0	0
2	0.31	–	0	0	0.29	–	60	20
3	0.32	–	40	15	1.01	+	1500	95
4	1.12	+	1500	95	1.16	+	1500	
5	1.29	+	2000		0.80	+	1500	
6	1.05	+	1500		0.87	+	2000	
7	0.94	+	2000		0.93	+	2000	
8	1.06	+	2000		0.91	+	2000	

Note: N, serial culture number, represents a 48-hr period of incubation. The first tube of broth (10 mℓ) was inoculated with a nonfimbriate colony on agar. Subsequent cultures were inoculated with cells from the surface of the preceding culture (ca. 0.1% of cell population). The two sets of data were derived from two separate colonies picked at random. Each unit of growth, G, is equivalent to 10^9 cells $m\ell^{-1}$. Pellicle, P, is indicated by –, if not observed, or +, if present. Hemagglutinating activity, H, is expressed in the units of Duguid and Gillies.[201] F, the percentage of fimbriate cells, was determined by electron microscopy. Data from Duguid and Wilkinson.[691]

A change in the conditions of the environment may just favor one phase at the expense of the other. Pellicle formation, as an alternative and alternate phase, is a response that enables the organism to make do in a rapidly deteriorating situation. It is obviously an advantage to an organism to be able to assume two forms, equally viable, one of which, however, is more viable than the other under a given set of conditions.

Fimbriate and nonfimbriate strains can be distinguished from each other by electron microscopy, electrophoresis, and colonial morphology. On agar plates, colonies of fimbriate cells are smaller and smoother compared to the nonfimbriate.[698] The electric charge density on the cell surface of cells with type 1 fimbriae is considerably less than that of the nonfimbriate cells.[707] The electrophoretic mobility of fimbriate cells is only about half.[698] Defimbriated cells behave like nonfimbriate cells and regain their characteristic electrophoretic mobility upon regeneration of fimbriae.[734] That indeed fimbriae make the difference in the behavior of cells is indicated by the similarity in electrophoretic mobility between fimbriate cells and a preparation of isolated fimbriae.[698]

As already mentioned, cultures of fimbriate strains usually contain both fimbriate and nonfimbriate cells, as well as scarcely fimbriate ones. In general, fimbriation is enhanced by serial culture in broth under aerobic, static conditions and inhibited by plating on agar. Fimbrial phase transition is particularly evident in *Shigella flexneri*. However, some strains do not exhibit phase transition and others do not quite lose all their fimbriae. Also, phase transition is sometimes too rapid to be explained by selection of mutants already in the culture.[691]

Nonfimbriate mutants, called Fim^-, of the fimbriate wild type (Fim^+) remain nonfimbriate under any circumstance. Mating with a Fim^+ or $Fim^{(+)}$ donor makes the Fim^- mutant fimbriate again.[735] Fim^+ is genotypically and phenotypically fimbriate, whereas $Fim^{(+)}$ is genotypically fimbriate but phenotypically nonfimbriate. Aside from these three classes,

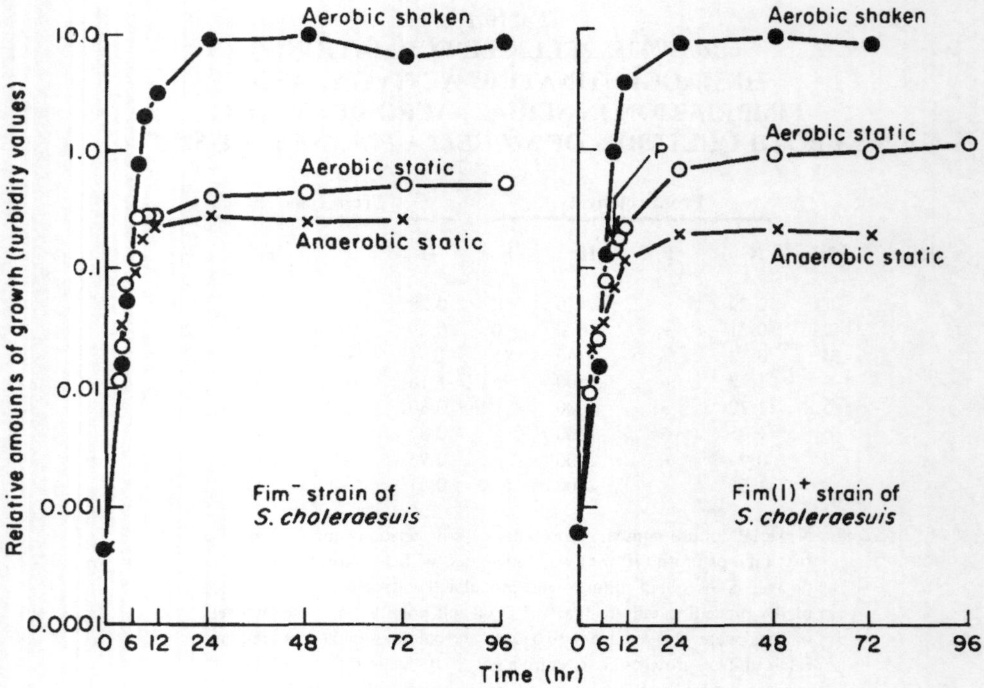

FIGURE 8. Growth curves of Fim$^-$ and Fim$^+$ strains of *Salmonella choleraesuis* in nutrient broth at 37°C. The inocula were derived from 48-hr broth cultures. Aerobic, static cultures (○) were grown in cotton-plugged test-tubes containing 10 mℓ of medium, incubated in air, and left undisturbed. Anaerobic, static cultures (x) were grown in similar tubes incubated in an anaerobic jar filled with hydrogen. Aerobic, shaken cultures (●) were grown in 40 mℓ of medium in 1-ℓ bottles shaken at 20 r/min. The time at which a pellicle in the aerobic, static culture was first observed is indicated by P. (From Old, D. C., Corneil, I., Gibson, L. F., Thomson, A. D., and Duguid, J. P., *J. Gen. Microbiol.*, 51, 1, 1968. With permission.)

there are mutants called Fim$^{\sigma+}$ (stable fimbriate) and Fim$^{\sigma-}$ (stable nonfimbriate). The relationships among these classes are as follows:[736]

The mutation from Fim$^+$ to Fim$^-$ is a rare event. The reversible Fim$^+$ $\rightleftharpoons$ Fim$^{(+)}$ interconversion corresponds to the commonly observed fimbrial phase variation. Fim$^{\sigma+}$ differs from Fim$^+$ in that it is stable and thereby not subject to fimbrial phase variation. Fim$^{\sigma-}$ differs from Fim$^-$ in that fimbriate recombinants can be obtained from crosses between Hfr Fim$^+$ and F$^-$ Fim$^{\sigma-}$. With Fim$^-$ as recipients, no recombinants are fimbriate. The results of such crosses suggest that Fim$^{\sigma-}$ retains some other component (perhaps a second chromosomal gene for fimbriation or a cytoplasmic factor) associated with the Fim$^{\sigma+}$ condition, whereas Fim$^-$ does not bear such an extra component.

Three models have been advanced to accomodate the observations.[736,737] All three postulate the existence of two genes, *A* and *B*, which are closely linked when functional. One model calls for *A* and *B* being chromosomal genes. *A* determines the synthesis of fimbriae, *B* the type of fimbriation. Thus, $Fim^{\sigma +}$ is genotypically *AB*, Fim^{+} is *Ab*, $Fim^{\sigma -}$ is *aB*, and Fim^{-} is *ab*. In the second model, both chromosomal genes are responsible for the synthesis of fimbriae, whereas the type of fimbriation is controlled by something else. Thus, both $Fim^{\sigma +}$ and Fim^{+} are genotypically *AB*, $Fim^{\sigma -}$ is *Ab*, and Fim^{-} is *aB*. In the third model, *A* is a chromosomal gene that controls the synthesis of fimbriae, *B* is an episome, which alternates between an autonomous state of existence and a state of integration into the host chromosome. *B* is assumed to determine the type of fimbriation. $Fim^{\sigma +}$, as in the other two models, is *AB*, when the episome is integrated. Fim^{+} corresponds to *A(B)*, when the episome is in its autonomous state. The spontaneous loss of *B* generates a nonfimbriate cell, which may be either *A* ($Fim^{\sigma -}$), if it is a simple loss, or *a* (Fim^{-}), if it is due to a mutation in *A*, upon which the maintenance of *B* depends. Fimbrial phase variation, if it is a mutational event rather than a repression-derepression system, may be analogous to flagellar phase variation in *Salmonella*, wherein the switch from one phase to the other seems to be due to site-specific inversion of a segment of DNA.[738-740]

The genetic information for the synthesis of fimbriae may be transferred from one species to another (from *E. coli* to *Salmonella*, for instance) by conjugation or by transduction (with the transducing bacteriophage P1, for instance).[741-743] A nonfimbriate *E. coli* mutant that failed to ferment arabinose and required threonine and leucine for growth was used as recipient. Fimbriate donor cells were infected with P1. Their resultant progeny were then used to infect the recipient cells. From the results of these experiments, plus those of interrupted matings between a fimbriate Hfr and a nonfimbriate recipient, was deduced the close linkage of the genetic information for the synthesis of fimbriae (*fim*) to the genes *thr* (for threonine requirement) and *leu* (for leucine requirement).

The *fim* site (now called *pil* for pilus), which determines the presence or absence of fimbriae, is found at 98 min on the *E. coli* chromosome, according to a recalibrated linkage map of *E. coli* K-12.[744] On the same closed circular chromosome of over 650 identified loci distributed over a normalized length of 100 min, *thr* is 2 min away clockwise at 100 min or 0 min, *leu* is a little farther clockwise, close to 1.5 min, and *azi* (for resistance to sodium azide or to phenethyl alcohol) is found at just a little over 2 min. Prior to the *fim* locus, *cyc C* (for resistance to cycloserine) maps at 94 min and *thi* (for thiamine thiazole requirement) is close to 89 min. In *Salmonella typhimurium*, the *fim* locus (also renamed *pil*) is assigned quite close to 14 min, where *lip* (for lipoic acid requirement) is also located, on the circular linkage map of more than 420 known genes.[745]

Just precisely how fimbriae mediate pellicle formation is not yet clear. The fimbriae on the surface of one cell may be envisaged to bind to the fimbriae of another cell, or, perhaps, to some surface components other than fimbriae. The cohesive forces in a pellicle may partly derive from hydrophobic interactions due to the predominance of nonpolar side chains of the protein subunits of fimbriae.[551] Hydrophobic interactions, however, would lack specificity. If the contacts are solely fimbrial, they may simply involve protein moieties. More likely, a pellicle is the consequence of the interactions between proteins and carbohydrates, the fimbriae of one cell being the proteins, the as yet unknown receptors on the surface of the other cell the carbohydrates. This judgment comes from the inhibitory effect of sugars on pellicle formation.

In *S. typhimurium*, pellicle formation is specifically inhibited by the presence of D-mannose in a broth culture.[203] Mannose derivatives, such as α-methyl-D-mannoside, a sugar not utilized by the organism, are also effective inhibitors, but not other sugars. The hemagglutinating activity of fimbriate salmonellae is likewise inhibited by D-mannose and its derivatives, indicating a similarity in specificity, if not in mechanism, between hemagglutination and pellicle formation. Modification of the hydroxyl group at the C-2, C-3, C-4, or C-6 position in the mannopyranosyl molecules results in the loss of inhibitory activity.[732] The α-linked mannopyranosides are better inhibitors than their β-linked anomers, demonstrating the importance of the C-1 position.

Although not obviously involved in conjugation and DNA transfer, pellicle formation may contribute to these processes — if there is any sex within the confines of the pellicle — simply by prolonging stable contacts among cells. Indeed, type 1 fimbriae have been reported to increase the rate of transfer of plasmids during conjugation.[746,747] Transfer of *col I* does occur in crosses between a Fim$^-$ donor and a Fim$^-$ recipient of *Shigella flexneri,* but is greatly enhanced if the recipient is fimbriate. Disrupting the pellicle by gentle shaking negates this enhancement. In *Salmonella typhimurium,* the presence of fimbriae in either donor or recipient increases the transfer of the colicinogenic factor. In either system, it can be surmised that the stabilizing influence of the pellicle prolongs the contact between conjugant pairs such that the transfer is assured.

Pellicle formation appears to serve a more obvious function. A selective advantage of pellicle formation is strongly suggested by the results of competition experiments with fimbriate and nonfimbriate strains of *S. typhimurium.*[205] In mixed cultures of a very large number of Fim$^-$ cells and a very small number of Fim$^+$ cells (the ratio of Fim$^-$ to Fim$^+$ was greater than 10^7 at the time of inoculation), the Fim$^+$ cells rapidly outgrew the Fim$^-$ cells, such that in 48 hr under aerobic, static conditions in the tubes, their number surpassed that of Fim$^-$ cells. Pellicles consisting of fimbriate cells were observed between 6 to 24 hr. The nonfimbriate cells remained in the oxygen-depleted depths of the tubes. Selective outgrowth of Fim$^-$ by Fim$^+$ was not observed under conditions that did not allow pellicle formation, i.e., in mixed cultures on agar plates, in static broth incubated anaerobically, or in broth aerated by continuous agitation. Furthermore, it was inhibited by α-methyl-D-mannoside, which prevented pellicle formation. Use of motile variants (*fla*$^+$) indicated that selective outgrowth was enhanced by motility. Drawn aerotactically to the surface of the broth, motile cells formed pellicles more rapidly than nonmotile cells, whose locomotion would be random rather than directed. Although motile Fim$^-$ cells outgrew their nonmotile counterparts, the competitive edge conferred by motility was not as strong as that due to fimbriation.

Because of the adhesive properties of fimbriate cells, a possible role of type 1 fimbriae for promoting infection has been considered. Some pathogenic strains of escherichiae, shigellae, and salmonellae turn out to be genotypically nonfimbriate, ruling out the essentiality of fimbriae for infection.[201,704,748] More recent experimental-infection studies with *S. typhimurium,* however, suggest that fimbriate strains are more pathogenic than nonfimbriate ones.[749] Inoculated into mice by the oral route, a Fim$^+$ strain caused more infections and deaths than the Fim$^-$ strain, from which the fimbriate strain was derived. The inoculated organism was also recovered more often from the feces of animals infected with Fim$^+$ than from the feces of those infected with Fim$^-$. Inoculated i.p. or by way of the conjunctiva, the strains produced about the same degree of infectivity.

The wide distribution of fimbriate species and their proven capacity for binding to cells of their own kind as well as to other kinds of cells suggest a similarly wide distribution of bacterial aggregation, the formation of pellicle, star, mating-aggregate, or other aggregation configurations. To find an aggregation system among other bacteria seems a matter of exploring the various fimbriate species and strains for aggregation under a wide variety of conditions. Whether pellicle formation contributes to survival, sex, pathogenicity, or some other function of selective advantage — or merely one of the many consequences of the fimbriate state that may or may not be advantageous, depending upon the environmental conditions the cells may find themselves — whatever its role, pellicle formation is a phenomenon that cannot be dismissed as trivial. If it makes for no other generalization, the phenomenon undeniably implicates surface appendages in cell-cell interaction.

V. AGGLUTINATION ASSOCIATED WITH BACTERIAL TRANSFORMATION

The sexual process in unicellular systems is essentially the transfer of genetic material from one cell to another. The transfer may be accomplished directly by cell-cell contact or indirectly at a distance. Direct transfer is accomplished by transient conjugation, as in enterobacteria, or by irreversible fusion of cells, as in yeasts. Indirect transfer, where donor and recipient are not together, is exemplified by virus-mediated transduction or by transformation.

Transformation is the process by which a naked fragment of foreign genetic material becomes a functional part of the genetic program of a cell. Unlike conjugation or transduction, transformation is a direct interaction between the cell surface and DNA, free and exposed. It involves the development of competence of the recipient, the uptake of DNA, its integration into the genetic mechanism of the recipient cell, and its expression in the form of an observable phenotype. First observed in the pneumococcus, the phenomenon is known in a number of bacterial genera, among them *Haemophilus, Neisseria, Rhizobium, Bacillus, Streptococcus, Moraxella, Micrococcus,* and *Escherichia,*[750-757] and recently in budding yeast.[758,759]

The phenomenon has not suffered from neglect since its discovery by Griffith in 1928.[760] Many reviews on the subject have since appeared, including the classical paper that ushered in molecular biology, the paper by Avery et al.[761] The reader is referred to a few later ones.[762-766,791] The discovery by Avery and colleagues that DNA is the transforming principle led directly to the current dogma that DNA is the genetic material, in cellular systems, anyway. More recently, transformation in *E. coli* has come to the fore as the basis for recombinant DNA technology and controversy.

Competence of a prospective transformant cell is a necessary condition for DNA uptake. It is operationally defined as the ability of the cell to irreversibly adsorb DNA from the immediate environment in such a manner that DNA becomes resistant to nuclease activity. Competence is heritable, for mutant strains remain incompetent. However, it is not constitutive. In some systems, competence develops only under certain conditions. The region of the growth curve of a batch culture when competence develops varies with the species. The time of peak competence is at the beginning of exponential growth in streptococci and *Rhizobium.*[754,767] It is late exponential or near-stationary in *Haemophilus* and *Bacillus subtilis.*[228,768,769] In *Neisseria* and *Micrococcus radiodurans,* it is not associated with any distinct phase of growth,[756,770,771] but a critical range of cell concentration is required for competence development in pneumococci.[772,773] The Wicky strain of streptococci requires the addition of extracts derived from competent cells.[774] All these show that competence development is inducible.

A number of nutritional conditions have been reported to inhibit or stimulate competence. Some of these conditions are found in Table 2, Chapter 3. These various effectors might be partly accommodated by a suggestion that competence develops when cells are released from catabolite repression.[775] Cyclic AMP, which is known to relieve catabolite repression, has been reported to cause noncompetent cells to become partially competent.[776] Some of the inhibitors, such as the inhibitors of RNA synthesis, may indirectly inhibit the synthesis of proteins. In *Haemophilus influenzae,* active growth is inimical to competence development, but conditions must be such as not to prevent protein synthesis.[769]

The literature is replete with attempts to correlate the activity of the cell surface with the development of competence. There obviously must be some alteration of the cell wall in order to facilitate the passage of a macromolecule that is the transforming DNA fragment. An increase in autolytic activity has been implicated in pneumococci, in streptococci, and in *Bacillus subtilis.*[777-783] The site of autolytic activity may be the growing region of the cell. Leakage of DNA from competent cells into the medium has been reported in pneumococci and in *B. subtilis*.[784,785] Other physiological events appear to be correlated with competence development. These include an increase in the negative charge of the cell surface,[786] an increase in heat sensitivity,[787] an enhancement of resistance to penicillin,[788] and a decrease in the synthesis of nucleic acids but not of proteins.[789-792] In *Moraxella*, there seems to be a close correlation between competence development and fimbriation.[207,717,793,794]

An antigenic alteration of the cell wall is detected during competence development.[29,795-799] Antibodies to formalin-killed competent cells inhibit transformation. Sera from animals immunized with noncompetent cells do not show any inhibitory activity. Most interesting, however, is the finding that a protein extract appears to be required for inducing the capacity of the cells to bind DNA. It is called competence factor or CF.

CF has been reported in pneumococci, streptococci, *Haemophilus*, *Bacillus subtilis*, and *B. stearothermophilus*.[213,774,800-805] It may be cell bound[806] or found free in the culture medium.[774] Apparently required for DNA uptake, CF is synthesized by competent cells. However, streptococcal CF, a protein of about 5×10^3 daltons,[807,808] may be synthesized also by noncompetent cells in synthetic medium and by nontransformable strains.[809,810] Pneumococcal CF is protease sensitive, requires mercaptoethanol for optimum activity, and has a molecular weight of about 10^4 daltons.[210] Insensitive to pancreatic RNase, phospholipase C, and lysozyme, it lacks activity as an autolysin, a hemolysin, an exonuclease, or a protease.

There is evidence for another component involved in competence: the cellular receptor for the competence factor.[799,811] The receptors are identified by their activity of physically binding and reversibly inactivating CF. They are presumed to be located on the plasma membrane, because purified CF is inactivated by membrane preparations. They may be located within the same region as the growing zone of the cell envelope, perhaps the site of DNA uptake.[814]

The synthesis of proteins other than CF may be required for the uptake of DNA.[211] Addition of inhibitors of protein and RNA synthesis prevents the appearance of DNA-binding capacity. Chloramphenicol, which does not prevent CF binding, impairs both transformation and DNA uptake by CF-treated cells. The synthesis of these other proteins must then occur subsequent to CF binding by the presumptive cellular receptors of CF.

Also associated with competence development is the tendency of competent cells to aggregate among themselves. One of the proteins synthesized during competence development could be an agglutinin, which presumably mediates the cell-cell interaction.

The tendency of competent cells to aggregate may have been noted by early workers, but the first report seems to be that of Javor and Tomasz.[812] They observed that competent pneumococcus cells tended to clump in the growth medium. Before competence development, the cells showed no signs of clumping. This observation clearly indicates that mural changes during the development of competence are gross enough as to be demonstrable in the form of cell aggregation.

Pakula and colleagues first demonstrated that competent cells of streptococci precipitated in dilute hydrochloric acid (10^{-3} to 10^{-2} *N*).[29] Noncompetent cells remained as a homogeneous suspension. Agglutination was observed with either Challis or Wicky strain of streptococci.

This demonstration of competence-associated agglutination was extended by Tomasz and Zanati to pneumococci.[30] A culture of pneumococci was split into two. The pH of both was adjusted to 8.0. To one was added CF derived from competent cells, to the other an extract of noncompetent cells. The CF-treated half, upon incubation at 30°C for 20 min, developed competence and exhibited agglutination in dilute acid (Figure 9). The other half remained noncompetent and nonagglutinative. Portions of either half were then subjected to spheroplasting treatment. Transferred to dilute acid, the CF-treated spheroplasts agglutinated. The control spheroplasts did not. The agglutinative activity was sensitive to trypsin digestion and hence was presumed to be a protein, an agglutinin.

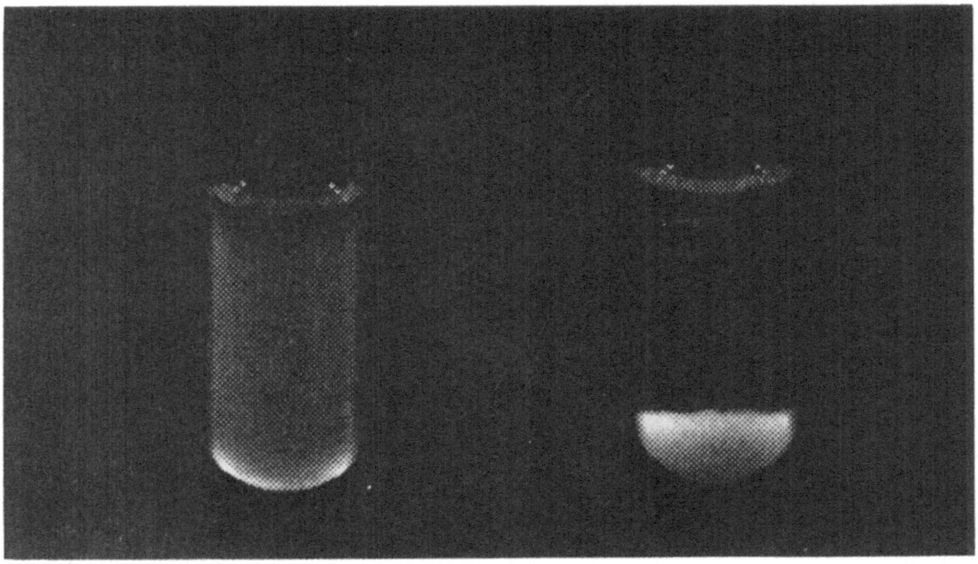

FIGURE 9. Agglutination of competent pneumococci in dilute acid. Left tube, cells treated with an extract of incompetent cells. Right tube, cells treated with CF from competent cells. (From Tomasz, A. and Zanati, E., *J. Bacteriol.*, 105, 1213, 1971. With permission.)

Agglutination was not due to CF molecules combining with their presumptive receptors on the cell surface.[30] Spheroplast membranes bound CF but did not give the agglutination reaction. Competent cells, presumably loaded with cell-bound CF, agglutinated in dilute acid; so did noncompetent cells to which CF had been added. These competent cells, incubated with chloramphenicol for 90 min at 37°C, lost competence and cell-bound CF, but their agglutinative property remained unaffected. Addition of chloramphenicol, however, before the cells became competent in the presence of CF, prevented agglutination, without affecting the binding of CF. In contrast, trypsin prevented both CF binding and agglutination. The induction to competence of noncompetent cells did not occur at pH 6.8; neither did the agglutinative property appear, although CF binding took place. Genetically incompetent cells did bind CF, but did not develop the agglutinative property.

Two models have been advanced to explain the above results.[30] The simpler one calls for the synthesis of the agglutinin being somehow triggered by the attachment of CF molecules to their presumptive receptors on the plasma membrane. The more complicated alternative is that agglutination is the result of an unmasking of sites on the plasma membrane by a conformational change due to the attachment of CF molecules, which behave as effectors. The inhibition of protein synthesis would lead to the inhibition of the synthesis of autolysins, which might be part of the unmasking process.

Because the phenomenon is demonstrated at rather low pH, it is not possible to state with confidence what role agglutination associated with competence development plays in genetic transformation. Nevertheless, the mural changes that occur during competence development are reflected in the tendency of competent cells to clump in the growth medium. The phenomenon has also been noted in *Bacillus subtilis*[31] and in *Haemophilus influenzae*.[1366] It remains to be seen how widespread the phenomenon is among the various transformation systems.

Transforming DNA is known to be released by cells into the culture medium.[784,785] The peak of extracellular DNA concentration is coincident with the peak of competence of the cell population. The clumps in the growth medium may include both recipients and donors. Tomasz has ventured to speculate that a clump may represent a rudimentary conjugation arrangement, whereby the release of DNA by a recipient is coordinated, temporally and spatially, with the uptake by a donor.[813] It is interesting that early workers on transformation regarded the presence of antiserum agglutinating the recipient cells as an environmental condition for competence.[761]

VI. FORMATION OF DENTAL PLAQUE AND AGGREGATION OF ORAL BACTERIA

Confluent microbial colonies found on tooth surfaces are called dental plaques. The Biblical skin of our teeth and an original source of Leeuwenhoek's animalcules, the dental plaque is so tenaciously attached to the tooth enamel such that it withstands the mechanical and chemical cleansing activities of the oral cavity. Although it is removable by routine brushing, it is resistant to rinsing with water. It may become mineralized into a calcareous deposit known as calculus. Much of the recent surge in research interest related to dental plaques derives from their likely causal relationship to dental caries and periodontal diseases. The subject has been amply reviewed.[37,815-831]

The microorganisms in dental plaque are mostly bacteria. It has been estimated from direct microscopic counts that there are about 2.5×10^{11} bacteria in a gram of plaque material.[832] However, only 10 to 20% of the total bacterial population is viable, anaerobic plating techniques favoring higher viable counts.

The bacterial community that constitutes the plaque is made up of diverse species. The cultivable microbiota includes the following groups: facultative streptococci (Gram-positive facultative cocci), facultative diphtheroids (Gram-positive facultative rods), anaerobic diphtheroids (Gram-positive anaerobic rods), peptostreptococci (Gram-positive anaerobic cocci), *Veillonella* (Gram-negative anaerobic cocci), *Bacteroides* and fusobacteria (Gram-negative anaerobic rods), *Neisseria* (Gram-negative facultative cocci), and vibrios.[832,833] The coccal forms *Streptococcus mutans, S. sanguis,* and *S. mitis* appear to be the predominant species on the tooth surface, but the catchall group of Gram-positive filaments is not far behind (Table 2). A definitive census of the bacterial population is hard to come by, not only because of the complexity of the oral flora, but because of its variability. The number and the proportional distribution of the various species vary from individual to individual, from tooth to tooth, and from time to time. Needless to underscore, they are dependent upon the state of health of the individual, his oral hygiene, and his nutritional habits. The dental plaque is indeed a very variable and very heterogeneous ecological microcosm. Add to that the cultural fastidiousness of some of its components, it is surprising that it is beginning to yield to the relentless probing by oral microbiologists.

The dental plaque is not just a mass of bacteria. It contains, in addition to its cellular components, acellular materials, the most important of which are polymers elaborated by bacteria and polymers, collectively referred to as mucins, found in the saliva. These polymers may mediate the aggregation of bacterial cells in the plaque by serving as matrix to hold the cells together and in place. Too, these polymers may be the active components in the formation of what is called the acquired pellicle, a thin, cell-free, amorphous film on the surface of the tooth.[818,834] It is commonly suggested that the bacterial colonies develop on this acquired pellicle, rather than directly on the enamel surface. Whether the pellicle is an accident or an absolute requirement for plaque formation is a moot issue. Bacterial cells absorb onto this acquired pellicle soon after a tooth is thoroughly cleaned. The microcolonies from this initial population develop in a few days, eventually coalescing to form a soft, firmly adherent mass that is the dental plaque. Coccal forms predominate during the early phase of development. But as the plaque matures, filamentous forms become more numerous.[835-837]

Table 2
DISTRIBUTION OF PROMINENT BACTERIA IN THE ORAL CAVITY[a]

Organism	Coronal plaque	Gingival crevice	Tongue dorsum	Buccal mucosa	Saliva
Streptococcus salivarious	<0.5	<0.5	20	11	20
S. mitis	15	8	8	60	20
S. sanguis	15	8	4	11	8
S. mutans	0—50	?	<1	<1	<1
Enterococci	<0.1	0—10	<0.01	<0.1	<0.1
Gram-positive filaments[b]	42	35	20	?	15
Lactobacilli	<0.005	<1	<0.1	<0.1	<1
Veillonella spp.	2	10	12	1	10
Neisseria spp.	<0.5	<0.5	<0.5	<0.5	<1
Bacteroides oralis	5	5	4	?	?
B. melaninogenicus	<1	6	<1	<1	<1
Vibrio sputorum	1	5	<0.5	<0.5	?
Spirochetes[c]	<0.1	2	<0.1	<0.1	<0.1
Fusobacterium spp.	4	3	1	?	<1
Haemophilus spp.	?	+	+	+	5

[a] Estimates are expressed as percentages of the total flora cultivable on anaerobically incubated blood agar.

[b] Include *Actinomyces, Nocardia, Rothia, Corynebacterium, Bacterionema, Leptotrichia*, and other Gram-positive filamentous forms.

[c] Include *Treponema* and *Borrelia* species.

From Gibbons, R. J. and van Houte, J., in *Textbook of Oral Biology*, Shaw, J. H., Sweeney, E. A., Cappuccino, C. C., and Meller, S. M., Eds., W. B. Saunders, Philadelphia, 1978, 684. With permission.

Both cell proliferation and surface interactions are overtly involved in the development of a dental plaque. Three surface-interaction processes may be discerned (Figure 10). Cells adsorb selectively onto the tooth surface by way of cell-substratum adherence (process 1). By homotypic cell-cell interaction, cells of one species may accumulate and maintain their positions in the growing plaque (process 2). The presence of other bacterial species in the plaque suggests the participation of heterotypic cell-cell interaction (process 3). Whereas these processes are experimentally separable, the proliferation of cells, the only effective way by which the total oral population increases, complicates the analysis. Moreover, the surface processes must be considered as temporally overlapping, initially reversible reactions.

Adherence, the first process in plaque formation, is clearly outside the confines of our definition of cell aggregation and will not be treated here to the same extent as the two other processes. It must be pointed out that the accumulation and retention of cells in a plaque can be accomplished by means other than cell aggregation. Cells may merely grow, divide, and retain contact with one another by chain formation. However, the evidence that will be presented below leaves no doubt that indeed aggregation processes take place and that plaque formation is not merely the failure of progeny to completely separate after cell division.

Aggregation of oral bacteria is a general term for a collection of cell-aggregation phenomena. Apart from adherence, there are both homotypic and heterotypic interactions involving various species of oral bacteria. Either may be mediated by polymers in the saliva or polymers synthesized by bacterial cells. Glucan-induced agglutination of *Streptococcus mutans*, the best studied of the bacterial aggregation systems, is an example of a homotypic interaction that appears to be mediated by bacterial polysaccharides. Saliva-induced aggregation of *S. sanguis* is a homotypic interaction that is possibly mediated by glycoproteins in the saliva. The co-aggregation of *S. sanguis* and *Actinomyces viscosus* is an intergeneric heterotypic system. Finally, there is acid-induced aggregation of plaque bacteria.

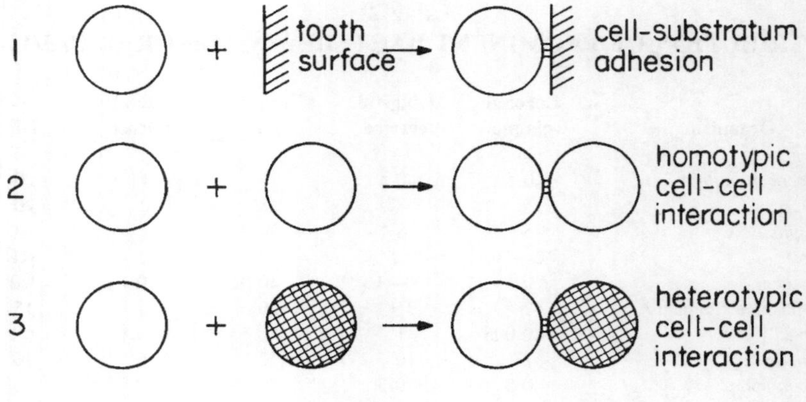

FIGURE 10. Surface interactions in dental-plaque formation. Circles represent cells. Adapted from Schachtele et al.[230]

A. Glucan-Induced Agglutination of *S. mutans*

S. mutans preferentially colonizes the tooth surface (Table 2). It was the first bacterial species to be associated with dental caries in humans.[838,839] The evidence and the arguments in favor of an important role for *S. mutans* in dental caries have been summarized in the literature.[37,822,825-827] Plaque formation in humans and experimental animals has been associated with *S. mutans* and dietary sucrose.[826,835,840,841] Consequently, the role of bacterial polymers on adherence of this streptococcal species has been intensively studied in vitro and in vivo. Suffice it to say that sucrose or dextran of high molecular weight promotes adherence. It is presumed that sucrose is converted to glucan, which mediates the interaction between the cell surface and the substratum. Adherence is highly selective in that only certain species of oral bacteria are adherent, *S. mutans* being particularly so. It has been shown, however, that adherence is separable from cell-cell interaction, although they share some common properties. Hence, although investigators tend to combine both processes, we shall confine the following account principally to glucan-induced agglutination.

It was Guggenheim and Schroeder who first reported glucan-induced agglutination of oral bacteria.[842] They observed that washed cells of glucose-grown *S. mutans* agglutinated within several seconds after the addition of sucrose. Agglutination was not observed when dextranase was added prior to sucrose addition. The agglutination reaction, demonstrated on a microslide, was devised as a convenient way of estimating the rate of glucan synthesis. The insoluble polysaccharide synthesized from sucrose was identified as a dextran branched at the 1:3 and 1:4 positions. Earlier reports had shown that dietary sucrose favored plaque formation[835] and that dextran-like polymers were synthesized by *S. mutans*.[843-845] Hence, the reported sucrose-induced agglutination was presumed to be dextran mediated. The enzyme system that catalyzes the formation of dextran from sucrose, dextransucrase, comes in two forms, soluble or cell bound.[845,846]

That dextran mediates agglutination of *S. mutans* was confirmed by the work of Gibbons and Fitzgerald.[82] They found that washed cells agglutinated upon the addition of dextran of high molecular weight (2×10^6 daltons) (Figure 11). Agglutination was strain specific. In some strains, it occurred within seconds after the addition of dextran. Other strains agglutinated within minutes or even hours, still others not at all. Among the nonreactive organisms were *S. sanguis, S. bovis,* and *Lactobacillus mesenteroides,* species known to synthesize dextrans. Agglutination was observed between pH 5 and 10. It was maximal at pH 8.5. At this pH, a concentration of about three molecules of dextran per cell, it was estimated, could cause agglutination. Linear or highly branched dextrans were effective about equally. On

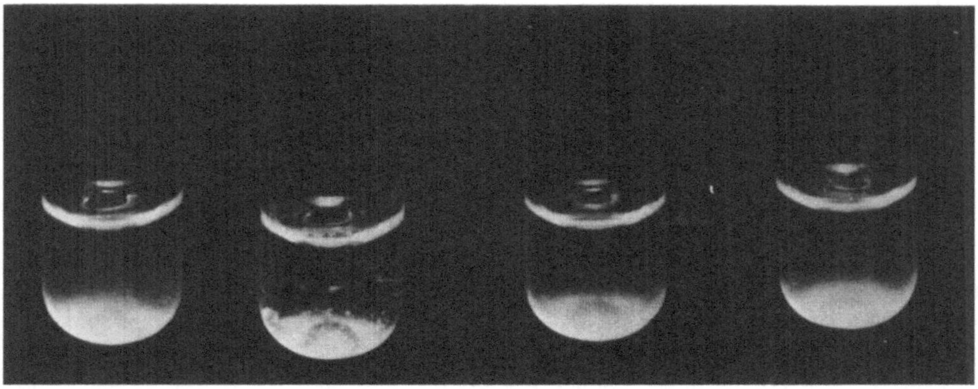

S. mutans 6715 control | *S. mutans* 6715 1 μg dextran | *S. sanguis* 34 control | *S. sanguis* 34 1 μg dextran

FIGURE 11. Agglutination of *S. mutans* by dextran of high molecular weight (2×10^6 daltons). (From Gibbons, R. J. and van Houte, J., *J. Periodontol.*, 44, 347, 1973. With permission.)

the other hand, dextran of low molecular weight (2×10^4 daltons), or such polysaccharides as starch, agar, levan, and inulin, were ineffective. Sucrose, but not glucose or fructose, also caused agglutination, which was inhibited by high concentrations of glucose, levan, or dextran of low molecular weight. Heat treatment (100°C, 10 min) caused the loss of agglutinability of the cells.

Cells that are agglutinated in the presence of sucrose are reversibly dispersed by urea, EDTA, and high temperature (up to 54°C).[847] Dispersed cells reagglutinate upon removal of urea, upon the addition of cations, or upon cooling. Cyclohexyl-*iso*cyanide and extremes of pH (<4.5 and >9.0) also cause dispersion. These observations implicate the participation of hydrogen bonds, involving perhaps the hydroxyl groups of the polysaccharides elaborated by the cells from sucrose.

The site of cell-cell interaction during dextran-induced agglutination has been examined.[225] Periodate oxidation of exogenous dextran destroys its agglutinative capacity, which is not restored upon subsequent reduction with borohydride. Competitive inhibition of dextran-induced agglutination by sugars and sugar alcohols suggests that the C-3 and C-1 positions of the sugar units in dextran play decisive roles. It does not appear likely that the dextran receptors on the surface of *S. mutans* are identical to the dextran-reactive site of the glucosyltransferase. However, the receptor sites may include a protein component, because papain or heat treatment abolishes agglutinability. Pretreatment of cells with urea or sodium dodecylsulfate, or of dextran with cyclohexyl-*iso*cyanide, prevents agglutination. These pretreatments presumably cause the denaturation of the receptor and the dextran, respectively. As in many other aggregation systems, divalent cations seem necessary, for EDTA-treated cells do not agglutinate, but do so in the presence of Ca^{2+} or Mg^{2+}. Sodium chloride does not disperse the agglutinates, thus excluding the significance of coulombic forces.

Mutants defective in sucrose/dextran-induced agglutination of *S. mutans* have been isolated. One mutant selected for its having lost the capacity to produce sticky polysaccharide in the presence of sucrose shows decreased cariogenicity.[848] Unlike the parent strain, it does not agglutinate in the presence of commercial dextran, streptococcal glucan, or sucrose. Although it synthesizes extracellular polysaccharides, it may be defective in the ability to synthesize cell-bound glucan or the receptor that binds glucan to the cell surface.

Similar mutants have been characterized as defective in the synthesis of extracellular polysaccharides.[849,850] Electron microscopy reveals that the parent strain synthesizes extracellular polysaccharide of a fibrillar nature, whereas the mutant material appears to be predominantly globular. One nonadhesive isolate has been shown to possess an altered glucosyltransferase activity with physical properties different from those of the wild-type enzyme.[851]

Adherence to solid surface and cell-cell interaction remain confused in the current literature. They are treated by some authors as if one is a measure of the other, by others as if they are the same in mechanism. The two processes, however, are experimentally separable. There are mutants that do not form plaque in vitro, but are only minimally altered in the ability to agglutinate in the presence of sucrose or glucans.[228] These mutants show a decrease in the synthesis of alkali-soluble, cell-bound glucans and an increase in water-soluble, extracellular glucans. Thus, they show that adherence and agglutination are separable characters and that the synthesis of alkali-soluble, cell-bound glucans is related to adherence. These mutants, which differ in colonial morphology from one another, exhibit attenuated virulence.[852] The morphology of their dextranase-sensitive, sucrose-dependent, extracellular glucans appears different by electron microscopy.[853] The wild-type glucan is predominantly fibrillar, the mutant more globular. A fuzzy coat, neither glucanase sensitive nor sucrose dependent, is found both in the wild type and in the mutant. The fibrillar extracellular glucan seems related to adherence, the cell surface fuzz to glucan-mediated agglutination. Surface fibrils have been previously observed in oral streptococci.[854-856]

Other mutants have been characterized to further substantiate the nonidentity of adherence and agglutination.[857] One mutant shows deficiencies in plaque formation, adherence to glass, cariogenicity, and synthesis of water-insoluble, adhesive glucans. It is not deficient in the capacity to agglutinate in sucrose or exogenous glucan. The deficiencies are ascribed to a defective glucosyltransferase activity. Another isolate shows just the reverse.[857] Compared to the wild type, it exhibits enhanced plaque formation, adherence, cariogenicity, and synthesis of glucans. A mutant with elevated levels of dextransucrase has been isolated.[858] It displays enhanced soluble and cell-bound enzyme activities as well as enhanced adherence. Still another kind of nonadhesive mutant, less virulent but fairly active in the synthesis of insoluble glucans, is able to agglutinate in exogenous dextran or sucrose.[857] This isolate may be deficient with respect to a cell-surface antigen required for adherence but not for agglutination.

Continual transfer of a strain of *S. mutans* results in the loss of cariogenic potential as well as the capacity to adhere to solid surfaces in the presence of sucrose and to be induced to agglutinate in sucrose or dextran.[82] This phenomenon is probably due to the selection of a mutant altered in the ability to bind glucan. The resultant variant strain, unable to agglutinate in sucrose or dextran nor to adhere to smooth surfaces, produces the same levels of cell-associated and extracellular glucosyltransferase activities as the original strain.[859] It seems no different from the original in many respects, but is markedly defective in binding exogenous glucan.

The more interesting class of mutants consists of mutants defective in glucan-induced agglutination, but not in adherence. These mutants have been selected by repeated agglutination of the wild type with exogenous glucan.[860] Characterized mutants are unaltered in, among other traits, dextranase activity, glucan and fructan synthesis, in vitro plaque formation, and virulence. They are, however, refractory to glucan-induced agglutination. These mutants decisively demonstrate that adherence and susceptibility to glucan-induced agglutination are genetically different traits. They also bring to doubt the importance of glucan-induced agglutination in plaque formation as well as cariogenesis. It is clear that without adherence, agglutination cannot result in the colonization of a tooth.

Another interesting mutant is that reported to agglutinate in dextran, but not in sucrose.[861] Deficient in cell-associated glucosyltransferase, it is nevertheless adherent to rat teeth and to saliva-coated hydroxyapatite. Thus, neither glucosyltransferase activity nor glucan synthesis from sucrose seems necessary for dextran-induced agglutination or adherence. The converse mutant has also been reported.[862] It does not agglutinate in the presence of exogenous glucan or dextran, but does so when incubated with sucrose. This mutant might lack the presumptive dextran-binding molecule specifically involved in agglutination.

Because of its high affinity for dextran, the glucan-synthesizing enzyme bound to the cell surface could conceivably serve as receptor in glucan-induced agglutination.[82] One model proposes the direct participation of dextransucrase as a ligand.[226] The enzyme is assumed to bind to dextran, which is anchored at one end to a protein receptor tightly bound to the cell surface. The aggregation of the dextransucrase molecules, which appear to contain a single binding site for dextran, enables the molecular aggregate to hold two or more dextran molecules and thus cause agglutination.

There is, however, much evidence against the direct participation of glucosyltransferase molecules. Immune sera directed against strains of *S. mutans* have been reported to inhibit sucrose/dextran-induced agglutination.[863] Cross-inhibition among strains does occur, but the antiserum to partially purified glucosyltransferase does not inhibit agglutination. Thus, it seems that the dextran–binding receptor involved in cell-cell interaction is a cell-bound molecule other than the glucan-synthesizing enzyme. That cell-bound dextransucrase activity is not directly involved has been concluded from other observations.[225,231,859-861,864-866] There are the agglutinationless mutants that are not deficient in dextransucrase activity. The mutant described above as defective in sucrose-induced agglutination lacks glucosyltransferase, but agglutinates in the presence of exogenous dextran. Blocking sucrose/dextran-induced agglutination by trypsin or EDTA does not bring about a corresponding loss of cell-bound dextransucrase activity. Conversely, an inactivation of cell-bound activity of the enzyme by various means, such as heat or protease or toluene, does not adversely affect dextran-induced agglutination. In media wherein the levels of dextransucrase are drastically reduced such that sucrose-induced agglutination does not occur, dextran-induced agglutination is nevertheless observed.[867] Thus, the function of dextransucrase extends no further than the conversion of sucrose to glucan. Whereas cell-free dextransucrase readily transfers glucose to soluble dextran, the cell-bound enzyme does not use soluble dextran as an acceptor.[868] Other acceptors such as maltose and α-methylglucoside are functional, but whereas dextrans of low molecular weight block dextran-induced agglutination, α-methylglucoside does not. It is clear that the dextran-receptor site does not reside in the enzyme molecule but in some other component on the cell surface.

An extracellular dextran-binding protein has been isolated from *S. mutans*.[869,870] Isolated from culture filtrates and purified by affinity chromatography followed by gel filtration, it is devoid of dextranase and dextransucrase activities. It binds dextran and causes inhibition of dextran-induced agglutination. The inhibition is counteracted by the addition of more dextran. The lectin is also recovered in low amounts from whole cells sonicated gently so as not to cause cell rupture. Other protein species with dextran-binding activity have been identified.[871] It seems that these are multiple forms of the same lectin activity.

A recent study on the role of dextran binding in glucan-induced agglutination allows us to see a more comprehensive picture.[862] Glucans and dextrans of either high or low molecular weight bind to cells of *S. mutans*. However, strains of different serotypes exhibit different binding activities, which do not correlate well with the capacity to agglutinate in dextran. Cross-reactions occur among various serotypes of *S. mutans*. Soluble glucan isolated from one serotype may bring about the agglutination of cells of another. One strain (serotype *d*) binds glucans of six other strains. A serotype may or may not react with its own glucan. Maximum agglutination in dextran (7 μg, 2 × 10^6 daltons) is observed at the end of 1 hr at 37°C. Dextrans of low molecular weight, although bound by cells, do not cause agglutination. Heat-treated cells (100°C, 10 min), as well as cells treated with protein denaturants (SDS or urea), enzymes (trypsin or dextranase), a chelator (EDTA), or with lectins (concanavalin A, RCA I, or RCA II), lose the ability to agglutinate in dextran. Their ability to bind dextran of high molecular weight is considerably reduced, but their ability to bind dextrans of low molecular weight remains unaffected. Pre-incubation of cells in dextrans of low molecular weight does not alter their agglutinative property nor their binding capacity for glucan. Conversely, pre-incubating cells in high-molecular-weight dextran or glucan does not impair the binding of dextrans of low molecular weight. Contrary to previous findings, dextran-induced agglutination is not inhibited by low-molecular-weight dextrans, but antisera directed against whole cells, against the type-specific *d* polysaccharide, or against glucan all inhibit agglutination. These observations point to the nonidentity of the binding sites for dextrans of low molecular weight and for dextrans of high molecular weight and to the involvement of polysaccharides as receptors and the participation of some protein component in the cell-cell interaction.

The agglutinative and dextran-binding properties of cells grown in a synthetic medium are quite different from those grown in a complex medium.[872] In contrast to cells grown in a complex medium, cells grown in a synthetic medium are less adherent to glass, do not bind exogenous glucosyltransferase, nor do they agglutinate in dextran/glucan of high molecular weight. They do bind dextrans of low molecular weight and glucan synthesized by the strain and readily agglutinate when incubated with sucrose. Both types of cells produce constitutive levels of glucosyltransferase activity, although cells in the synthetic medium synthesize less glucan. The type-specific polysaccharide antigens are nonetheless present, thus excluding the direct involvement of such antigens in glucan-induced agglutination. The influence of cultural conditions on the glucosyltransferase activity of *S. mutans* has been described in other reports.[867,873-876] It has been noted also in other species of streptococci.[877-879]

As mentioned earlier, dietary sucrose promotes the accumulation of *S. mutans* cells on teeth of humans and experimental animals. This accumulation is the consequence of the ability of the species to form extracellular polysaccharides (glucans and fructans) from sucrose.[880,881] The fructans, polymers of fructose units, are similar to levan. The glucans consist of polymers of glucose units mainly α-1,6-linked in the manner of classical dextran and of polymers, called mutan, of glucose units mainly α-1,3-linked.[842,844,881,882] Because dextranase and mutanase, as well as dextrans of low molecular weight, inhibit plaque formation, the glucans seem to be of greater significance than the fructans.[883-886] The glucosyltransferase mutants encountered earlier also point to the more important role of glucan in plaque formation. Furthermore, because they tend to be more insoluble than the fructans and more resistant to depolymerization by oral bacteria, they are more likely to be components of the matrix that holds the plaque together.

For the synthesis of the extracellular polysaccharides, intact molecules of sucrose are necessary. The synthetic reaction catalysed by glucosyltransferases is driven by the energy released from the glucosidic bond:

$$\text{Sucrose} + (\text{Glucan})_n \xrightarrow[\text{Dextransucrase}]{} (\text{Glucan})_{n+1} + \text{Fructose.}$$

The enzyme transfers the glucose moiety of the disaccharide molecule to a glucan acceptor, thereby increasing the length of the acceptor polymer by one glucose unit. The glucosyltransferases of *S. mutans,* possibly made up of a number of molecular species, have been reviewed by Ciardi.[887] The properties of the enzyme preparations from various laboratories are compared in Table 3. Glucosyltransferase activity may be recovered from the culture filtrate as an extracellular material or may be cell-associated, either intracellular or avidly bound to the cell surface. The intracellular forms are the precursors of the aggregated species on the cell surface and the enzyme activity found in the culture medium. As insoluble glucan accumulates, the glucosyltransferase activity of a culture tends to diminish, presumably because of the entrapment of the enzyme molecules in the growing glucan matrix.[873,875]

Table 3
CHARACTERISTICS OF GLUCOSYLTRANSFERASES PURIFIED FROM *STREPTOCOCCUS MUTANS*

Strain	(serotype)	Product	Mol wt (× 10^3)	K_m (m*M*)	IP	CHO binding	FTF	Ref.
OMZ-176	(d)	Ins		0.98—7.23	4.24—6.03		No	846
6715	(d)		94	3.0	4.0	Yes	No	888, 889
HS-6	(a)	Ins						890
		Sol	170	2.0			No	
HS-6	(a)	Ins	400—2000			Yes	Yes	891
		Sol				Yes	Yes	
GS-5	(c)	Ins	190 + 235	1.7	4.3, 6.2	Yes	Yes	892
		Sol	45	2.1	4.3, 6.5		No	
6715	(d)	Unf		5.0		Yes	No	893, 894
		Ins	150		4.3			
		Sol	175		3.9			

Note: Abbreviations: IP, isoelectric point; CHO, carbohydrate; FTF, fructosyltransferase; Ins, water-insoluble glucans; Sol, water-soluble glucans; Unf, unfractionated system.

The evidence from various sources makes it improbable that the glucosyltransferase molecules act as receptors for dextran in the cell-cell interaction. Rather, there are more important cell-bound protein receptors that by specifically binding dextran bring about dextran-induced agglutination. Doubtless, glucosyltransferase does bind dextran and its aggregate can serve as a ligand, as envisaged in one model, but its direct contribution is probably minor, because dextran-induced agglutination is not dependent on the active enzyme. The primary role of glucosyltransferase is simply the synthesis of glucan, which is the indigenous mediator, from sucrose. Mutants that agglutinate in the presence of exogenous dextran but not in sucrose are presumed to be deficient in dextransucrase. Mutants that agglutinate in the presence of sucrose but not in exogenous dextran may be thought of as deficient in cell-bound protein receptors specific for dextran. The glucan synthesized *de novo* by cell-bound dextransucrase might be bound to the cell surface by a receptor that is specific for indigenous glucan but unable to bind avidly to exogenous dextran.

Although adherence can be dissociated physiologically and genetically from cell-cell interaction, their mechanisms may be analogous if glucan is first deposited on a solid surface. In the oral cavity, that solid surface is the tooth enamel. If there are present, as the current evidence shows, bacterial-glucan polymers in the acquired pellicle, then the binding of cells to the acquired pellicle would constitute a mechanism that is half of that of cell-cell interaction, that is, the glucan-coated enamel may be viewed as if it were one gigantic bacterial surface. Otherwise, if cell-cell interaction were heard as the clapping of hands, adherence would be the subliminal sound of one hand clapping.

Glucan-induced agglutination is manifested also by other oral bacteria, notably species of *Actinomyces*. A discussion of that system is reserved, for the sake of convenience, for a later subsection (see Section VI.C below).

B. Saliva-Induced Aggregation of Oral Bacteria

The aggregation of plaque-forming bacteria, as well as the sorption of the organisms to human enamel, has been shown to be induced by saliva.[895-897] *Streptococcus mitis* and *S. sanguis*, but not *S. mutans*, form aggregates in whole saliva or in parotid and submaxillary secretions (Figure 12). Aggregation occurs between pH values 5 and 7.5. The inducing factor in the saliva, not likely to be gamma globulin but presumed to be a large molecule for it is nondialysable, can withstand boiling for 20 min. The cell surface, which binds the

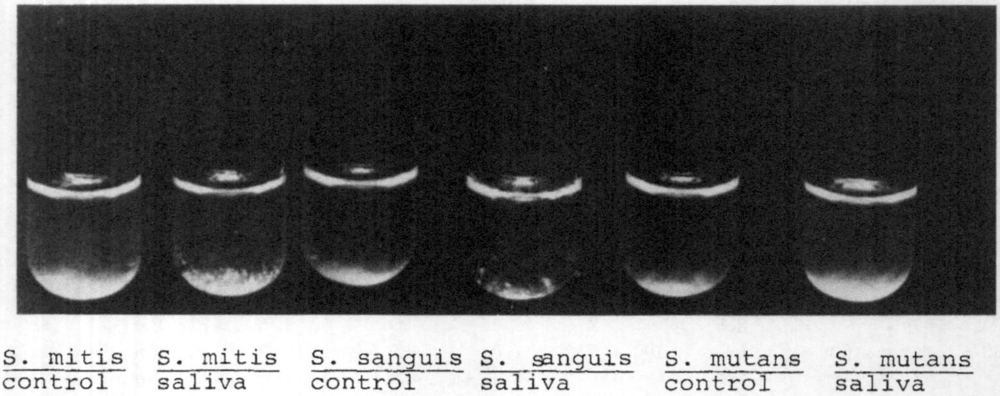

FIGURE 12. Saliva-induced aggregation of *Streptococcus mitis* and *S. sanguis*. (From Gibbons, R. J. and van Houte, J., *J. Peridontol.*, 44, 347, 1973. With permission.)

inducing factor, is heat sensitive. Although the aggregation reaction is strain specific, many plaque isolates, including *Actinomyces viscosus* and *A. naeslundii*, form cell aggregates in saliva.

Saliva-induced aggregates of *Streptococcus sanguis* or *S. mitis* are large enough to be observed by the unaided eye. Such aggregates, however, are fragile and readily break apart upon gentle agitation. They can be dissociated into more stable particles with a median diameter of about 4.5 μm and comprising about 100 cells.[898] These stable particles may constitute core aggregates, to which free cells attach themselves to form the larger aggregates. Aggregation is dependent on the concentration of cells and of saliva. It is unaffected by physiological age of the culture, log-phase cells being as inducible as those in early stationary phase, but as the cells age in stationary phase, their aggregating activity decreases. Aggregation of *S. sanguis* is most pronounced at 10°C. At 40°C, activity is reduced to about 15% of the maximum, whereas about 65% of the activity is retained at 0°C. Inhibited by EDTA, aggregation is maximal in the presence of 1 meq of Ca^{2+} per liter. The number of core aggregates attains a maximum within 10 to 20 min and decreases thereafter. Formalin-treated cells yield fairly stable aggregates. Saliva-induced aggregation of formalin-treated cells, however, is independent of Ca^{2+} and is maximal between 20 and 30°C.

The aggregating activity of oral secretions is now well established.[895-901] Different factors may bring about aggregation of cells of one type or of several species. Evidence for separate factors for *S. sanguis* and *S. mitis* has been presented.[898] The factors can be separated by differential adsorption to intact cells and to hydroxyapatite and by differential stability to prolonged dialysis. A number of salivary constituents are known to interact with the surface of oral bacteria to cause aggregation. They include antibodies, salivary glycoproteins, and blood-group-reactive substances.[897,902-906]

Agglutinin components of saliva have been partially characterized. From whole saliva has been isolated a high-molecular-weight glycoprotein containing substantial quantities of hexosamine and causing the aggregation of *S. sanguis*.[897] The agglutinin activity derived from resting parotid saliva has been isolated by high-speed centrifugation and further purified by chromatography in DEAE-Sephadex® A 50 and hydroxyapatite.[901] The partially purified material appears to be a glycoprotein with a molecular weight of 10^6 to 10^7 daltons. The agglutinin is active in inducing the aggregation of *S. mutans*. It contains a good amount of carbohydrates, prominent of which are fucose, hexoses, and hexosamines, and a small amount of sialic acid. Because parotid or submaxillary saliva is not uniformly active for different species and strains of streptococci,[897,898] it is likely that there are a number of salivary agglutinins with different specificities for different bacteria. Furthermore, different individuals secrete saliva of different sensitivity to sulfhydryl reagents.[907] Thus, there may be a wide variety of agglutinins, each specifically effective for one strain by way of specific oligosaccharide side chains, or there may be a small number of agglutinins, some of which are singly responsible for causing aggregation of a number of strains.[901]

Saliva-induced aggregation of *S. sanguis* or of *S. salivarius* is inhibited by lactose, galactose, and fucose.[869] This observation suggests the contribution of carbohydrates to the receptor sites.

One of the components of saliva is the cell-lytic enzyme lysozyme. In various microbial systems, including both bacteria and yeasts, lysozyme is known to bind to cells and thereby cause cell aggregation.[908-911] Lysozyme-mediated aggregation of streptococci has been studied recently.[912] Cell aggregation and lysozyme-binding activity are a function of enzyme concentration. Binding is inhibited by sugars such as D-mannose, *N*-acetyl-D-glucosamine, and *N*-formyl-D-glucosamine. Various serotypes differ in their capacity for binding of the enzyme as measured by maximum capacity and kinetics of binding and by their susceptibility to binding inhibition by sugars. Antiserum directed against human leukemic lysozyme inhibits cell aggregation in the presence of human parotid saliva.

The role of sialic acid in saliva-induced aggregation has been examined.[913] Saliva treated with protease or neuraminidase loses its activity to aggregate strains of *S. sanguis,* as does saliva mildly acid hydrolyzed. The loss correlates well with the release of sialic acid. Protease or neuraminidase also causes dispersal of cells in aggregates. Saliva-induced aggregation is not inhibited by sugars such as glucose, glucosamine, and galactose, nor by dextran, nor by concanavalin A. Lecithin exhibits some inhibitory activity. Inhibition is most marked when the streptococcal cells are incubated with a preparation of gangliosides. Acid-hydrolyzed gangliosides are less effective. The aggregating activity of *S. sanguis* is not related to antigens *a, b, c, d,* and *e,* nor to the rhamnose or phosphate content of the cell wall. A neuraminidase activity in clarified saliva is suggested by the observation that saliva loses aggregation activity at room temperature. The loss is related to the release of sialic acid.

C. Heterotypic Aggregation (Interspecies Interaction) of Oral Bacteria

Among the many species of bacteria found in dental plaque is the filamentous form *Actinomyces viscosus.* It synthesizes levan and other extracellular carbohydrate polymers.[914-916] High-molecular-weight dextran isolated from *Leuconostoc, Streptococcus sanguis,* or *S. mutans* causes rapid agglutination of *A. viscosus.*[917] As few as three dextran molecules per cell can cause agglutination, as has been estimated for dextran-induced agglutination of *S. mutans.* Other simple and complex carbohydrates, such as sucrose, glucose, mannose, cellobiose, agar, starch, and inulin, are ineffective. Maximal agglutinating activity is shown by dextran with a molecular weight of 2×10^6 daltons. Dextrans with a molecular weight of less than 10^5 daltons are without activity. Agglutination is ion dependent, divalent ions being much more efficient than monovalent ones. It is dependent on the physiological age of the cells. Agglutinating activity, greatest during late-log or early-stationary phase of a batch culture, is gradually lost upon aging of cells in stationary phase. Dextran-induced agglutination is sensitive to heat, proteases, and glucosidases. It is inhibited by concanavalin A but not by phytohemagglutinin. Thus, both proteins and carbohydrates are implicated as receptor components.

There has been no report as to the dextran-synthesizing ability of *Actinomyces viscosus.* Thus, the reported lack of effect of sucrose as an inducer of agglutination. For it to be an active component of dental plaque, *A. viscosus* has to utilize glucan synthesized by other oral bacteria, such as the streptococci. If it interacts with the cell-associated polymer of other species, heterotypic aggregation must be the consequence.

Indeed, dextran-mediated intergeneric aggregation between *A. viscosus* and dextran-synthesizing streptococci has been described.[918] When cells of *A. viscosus* and *S. sanguis* are mixed together, stable aggregates containing both species are rapidly formed. *S. mutans* likewise co-aggregates with the actinomycete. Like the homotypic dextran-induced agglutination of the actinomycete, co-aggregation between *A. viscosus* and *S. sanguis* is ion dependent. It is not affected by changes in pH between 5.0 and 8.5.

Only streptococci grown in sucrose, and not those grown in other sugars, co-aggregate intensely with *A. viscosus* in the absence of exogenous dextran.[918] In contrast, the actinomycete is not affected by sucrose in the growth medium. An incubation period of 4 hr in sucrose is sufficient for streptococci to become competent to co-aggregate with the filamentous form. The physiological age of the streptococcal cells is not a factor in the co-aggregation. Neither is that of the actinomycete when grown in brain-heart-infusion broth, but in trypticase-soy broth, aggregating activity of the actinomycete decreases rapidly in stationary phase and is best demonstrated during the late-logarithmic phase of a culture. Although the ratio of the two components in the mixture can be effectively varied up to a 100-fold, the most effective input ratio appears close to unity.

Co-aggregation may be provoked by exogenous dextran.[918] Glucose-grown cells of *S. sanguis* are incubated with dextran for 30 min. After the unbound dextran is removed, the dextran-treated cells are mixed with untreated cells of *A. viscosus*. Dextran of high molecular weight (2×10^6 daltons) or glucan isolated from *S. sanguis* promotes co-aggregation, whereas dextrans of low molecular weight (1×10^4 daltons) do not. Untreated cells do not co-aggregate with the actinomycete. *S. mutans*, in contrast to *S. sanguis*, forms homotypic aggregates due to dextran-induced agglutination. Hence, it is necessary to disperse these agglutinated cells before mixing with *A. viscosus*. Co-aggregation of *S. mutans* with *A. viscosus* is indicated by a decrease in the number of colony-forming units of the reaction mixture. Pre-incubation of cells of *A. viscosus* with dextran likewise results in heterotypic aggregation with the streptococci.

Modification of the cell surface of *A. viscosus* by heat or protease destroys the ability of the cells to agglutinate in dextran and to co-aggregate with *S. sanguis*.[918] Dextranase destroys the ability of *S. sanguis* to co-aggregate with *A. viscosus*, but heat or protease does not. Hence, the receptors on the actinomycete cell surface are probably proteinaceous in nature, whereas those on the streptococci are nonproteinaceous.

The "corn-cob" structures on the superficial layer of dental plaque are due to coccal forms co-aggregating with filamentous bacteria (Figure 13). First observed in stunning detail with the scanning electron microscope by Jones, they seem to have been known to previous workers as early as the turn of the century.[919,920] These obviously heterotypic arrangements, which first appear on the third day of dental plaque development, might well be made up of streptococci bound by their glucan to actinomycetes.[921] Isolates of *S. sanguis* from corn-cob structures have been demonstrated to form similar structures in vitro with *Bacterionema matruchotii*.[922] Heat (100°C, 10 min) or proteases inactivate the coccal isolates but not the filamentous organism.

Intergeneric aggregation between *Actinomyces naeslundii* and dental-plaque streptococci has been reported.[923] Approximately equal numbers of each component are required for maximal co-aggregation. Strains of *A. naeslundii* aggregate more strongly with strains of *S. sanguis* and *S. mitis* than with *S. mutans* strains.[924] Pretreatment of *A. naeslundii* with papain, trypsin, or heat impairs its ability to co-aggregate with *S. sanguis*. The same treatments do not affect the ability of *S. sanguis* to co-aggregate with *A. naeslundii*. Electron microscopy reveals that contacts are mediated by a tuft of electron-dense fuzz.

The mechanism of co-aggregation between actinomycetes and streptococci has been examined in more detail recently.[925] The co-aggregation between *A. viscosus* T14V and *S. sanguis* 34 is inhibited by EDTA and not by high concentrations of NaCl. It is best demonstrated between pH 8.0 and 8.5. It appears to require the interaction of a protein or glycoprotein on the surface of the actinomycete and a carbohydrate component on *S. sanguis*.

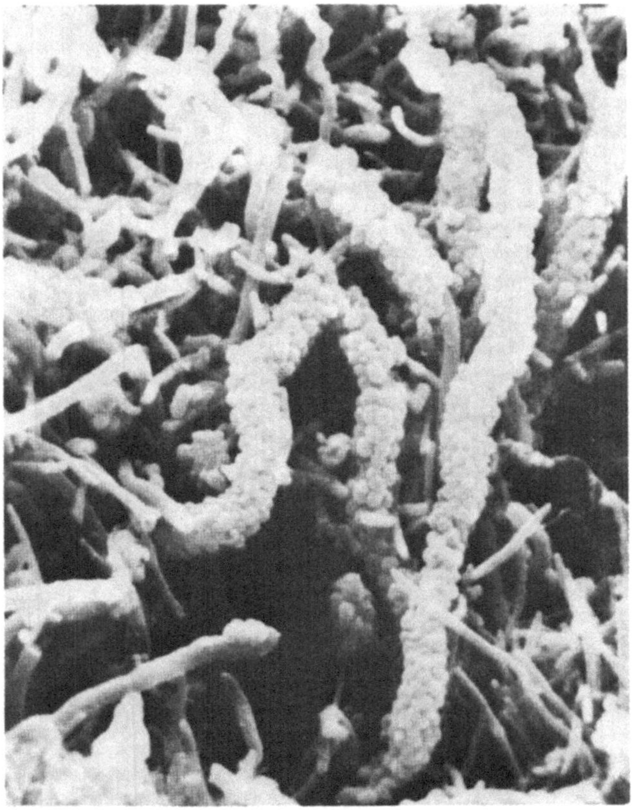

FIGURE 13. "Corn-cob" structures in human dental plaque. Scanning electron micrograph shows coccal forms attached to filamentous forms. (From Gibbons, R. J. and van Houte, J., in *Textbook of Oral Biology*, Shaw, J. H., Sweeney, E. A., Cappuccino, C. C., and Meller, S. M., Eds., W. B. Saunders, Philadelphia, 1978, 684. With permission.)

Contrary to earlier reports, it is not dependent on dextran, because it is resistant to dextranase. However, this does not exclude the participation of an indigenous glucan different from dextran. The avirulent strain of the actinomycete is nonaggregative. The co-aggregation is inhibited by lactose, D-galactose, β-methyl-D-galactoside, and *N*-acetyl-galactosamine, but not to any considerable extent by other sugars, such as α-methyl-D-galactoside, melibiose, sucrose, D-glucose, and D-mannose. The co-aggregates already formed are dispersible into free cells by the inhibitory sugars. Electron-microscopic examinations of co-aggregates labeled immunochemically with antibodies to *A. viscosus* suggest that co-aggregation is mediated by fibrils on the surface of the actinomycete.

A fibrillar layer on the cell wall of *A. viscosus* and *A. naeslundii* has been observed earlier.[926] The virulence-associated antigen V of *A. viscosus* T14V, first identified by Hammond et al.,[927] has been correlated with fimbria-like appendages and the ability to co-aggregate with *S. sanguis* 34.[928] The surface components responsible for co-aggregation between the two strains appear to be fimbriae on the actinomycete.

The specificity of the co-aggregation reaction between various strains of *A. viscosus* and *A. naeslundii*, on one hand, and various strains of oral streptococci, on the other, allows the sorting of actinomycetes into two groups and of streptococci into four.[929] The groups are identified by their combination patterns and the effect of lactose on their co-aggregation. Three types of cell-cell interactions are defined: (1) co-aggregation, not reversed by lactose, between a heat-sensitive streptococcus and a heat-resistant actinomycete; (2) co-aggregation, reversed by lactose, between a heat-sensitive actinomycete and a heat-resistant streptococcus; and (3) co-aggregation blocked only by heat treatment of both species. The third type of interaction is a combination of the other two, because lactose reverses co-aggregation between heat-treated streptococcus and untreated actinomycete, but not co-aggregation between untreated streptococcus and heat-treated actinomycete. Heat-sensitive cells can be inactivated also by protease digestion or amino-group acylation, but those cells unimpaired by heat are insensitive to these treatments.

Co-aggregation is independent of physiological age, except that involving *A. viscosus* T14V, which shows reduced ability when allowed to incubate long past the start of stationary phase.[929] Save for certain strains of *S. mutans* and rodent strains of *A. viscosus*, glucose-grown cells of streptococci and actinomycetes may be caused to aggregate by dextran or levan of high molecular weight. Co-aggregation is not inhibited by dextran of low molecular weight nor by dextranase. At low pH (4.5), co-aggregation with most strains is greatly diminished, but addition of Ca^{2+} enhances this aggregation at low pH.

It would seem that cell-cell interaction involves protein or glycoprotein receptors on the cell surface of actinomycetes and carbohydrate receptors on the cell surface of streptococci. The protein on the actinomycete cell surface might well be a fimbrial component possessed with a specific affinity for β-linked galactosyl residues, whereas the streptococcal carbohydrate might well be a tightly bound component of the murein. The interaction would then be classifiable as lectin mediated.

Intrageneric co-aggregation is found among the oral streptocci. *S. mutans* and *S. sanguis* form heterotypic cell-cell interactions, which are stimulated by sucrose and sensitive to dextranase.[230] Furthermore, *S. mitis* competitively inhibits co-aggregation between *S. mutans* and *S. sanguis*.

Another heterotypic system involves Gram-positive filaments and Gram-negative cocci. Cells of *Veillonella* species have been reported to co-aggregate with actinomycetes.[923] They do not by themselves adhere to hard surfaces, but instead attach to preformed plaques of Gram-positive diphtheroids.[930] It is safe to say that many more heterotypic interaction systems among oral bacteria simply await detailed documentation.

D. Acid-Induced Aggregation of Oral Bacteria

Bacterial cells derived from dental plaque that has been disaggregated by dilute NaOH have been shown to aggregate again as the suspension is made acidic.[931] We shall call this phenomenon, to distinguish it from the other systems, acid-induced aggregation of oral bacteria.

Acid-induced aggregation is dependent on cell concentration, being best demonstrated at high concentrations of its components, as would be expected of a colloidal instability system.[931] Likewise expected is its dependence on ionic strength.The higher the ionic strength, the more readily demonstrable the phenomenon. Also to be expected is the favorable effect of divalent cations. The favorable effects of high cell concentration, high ionic strength, and high divalent-ion concentration are reflected not only in the greater amount of precipitated mass, but also in an increase in pH at which the phenomenon becomes demonstrable. Aggregation of alkali-dispersed plaque bacteria occurs at about pH 4.0 when the cell concentration is about 0.13% (v/v). At higher cell concentrations, that critical pH may be raised to 4.5 and to pH 5 to 6 in the presence of divalent cations. The acellular fractions of dental plaque inhibit aggregation, except for one fraction that seems to enhance it.

The filamentous form *A. naeslundii* has been studied in detail with regard to acid-induced aggregation.[932] When the pH of a culture in a rich medium supplemented with glucose falls below 5.5, cells aggregate to form bacterial masses adherent to the culture vessel. In the absence of exogenous glucose in the medium, maximum cell population density is reduced, the pH remains above 6.5, and the cells remain nonadherent and nonaggregative. Adjusting the final pH of the culture to below 5.5 causes cell aggregation. The aggregation reaction of a washed-cell suspension containing greater than 5×10^8 cells $m\ell^{-1}$ takes about 90 min for completion. Acid-induced aggregation is observed also in strains of *A. viscosus, Streptococcus mutans, S. sanguis,* and *S. salivarius,* but not in strains of *S. faecalis, S. mitis, Lactobacillus casei, Staphylococcus aureus, Escherichia coli,* and *Proteus vulgaris.* Recall the similarity of this phenomenon to agglutination associated with transformation competence (see Section V of this chapter).

Acid-induced aggregation of oral bacteria is doubtless analogous to colloidal instability systems. The zeta-potential of various oral streptococci is pH dependent, species specific, sensitive to counter-ions, and directly related to acid-induced aggregation.[933]

The vigorous interest in the many cell-aggregation systems of oral microbiology stems largely from the unabashed optimism that through them, a full understanding of dental-plaque formation will be achieved and thereby insight into the control of dental caries and periodontal diseases. Save for the attraction of aggregation in cellular slime molds, there is hardly anything comparable among the cell-aggregation systems to the drawing power of dental-plaque development. Much of the optimism can be discerned in a recent meeting of the International Association for Dental Research. A good number of the papers presented at that meeting bear directly or indirectly on the problems of aggregation of oral bacteria. A few are worth noting. One paper concludes that cyclic AMP and cyclic GMP promote agglutination of *S. mutans* in sucrose.[934] There is a report on the preparation from *S. sanguis* of a cell-wall protein that mediates adherence to hydroxyapatite.[935] Another paper reports that hair-like structures (pili), approximately 0.4 μm in length, are peritrichously arranged on the cell surface of *S. sanguis,* but that piliation is not related to agglutination.[936] There is a further study on the co-aggregation between actinomycetes (*A. viscosus* and *A. naeslundii*) and streptococci (*S. sanguis* and *S. mitis*).[937] Preliminary results are presented on the effects of Li, Sr, and F.[938] These elements in trace amounts appear to reduce both the number of saliva-induced aggregates of *S. mutans* and the size of the aggregates. They increase the number of saliva-induced aggregates of *A. viscosus* while decreasing their average size. There is a report on the calcium dependence of saliva-induced aggregation of *S. mutans,*[939] another on the agglutination of *A. naeslundii* by a contaminating lectin found in commercial mannose.[940]

The many aggregation systems of the oral flora are unified not only by their ecological relation, but also by the role of polymers in the seemingly varied processes. They afford the student of cell-cell interaction a close-up view of the significance of microbial aggregation in the genesis of infection. There is much to expect from the work of the many laboratories that have been drawn to the study of dental-plaque development and aggregation of oral bacteria. Aside from more and more papers, there is the prospect of a near future wherein dental caries shall have ceased to be a universal source of pain and inconvenience.

VII. CELL AGGREGATION IN *STREPTOCOCCUS FAECALIS*

Somewhat related to aggregation of oral bacteria is the recently reported cell aggregation in *Streptococcus faecalis.*[941] It is, however, more related to mating-aggregate formation during conjugation in *Escherichia coli* or to sexual agglutination in eukaryotes. The phenomenon appears to be plasmid controlled and associated with mating and gene transfer.

Clinical isolates of *S. faecalis,* as well as isolates from other sources, are known to harbor plasmids coding for antibiotic resistance, production of bacteriocins and hemolysins, or bacteriocin sensitivity.[942-949] Gene transfer mediated by these plasmids has been demonstrated.[943,945-953] Certain plasmids bearing determinants for antibiotic resistance or bacteriocin/hemolysin are capable of self-transfer. In addition, self-transferable plasmids have been shown to mobilize other plasmids, which are incapable of self-transfer, as well as chromosomal segments.[947,949,952] The process of gene transfer seems analogous to sexual conjugation, for intercellular contact between donor and recipient is necessary. There is so far, however, no evidence for pilus-like structures, which may mediate cell-cell interaction, in *S. faecalis*. Neither transduction nor transformation has been reported in this streptococcal species.

Mating mixtures of *S. faecalis* strains grown in nutrient broth form clumps.[941] In contrast, recipient cells and donor cells separately incubated remain in homogeneous suspension. Virtually all of the cells in a mating mixture are found in large masses, which settle to the bottom of the mixing vessel on short standing. An incubation period of 40 to 120 min after mixing at 37°C is required for clumps to become visible to the unaided eye. The clumping reaction is observed in various combinations of different strains and recipients, but is most intense in mixtures containing a donor strain that bears the plasmid pPD1 and three cryptic plasmids.

The donor cells themselves, in the absence of recipient cells, form aggregates when incubated in filtrates of the culture medium in which recipient cells have been grown.[941] The reverse manipulation does not work: the recipient cells do not clump in the presence of filtrate derived from the donor cells. Thus, there is a factor, excreted by the recipient cells into the culture medium, which induces the donor cells to clump, either with the recipient cells or even among themselves. This inducing factor has been designated CIA, for clumping-inducing agent. CIA possesses the characteristics of a relatively small protein or a peptide. It is resistant to boiling for 10 min. It is likewise resistant to DNase and RNase. Its activity, however, is destroyed by trypsin in short time. This inactivation by trypsin is counteracted by heat or the presence of trypsin inhibitor. CIA is retained by a filter with an exclusion limit of 10^3 daltons but passes through a filter that retains molecules of up to 10^4 daltons.

Regardless of the concentration of CIA or of cells, an incubation time of at least 30 min is required to demonstrate donor-cell aggregation due to recipient-derived CIA.[941] Chloramphenicol or rifampicin, added along with CIA, prevents clumping of donor cells, but 6-parahydroxyphenylazouracil, which inhibits DNA synthesis and cell growth, does not. Neither does chloramphenicol or rifampicin if added 15 min after CIA addition. Thus, whereas DNA synthesis is dispensible, RNA and protein synthesis must take place during induction prior to visible clumping. Once the required proteins are made, the donor cells become competent to aggregate among themselves and become indifferent to inhibitors of RNA and protein synthesis.

Plasmid-free isolates of *S. faecalis* produce CIA that induces the clumping of plasmid-bearing strains.[941] The acquisition of a conjugative plasmid (pAMγ1, for instance, which determines hemolytic activity) by an originally plasmid-free strain results in a drastic reduction in CIA activity of its filtrate. Moreover, the newly acquired plasmid confers on the cells the ability to clump among themselves in the presence of exogenous CIA. A nonconjugative plasmid (pAMα1, which codes for tetracycline resistance) is ineffectual in eliciting CIA-synthesis inhibition or response to CIA activity.

Do clumping and CIA influence mating and gene transfer? Indeed, they do, quite dramatically. Donor cells are pre-incubated with recipient filtrate (for a period of time short of visible clumping) and then incubated with the recipient cells in a mating-reaction mixture. CIA treatment of donor strains increases gene-transfer efficiency 10^2- to 10^6-fold.[941,954] The enhancement of gene transfer by CIA is sensitive to inhibition by rifampicin. The filtrate of donors does not cause clumping nor does it stimulate gene transfer. Strains that do not clump do not exhibit enhanced donor capacity in the presence of CIA.

The extracellular, trypsin-sensitive, heat-stable activity that is CIA qualifies as a bacterial sex hormone, a pheromone, functionally similar to pheromones found in a number of sexual systems including yeasts and algae. It serves as a mating signal, rather than a molecular component of the aggregation reaction. What it does is provide the donor cells the message and the opportunity to make alterations on their walls so as to make them more aggregative. The formation of aggregates facilitates the mating process by favoring the required donor-recipient contact. The consequence is a dramatic enhancement of gene-transfer efficiency in the cell population.

Cells aggregated by CIA can be dispersed in the presence of EDTA.[954,955] Reaggregation takes place only in the presence of Ca^{2+} and phosphate. By exposure of the dispersed cells to proteases (Pronase or trypsin), sodium dodecylsulfate, or low pH, reaggregation is prevented. These observations implicate proteinaceous components on the cell wall and rule out the involvement of bifunctional ligands that can be eliminated by disaggregation.

The CIA activity of the culture medium increases with the cell population.[955] In one recipient strain, it attains a maximum at the start of stationary phase and decays rapidly thereafter. In another, maximal activity is maintained for many hours. The CIA titer of a filtrate is reflected in the increase of mating efficiency. The decrease in titer in the case of one strain is presumably due to proteolytic activity during stationary phase. Because CIA activity is heat stable, a filtrate may be autoclaved for 20 min. Autoclaved CIA preparations retain their activity on storage at 4°C for many weeks.

To a CIA preparation from a single strain, different clinical isolates respond with varied effectiveness.[955] Those without any plasmid do not respond at all. Strains that are hosts to the plasmid pPD1 or pAMγ1 are most responsive to CIA from a strain that is plasmid free. Strains bearing the plasmid pOB or pAD1 are poor responders. However, when these plasmids are eliminated, the cured derivatives produce CIA to which the parent strains now respond. It is revealing that a strain bearing pPD1 responds both to cured derivatives and to parent strains containing pOB or pAD1. It appears then that the presence of a plasmid shuts off the production of (or inactivates) the CIA to which a prospective donor with that plasmid responds, without prejudice to the production of a different kind of CIA specific for a donor with a different plasmid. Indeed, by cross-reacting filtrates of various recipients with various donor cells, it can be concluded that there is a variety of pheromones produced by recipient cells. Each of these is specific for a donor strain bearing a plasmid of a particular class.

One molecular species, the CIA which induces cells bearing the conjugative plasmid pPD1, retains full activity after precipitation with trichloroacetic acid.[956,957] Designated as cPD1, the pheromone can be separated from other CIA activities by chromatography in a polyacrylamide column. Its apparent molecular weight is about 700 daltons.

"Transposable elements are specific DNA segments that can repeatedly insert into a few or many sites in a genome."[958] The simple sequences, generally smaller than 2 kb, bear the insertion function and no other function unrelated to it. They are called IS elements. The more complex sequences are called Tn elements or transposons. Often containing two IS elements and generally larger than 2 kb, they contain functions other than insertion. Episomes, often containing IS and Tn elements, are much larger, self-replicating units.

Transposable elements have been identified in *S. faecalis*.[959] A Tn element carrying a chromosomal tetracycline-resistance marker has been found to transpose to the conjugative plasmid pAMγ1 or pAD1.[959,960] An analysis of the new plasmid construct suggests that the Tn element itself codes for a response to a unique CIA. The cell aggregation that is induced is related to an enhancement of mating frequency.

The clumping phenomenon and the synthesis of CIA are fairly common among isolates of *S. faecalis*. A third of clinical isolates form clumps when challenged with the filtrate of one strain, whereas about two thirds of them produce CIA that induces clumping of either one of two strains used as donors.[954] Production of CIA or response to it is more frequent among drug-resistant isolates.

The above observations have been generalized into a model that relates conjugative plasmids and pheromones with cell aggregation (Figure 14).[954] The model assumes that cell aggregation is mediated by complementary substances on the surface of the cells. The binding substance (BS) on the surface of the recipient binds specifically to the aggregation substance (AS) on the surface of the donor. BS is found also on the donor surface. Consequently, the model also accommodates donor-donor interaction, which accounts for CIA-induced aggregation of donor cells. BS is determined by a chromosomal gene. Also on the chromosome are the determinants for pheromones, which are excreted by the recipient cells into the medium, where they are taken up by the plasmid-bearing donors. A particular species of pheromone interacts with a specific segment of a particular plasmid. This specific segment codes for a regulatory protein that somehow turns on another determinant (shown in the model as located on the plasmid, but may be chromosomally located) that brings about the synthesis of the AS. Located on the plasmid is another determinant for a substance which inactivates the indigenous pheromone or represses its synthesis. The plasmid determines the donor's specific response to a given species of pheromone. A plasmid-bearing cell responds only to a pheromone that is not indigenously present in that cell.

The mechanisms for mating, mating-pair formation, and gene transfer remain unresolved. It seems obvious nevertheless that cell aggregation is a prelude to sex, the plasmid-mediated transfer of genes from a donor to a recipient. The same conditions that bring about cell aggregation favor efficiency of plasmid transfer in the cell population. The formation of mating aggregates offers a more effective alternative to transient encounter by random collision in an aqueous environment. Cell aggregation could promote the spread of drug resistance in populations of *S. faecalis* or of other plasmid-bearing pathogens.

VIII. CELL AGGREGATION IN MYXOBACTERIA

The myxobacteria are a very interesting group. Mysteriously fascinating in their complexity and beauty, they cast a spell of confusion on both the systematist and the would-be molecular biologist. It is not quite correct that they are primitive forms of the cellular slime molds, similar though they may be in gross appearance and behavior. Nor are they unyielding and inscrutable.

As a group, the myxobacteria have been placed in various taxonomic pigeonholes, because of their aggregative nature and their macroscopic fruiting bodies, which are easily mistaken for fungal structures. They are unique in a number of ways, but no more so in that their kind alone among prokaryotes are represented as herbarium specimens rather than type cultures. Indeed, to many a cryptogamic botanist, they are better known for their colorful fruiting bodies than for their prokaryotic character. Nevertheless, to true believers, they are bacteria, no less true than the eubacteria.

The earliest known member of the group was described in 1809 by Link.[961] Identified as a group in 1892 by Thaxter, who saw in them the essence of multicellularity,[44] the myxobacteria have not, until recently, had the attention they deserve. The early literature attests to their being more of a collector's delight than an experimental tool. The past decade or so, however, has seen a turn about to correct this oversight. About time, for indeed there is much going for them as models for morphogenesis, differentiation, and development. Complex enough to be intriguing and yet simple enough for cold analysis, they exhibit the most elaborate life cycle among the prokaryotes. The typical myxobacter is very like *E. coli*, but a very different kind of *E. coli* indeed. And, yet, it is not quite a whale.

There is much work being undertaken in many laboratories devoted to dissecting the intricacies of these fascinating organisms, so much so that a number of well-thought-out reviews have been generated.[962-974] Methods for their developmental study used to be limited when they were mere laboratory curiosities. Now that they are domesticated, simple tech-

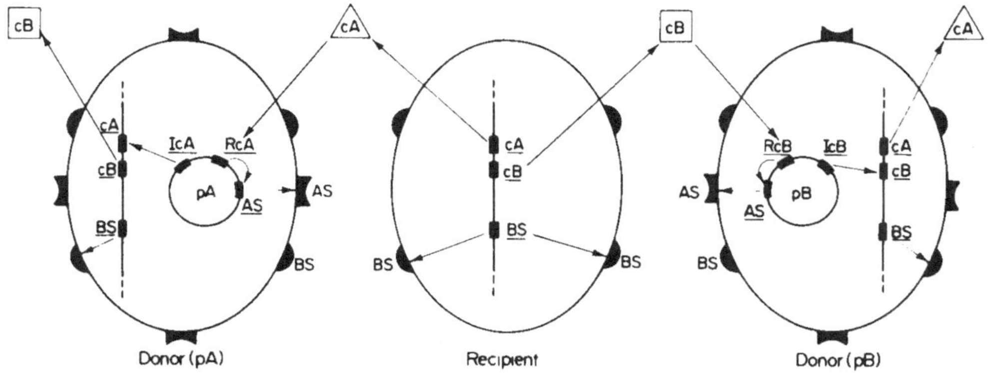

FIGURE 14. A model showing the relationships among conjugative plasmids, pheromones, and cell aggregation in *S. faecalis*. Oval figures represent cells. Segments of chromosomes are shown as vertical lines. The plasmids pA and pB are represented as circles in the donor cells. Underscored symbols are genetic determinants for corresponding gene products, which are represented by symbols not underscored. The chromosomal genes *cA* and *cB* code for the pheromones cA and cB, whereas *BS* codes for the binding substance BS, which eventually gets bound to the cell surface. The plasmid-located determinants *IcA* and *IcB* code for molecules that repress or inactivate the indigenous pheromones. *RcA* and *RcB* code for regulatory proteins that interact with the pheromones to activate AS, which in turn codes for the aggregation substance AS, which finds its way eventually to the cell surface. The BS (found on the cell surface of both the recipient and the donor) and the AS (found only on the cell surface of the donor) are complementary. (From Dunny, G. M., Craig, R. A., Carron, R. L., and Clewell, D. B., *Plasmid*, 2, 454, 1979. With permission.)

niques for their isolation, culture, characterization, and experimental manipulation are available.[240-242,974-979] Time-lapse movies on the dynamics of cell movement and the fruiting process have been produced by Reichenbach, Kühlwein, and Grimm in cooperation with *Encyclopedia Cinematographica*.[980-985]

Our present task is relatively simple compared to that of a general reviewer of the subject. It is simply to introduce the myxobacters to those who have not been fortunate enough to be acquainted with their aggregative properties. It is clear that fruiting-body formation of cells in an initial state of unicellular dispersion denotes the prior occurrence of cell-cell interaction and cell aggregation.

The current taxonomic position of myxobacteria has been arranged by McCurdy et al. in the latest edition of *Bergey's Manual*.[970,986-989] The fruiting myxobacteria are collected under the larger group of gliding bacteria. The group includes two orders: *Myxobacterales* Thaxter *emend.* Stanier 1957 and *Cytophagales* Leadbetter 1974.[962,990] In the first order are retained the fruiting myxobacters, with which we shall be concerned. The nonfruiting gliders, such as *Cytophaga* and *Sporocytophaga*, which used to be grouped with the myxobacteria in the older schemes of classification, are now placed in the new order, along with *Flexibacter, Beggiatoa, Thioploca,* and *Thiothrix* among others. Although sharing certain common characteristics with *Myxobacterales*, what used to be called nonfruiting myxobacters are sufficiently different from an evolutionary standpoint.[991] The molar G + C content of myxobacters ranges from 67 to 71%, that of *Cytophaga* and *Sporocytophaga* 33 to 44%.[992-994] Adansonian analysis further justifies the taxonomic separation and the creation of a new order from the monolithic taxon of an earlier day.[995-997]

The order *Myxobacterales* comprises eight genera in four families.[970,986-989] The families are Myxococcaceae, Archangiaceae, Cystobacteraceae, and Polyangiaceae. *Myxococcus*, the only genus in the first family, is the best studied of the genera. Consequently, the following discussion is mostly about this genus. Archangiaceae likewise covers a single genus, *Archangium*. Cystobacteraceae consists of the genera *Cystobacter, Melittangium,* and *Stigmatella*. Under Polyangiaceae are the genera *Polyangium, Nannocystis,* and *Chondromyces*.

Myxobacteria are nonphotosynthetic, heterotrophic, aerobic, unicellular, Gram-negative rods, no more than 1.5 μm in diameter. They may be spindle shaped (represented in the first three families) or uniformly cylindrical with rounded ends (Polyangiaceae). They multiply by binary transverse fission. They are found in the soil, on dung, decaying plant materials, and the barks of trees. Some have been isolated from aquatic habitats.[998-1000]

Though not very different from other Gram-negative bacteria, the myxobacters are noteworthy in a number of ways. To start with, they are colorful organisms. Their characteristic colors, red, orange, and yellow, are due to membrane-bound carotenoids.[1001,1002] Compared to enteric bacteria, they are considerably more sensitive to hydrophobic antibiotics.[972,1003] They lack flagella but are motile nevertheless. In common with the other gliding bacteria, they move from one area to another on a solid surface by slow, apparently directed, gliding motion. They are moderately flexible rods. Their flexibility may be accountable to lower amounts of peptidoglycan, qualitatively similar to that of enterobacteria but located in patches on their cell walls.[1004] The cells are typically enmeshed in slime, a cellular exudate also found in the tracks of gliding cells, in the mat left by a colony on agar, and in the sheaths of fruiting bodies.[1005, 1006] This adhesive slime is made up of complexes of proteins, polysaccharides, and lipids.[1006,1007] Myxobacters feed on other bacteria, dead or live, and other cells. They can hydrolyze insoluble macromolecules, such as cell walls, proteins, chitin, agar, and cellulose. The vegetative cells when foraging for food may travel as a swarm, displaying much cooperativity as a community of cells. When starved, they aggregate to form those colorful fruiting bodies, which contain resting cells called microcysts or myxospores. It is this developmental program which has intrigued the developmental biologist in search of a model organism and therein lies the uniqueness of the group as an experimental tool among prokaryotes.

The life cycle of the myxobacteria is simplified and generalized in Figure 15. The diagram is made up of two interlocking cycles: vegetative and developmental. In the vegetative cycle, the rods lengthen and then divide. The progeny are constrained to start the cycle repeatedly, unless the cycle is broken, when the vegetative cells are diverted to the developmental program. When the nutrients are depleted, the vegetative cells stream toward aggregation centers, where hundreds of thousands of cells gather and pile up. When a sufficient number of cells are gathered in a particular center, the resultant mound rounds off into a knob. Then follows a stage in the fruiting process when the knob is lifted by an emerging stalk into the air (Figure 16). Inside the fruiting body, the cells are converted into myxospores. These resting cells may for a long time remain dormant, resisting stresses due to an hostile environment. Under the proper conditions, the spores are released from the fruiting body to germinate into vegetative rods.

In the laboratory, the full splendor of the developmental process may be short-circuited. Vegetative cells may be induced to convert to spores without going through the rituals of aggregation and fruiting-body formation.[1008] Quantitative conversion of logarithmically growing cells is induced by the addition of 0.5 *M* glycerol to liquid cultures of *Myxococcus xanthus*.[1009] The technique applies to other myxobacterial species, such as *Stigmatella aurantiaca, S. brunnea*, species of *Archangium* and *Cystobacter*, and other species of *Myxococcus*.[968,978,1010,1011] Induction can be effected also by a variety of alcohols (such as propanol, propanediol, butanediol, phenethyl alcohol, ethylene glycol, threitol, and erythritol), amino acids, dimethyl sulfoxide, monovalent cations (Li, Na, NH_4, K), methionine starvation, incubation at 41°C, and a variety of other conditions.[1012-1015] Spores formed within the fruiting body and glycerol-induced spores are fairly similar.[1016, 1017] Both kinds are more resistant than vegetative cells to desiccation, extremes of temperature, UV irradiation, prolonged storage, sonication, and physical rupture. The spores of *Myxococcus* are spherical, optically refractile, and encapsulated. In some myxobacterial genera, the spores are merely shortened versions of the vegetative cells. The myxospores have been compared to other prokaryotic resting cells in a review by Sudo and Dworkin.[1017]

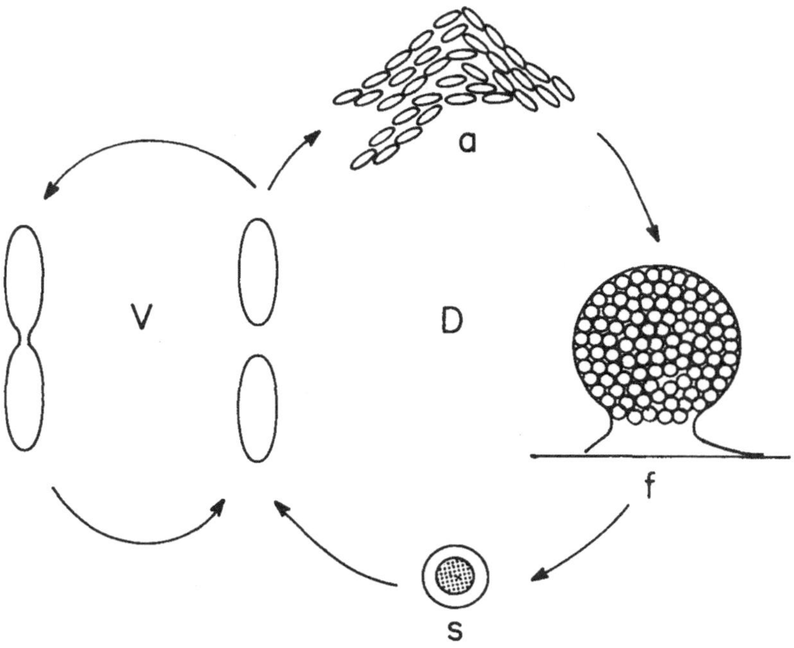

FIGURE 15. The vegetative (V) and developmental (D) cycles of myxobacteria. Aggregation, a; fruiting body, f; myxospore, s. The fruiting body depicted is that of *Myxococcus xanthus*. Adapted from Kaiser et al.[972]

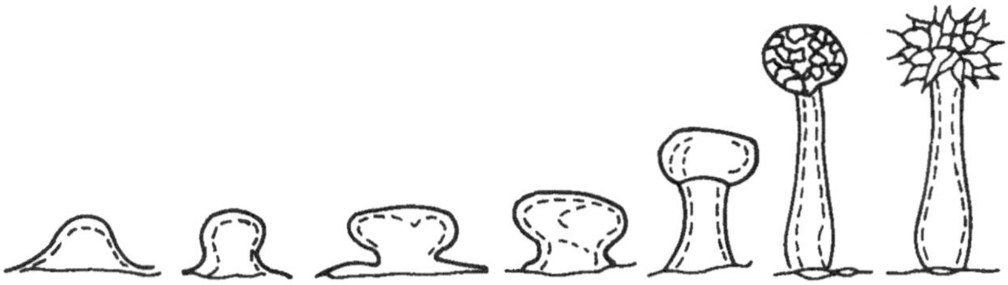

FIGURE 16. Fruiting-body formation in *Chondromyces apiculatus*. Adapted from Kühlwein.[237]

Clumping of cells during induction of sporulation has been reported in *Stigmatella aurantiaca.*[1015] In Casitone liquid medium plus inducer and vigorous aeration, vegetative cells are converted to myxospores in about 150 min. Induction is recognized visually by the clumping of differentiating cells about 30 to 60 min after the addition of inducer. In tests for various inducing conditions, clumping is considered to be diagnostic of positive induction. The phenomenon can be prevented from occurring by monovalent cations. The rate of clumping induced by glycerol rises with the temperature up to 32°C. It is measurable between 15 and 40°C. Between 20 and 30°C, the Q_{10} is about 2.3.

Sporulation induction with high concentrations of glycerol may not be physiological. For one thing, the starting cells are logarithmically growing. For another, the cell population density is quite high. Consequently, induction is fast and rather synchronous. Nevertheless, the clumping phenomenon that precedes the conversion of vegetative cells to myxospores suggests that changes occur on the cell surface to render the cells aggregative.

In nature (and on a solid surface), spore formation takes place in fruiting bodies in a number of days. Genetically determined, the shape of the spore-bearing structures is characteristic of the species. It ranges from the irregular masses of *Archangium* to the globular mounds of *Myxococcus xanthus*, to the coral-like arrangements of *M. corraloides*, to the mushroom caps of *Melittangium boletus*, to the more elaborate structures of *Stigmatella* and *Chondromyces* (Figures 17 to 19). The more elaborate fruiting body consists of several macrocysts or sporangia attached to the terminus of a stalk by their pedicels. The resting cells are found exclusively in the sporangia. The whole structure is covered by a layer of slime. In some species, the fruiting body is quite variable in form. Nonetheless, the systematics of *Myxobacterales* remains founded largely on the gross morphology of fruiting bodies. The central problem in the experimental study of these prokaryotic developmental systems is the description of the mechanisms by which these structures are constructed from a population of unicellular, undifferentiated rods.

Fruiting-body formation normally occurs on a solid surface. Cells of *Myxococcus xanthus* may be grown as a disperse population in liquid culture. They do not sporulate when starved under this condition, but do so when allowed to fruit on agar. A mutant that grows in a nondispersive state has been isolated by Burchard.[1018] This mutant forms macroscopic, multicellular spheres in a shake culture. Radially arranged within these spheres, the cells appear to be held together by slime fibrils. Most of them are converted to spores within the spheres when shaken in water or buffer solution. The spheres are resistant to disaggregation by physical and enzymic methods. When deposited on growth agar, they gradually disintegrate into a swarm of vegetative cells, but on fruiting agar, fruiting bodies are produced. Under the same conditions, nondispersive strains of *Myxococcus fulvus* behave similarly. These observations suggest that packing cells together in a compact mass may be conducive to differentiation. Indeed, cells of a nonmotile strain of *M. xanthus* (unable to form fruiting bodies on agar[1019]) convert to myxospores when pelleted in a centrifugal field and then spotted as a pellet on fruiting agar.

The early stages of development may occur in an aqueous milieu. When dispersely grown cells of *Stigmatella aurantiaca* are starved in liquid culture, they form clumps and take on the orange color of fruiting bodies.[1020] Clumps deposited on the surface of fruiting agar develop earlier than disperse cells, indicating that the aggregation process in liquid culture has shortened the period of development.

The coming together of cells in a shaken liquid culture is achieved by random collision. On a solid surface, in contrast, movement is clearly directed, toward aggregation centers, from which eventually arise fruiting bodies. Vegetative cells translocate themselves on a solid surface by gliding.[1021,1022] This mode of locomotion is characterized by continuous progression in the direction of the long axis of the cell. Occasionally, a gliding cell may stop or change direction. The rate of movement is low. For species of *Myxococcus* and *Archangium*, it is 5 to 10 μm min^{-1} on agar.[1023-1025]

Gliding motility is a cause for wonderment. As already noted, myxobacterial cells are bereft of flagella. They do not rotate nor do they take a sinusoidal path when they are in motion. Several theories have been proposed to account for gliding. One theory suggests that fimbriae, recently demonstrated in *Chondromyces crocatus* and in several other myxobacterial species, serve as locomotor organelles.[1022,1026-1030] A cell glides forward supposedly by extension and contraction of its fimbriae. Nonmotile mutants of *Myxococcus xanthus* have been found to lack fimbriae.[1028] A more recent report, however, shows that these isolates are nonmotile because of multiple mutations.[1031] But more disastrous to the theory is the fact that a number of strains without fimbriae are nonetheless actively motile.[972] Another theory assumes that a subsurface filament extending from one pole of the cell to the other causes cellular movement by its rhythmic contraction.[1032,1033] Supporting this proposal is the isolation of a nonmotile mutant with an altered filament.[1033] Still another theory assigns the directed extrusion of slime as a mechanism of propulsion.[237,245,1022,1034,1035]

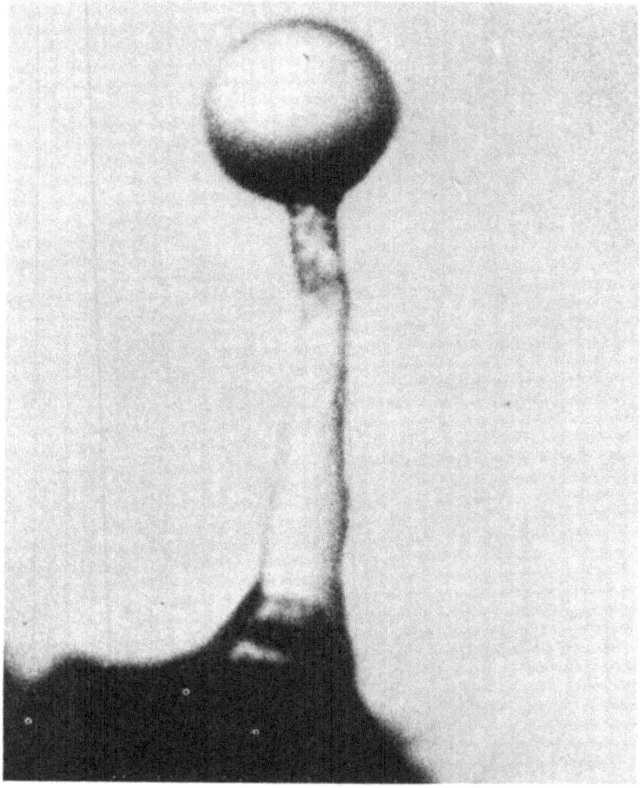

FIGURE 17. Fruiting body of *Myxococcus stipitatus*.[968] (Original photograph kindly furnished by Hans Reichenbach.)

A more attractive proposal for gliding motility has appeared in print recently.[1036] The locomotive power that drives flagellate bacteria resides in the rotary motor found at the base of the flagellum.[1037,1038] Whereas gliders lack flagella, they might well be equipped with these flagellar motors located in the cell envelope. Ring-like structures morphologically similar to flagellum-base elements have been identified in the nonfruiting gliders *Flexibacter columnaris* and *Cytophaga johnsonae*.[1036] These structures are absent in a nonmotile mutant of *F. columnaris*. The often-observed spinning action of a glider cell when tethered by one pole on a glass slide could be explained if these structures were rotary assemblies on the cell surface. The evidence that they may in fact be flagellar motors *sans* flagella rests on experiments with the active movement of latex spheres on the cell surface. Latex spheres, added to a suspension of actively motile cells, become attached to the cell surface, where they are observed to move about rapidly, sometimes as though they were being spun. Nonmotile mutants are incapable of moving latex spheres. In addition, the effects of metabolic inhibitors on gliding motility suggest that the generation of proton-motive force, rather than ATP, is the source for gliding, as it is for flagellar motility.[1039-1041]

Vegetative cells glide singly and in concert with hundreds to many thousands of its kind in loose social units called swarms.[1005,1019,1023,1025,1031-1033] A swarm may comprise a microcolony or the whole population derived from a single sporangium. The cells at the monolayer edge of a macrocolony make up swarms of various sizes. Smaller swarms may merge, larger ones break up, even as new ones form. Single cells may break away and wander about, only to rejoin the seething mass, as it seems in speeded-up time-lapse motion pictures. The

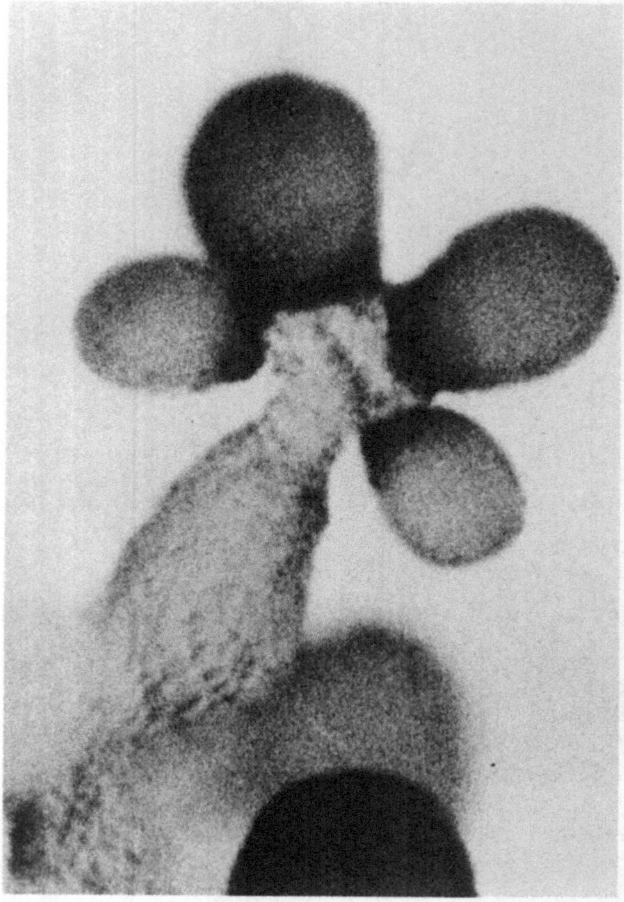

FIGURE 18. Fruiting body of *Stigmatella aurantiaca*.[968] (Original photograph kindly furnished by Hans Reichenbach.)

leading edge of a swarm consists of tongues of rods feeling their way outward. In certain cases, a swarm may expand radially. All the while that the swarm is on the move, the component cells increase in number by dividing. The individual cell in the reproductive act freezes in its tracks and moves on after cell division.[982] Rhythmic oscillation with regard to the outward movement of a swarm has been described.[1005] This is observed as ripples, which may represent a synchrony of fission among cohorts or a pulsatile emission of signals. However one looks at it, a swarm is a highly coordinated mass of interacting, migrating cells. Intercellular communication must exist for the expression of such a pattern of social behavior at a most elemental level.

The directedness of the movement of a cell population is evident in the phenomenon termed elasticotaxis.[1042] Myxobacterial cells streaked on an agar surface collect into fruiting bodies that are not, contrary to what is expected, stochastically distributed. Instead, they are arranged in rows approximately parallel to one another and perpendicular to the path of inoculation. This rather peculiar arrangement is the result of the oriented movement of swarms. The directive pressure appears to stem from the mechanical stress on the agar surface. Lines of tension and compression could force the gliding rods to move in line.

Myxobacterial cells on the move leave slime tracks behind them. Rather than strike out on their own, swarms tend to follow the tracks left by others (Figure 20).[1043] Occasionally,

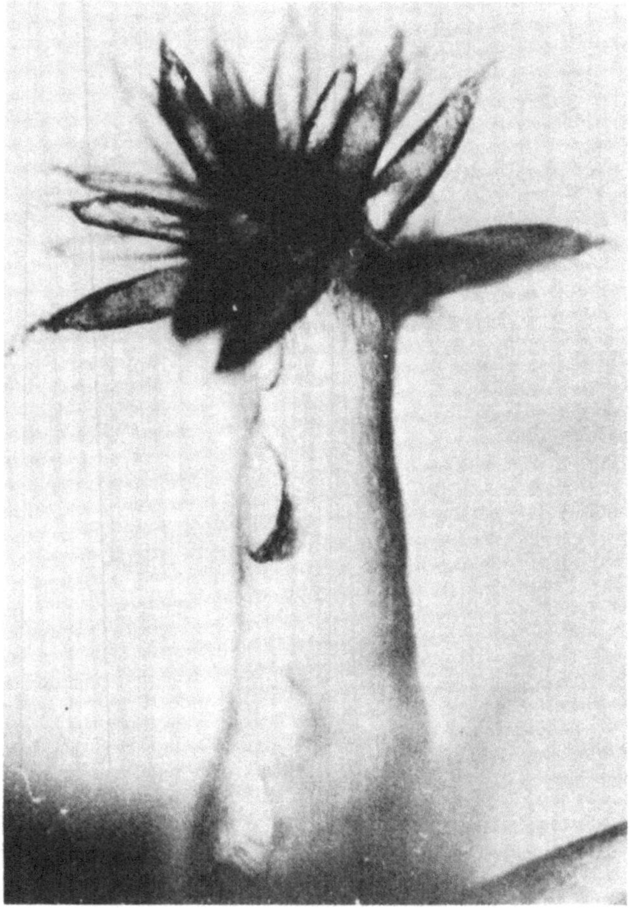

FIGURE 19. Fruiting body of *Chondromyces apiculatus*.[968] (Original photograph kindly furnished by Hans Reichenbach.)

however, they may be observed to cross a track. Also, it is possible to show experimentally the avoidance of slime tracks by swarms determined to reach aggregation centers. If the slime track between an aggregation center and a swarm is cut off by a gap or a barrier on the agar surface, the swarm may jump off the track, split into two, and go around the obstacle.

It would seem then that there is more to directed taxis than stress lines and slime tracks. Another homing mechanism could be chemotaxis. There are two obvious sources of chemical attraction for myxobacteria. Swarms of actively dividing cells are attracted to other bacteria, bacteria they can lyse efficiently and use up for food.[982] During the switch to the developmental program, when they have stopped dividing, cells are directed inexorably to aggregation centers. With the sources identified, one would think that proof positive for chemotaxis in myxobacters would have been obtained readily and the identification of chemotactic agents would pose no unusual difficulty, but until lately, evidence for chemical attraction has been limited to demonstrating the stimulatory activity of exudates of fruiting bodies on fruiting-body formation. The identification of chemoattractants for myxobacteria has been hampered by the slowness of their cellular movement, which comes close to the rate of diffusion of prospective tactic substances. Recently, it has been shown quite positively

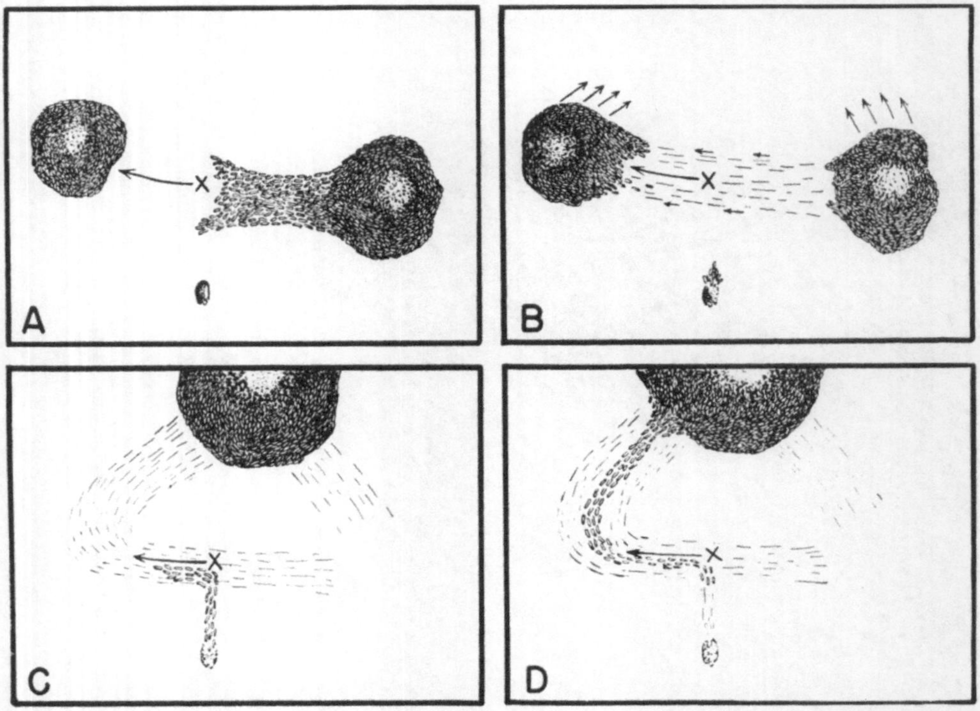

FIGURE 20. A demonstration of the propensity of *Chondromyces* cells to follow slime tracks. (A) An aggregation center is manipulated in the direction of the arrow from position X to the left. A macrocyst is found below X. (B) The mass of cells between the two centers have joined the translocated center to the left. Both centers now move upward in the direction of a common point. The macrocyst has ruptured, releasing its component cells. (C and D) The two original centers are now fused. The cells from the ruptured macrocyst join the new center, not by the shortest route but by following the more roundabout slime tracks. (From Bonner, J. T., *Morphogenesis: an Essay on Development*, Princeton University Press, Princeton, N.J., 1952, 165. With permission.)

that adenosine-5′-phosphate and cyclic GMP are chemoattractants for *Myxococcus xanthus.*[1044] Other nucleotides, including ADP, GMP, and the much-touted cyclic AMP, are ineffectual.

In addition to swarming movements and aggregation toward centers, there are the morphogenetic movements that lead ultimately to the realization of a fruiting body: the piling up, the lifting of the knob, the erection of the stalk, the compartmentalization of the head into sporangia. These gross morphological changes are accompanied no doubt by intracellular changes, including macromolecular synthesis, until the once vegetative cells are converted to myxospores within the sporangia of the fruiting body. They do not properly belong to cell aggregation, for they constitute a mere rearrangement of a structure that has become essentially multicellular. The initial events of development are aggregational.

Just what triggers the start of the developmental program? The stimulatory effect of starvation on fruiting-body formation was known to early workers.[234,235] There is now ample evidence to assure us that nutrient deprivation, or some event associated with it, triggers the initiation of development. For so long as the vegetative cycle is operative, the developmental cycle remains merely potential. When vegetative cells face starvation, they start making provisions for the time when food will not be conveniently available within the immediate environment and thus the program for converting themselves to resting cells that can persist under adverse conditions. The details of the turning on of the program remain obscure.

The formulation of chemically defined growth media has somewhat eased the problems of looking for conditions that constitute the initiation of fruiting-body formation. Those conditions that bring about the formation of fruiting bodies, because cell aggregation precedes it, may be construed as having triggered aggregation. On a defined agar medium with high concentrations of amino acids, it has been shown that limiting phenylalanine and tryptophan results in fruiting.[233] This initial observation has now been generalized. Fruiting is promoted by lowering the levels of any one of some ten amino acids in the medium, without altogether removing it, and is suppressed by increasing its concentration.[239,241,1045] Further studies with even lower levels of amino acids, and of identified sources of energy, carbon, and nitrogen, have led to the conclusion that limiting the concentrations of essential amino acids (isoleucine, leucine, valine, and methionine) or of growth-limiting phenylalanine induces fruiting-body formation in *M. xanthus*;[972,1046] so does limiting the concentrations of sources of energy, carbon, and nitrogen. Moreover, auxotrophic mutants can be induced to fruit by limiting the concentration of the nutrient required for growth. In general, the approaching state of starvation serves as a signal for the cells to start the process of fruiting.

The possible connection between metabolic regulation and the induction of development has been investigated.[1045,1047,1048] Aspartokinase catalyzes the first step in the metabolic pathway that leads to the synthesis of lysine, isoleucine, methionine, threonine, and diaminopimelic acid. The aspartokinase activity of extracts of *M. xanthus* is subject to feedback repression and feedback inhibition by threonine and lysine. The required amino acids isoleucine and methionine stimulate aspartokinase activity and inhibit fruiting-body formation. Threonine, as well as spermidine, reverses both effects. The level of aspartokinase activity drops significantly during the first hour of myxospore induction by glycerol. These observations suggest that a drop in aspartokinase activity may lead to starvation for diaminopimelic acid, which is an important component of the cell wall. An inhibition of cell-wall synthesis might be a requisite condition for the start of the developmental program. The explanation is obviously just one of many possible control mechanisms.

Nutrient depletion is a negative environmental stimulus. A positive one requires the enrichment of a medium for a substance. The addition of cyclic AMP to a medium of low nutrient content has been reported to enhance fruiting-body formation in *M. xanthus*.[1049] Presumably then an increase in intracellular cyclic AMP may trigger the initial events of development, particularly so because it has been shown that there is indeed a small but significant increase in intracellular cyclic AMP during aggregation, fruiting-body formation, and glycerol-induced myxosporulation.[1050,1051] Although displaying sharp concentration optima, the stimulatory effect is less than agent specific.[1049] Also effective are other adenine derivatives, such as isomers of AMP, ATP, adenosine, deoxyadenosine, and even adenine itself. Indeed, ADP is more potent than cyclic AMP. Other nucleotides tested have no stimulatory activity on fruiting-body formation.

There is a recent report that exogenous cyclic-AMP phosphodiesterase enlarges the territory of aggregates and accelerates the formation of fruiting bodies in *M. xanthus*.[1048] Theophylline and caffeine, both inhibitors of phosphodiesterase, counteract the stimulatory effect of the added enzyme, as do GMP and cyclic GMP. In the absence of the added enzyme, cyclic AMP increases the number of fruiting bodies, without shortening the time at which they are first observed. During myxosporulation induced by glycerol, there is a transient increase, followed shortly by a decrease, in the intracellular levels of cyclic AMP. The stimulatory activity of cyclic nucleotide phosphodiesterase is observed in a variety of other myxobacteria, including *Myxococcus coralloides*, *Stigmatella aurantiaca*, *Cystobacter fuscus*, and *Polyangium cellulosum*.

Cyclic AMP, which does not hasten the formation of fruiting bodies while increasing their numbers, may not be a factor in the migration of cells toward aggregation centers.[1049] The chemotactic signals may be supplied by some other compound, cyclic GMP, perhaps, which is hydrolyzed too by cyclic AMP phosphodiesterase and which is claimed to possess chemotactic activity.[1044] Cyclic AMP does not relieve fruiting inhibition by isoleucine or methionine, but does act synergistically with threonine to stimulate fruiting.[1049] It may serve as an agent for releasing the cells from catabolite repression, which enthralls them to a repetitive cycle of vegetative existence. One can find here further analogies to the induction systems in enteric bacteria. The accumulation of guanosine polyphosphates (tetra- and pentaphosphates), a response to amino-acid starvation, has been reported in *E. coli* to inhibit functions related to growth, such as the synthesis of peptidoglycans and ribosomal RNA.[1052-1054] In myxobacteria, they may serve to break the vegetative cycle and deflect the cells to the developmental sequence.[972]

By inhibition of cyclic-AMP phosphodiesterase, cytokinins may interact with the cyclic-AMP metabolism of mammalian systems.[1055] They have been shown in *E. coli* to counteract the growth-inhibitory effect of exogenous cyclic AMP.[1056] In *Myxococcus xanthus,* a recent report claims that N^6-(Δ^2-isopentenyl)adenine and, less effectively, two other cytokinins (6-benzylaminopurine and kinetin) inhibit fruiting-body formation.[1057] Because they concomitantly reduce the area of a swarm without affecting cell growth, cytokinins may interfere with the processes of movement and aggregation. It is not likely that they merely offset the action of cyclic AMP, because exogenous cyclic AMP or ADP fails to reverse their inhibitory effect.

It is reasonable to suppose that chemical signals are relayed from aggregation centers to the surrounding cells. None so far has been definitely identified, although diffusible factors have long been postulated. Fruiting bodies, which of course arise from aggregation centers, were first demonstrated by Lev to excrete diffusible substances that induce fruiting.[1058] On agar squares that allow growth but not fruiting, cells of *M. xanthus* or *Chondromyces exiguus* form fruiting bodies when the squares are laid over preexisting fruiting bodies. The induction appears to be specific at least at the generic level, for no cross-stimulation between the two species is observed. An aqueous extract of fruiting bodies of *Myxococcus virescens* has been claimed to stimulate development.[243] This extract yields, on paper chromatogram, a pigmented spot that cannot be found in extracts of vegetative cells. Because it reacts with Liebermann's reagent, the spot is presumed to contain a sterol and could be the pigment of the fruiting bodies. When the paper chromatogram is set on a plate of cells, fruiting bodies are subsequently observed to be concentrated in the vicinity of the spot.

The stimulation of fruiting by fruiting bodies does not necessarily denote chemotaxis. It might indicate merely the presence of substances that somehow bring about development. More indicative of chemotactic substances that influence aggregation is the evidence from other experiments. When vegetative cells are spread on a sheet of cellophane, which is placed on top of fruiting bodies, the location of most of the fruiting bodies found on the cellophane corresponds spatially to that of those underneath.[238,245,1059] Bonner's method for showing chemotactic response in cellular slime molds has been adapted for *Myxococcus fulvus.*[245,1060] On a shallow liquid culture of cells in a petri plate, liquid flow is generated by slightly raising the temperature at one spot. Under this condition, fruiting bodies are observed to have lined themselves up in the direction of the flow. The orientation arises presumably because, once a fruiting body breaks through the slime mat, cells tend to collect on the lee side of the aggregation center, leaving a clear zone upstream. This sort of experiment can be interpreted several ways, one of which is that the cells mechanically avoid the flow and take cover behind the emerging mound, but together with the other pieces of evidence, it can mean that there are in fruiting bodies certain diffusible substances to which aggregating cells tend to respond positively.

Chemotactic substances may simply direct movement, or may stimulate the rate of movement as well. Preparations from fruiting bodies of *M. xanthus* have been reported to increase the rate of migration of aggregating streams (from a mean rate of 0.05 mm hr^{-1} for the control to as much as 0.19 mm hr^{-1} for the treated).[1061] Cyclic AMP, AMP, ADP, and folic acid have not proved to increase aggregation rate. Vegetative cells do not respond to the rate-promoting preparations. Dependent on temperature and agar concentration, the rate of movement is unaffected by light or gravity.[1061,1062] The temperature optimum is close to 30°C, the optimum agar concentration about 1%.

To bring about starvation and consequently fruiting, one need only free the cells from the growth medium. Cells of *M. fulvus* and other myxobacterial species may be induced to fruit in static liquid cultures.[244] In a complex liquid medium, cells form a nonfruiting layer, several-cells deep, at the bottom of a petri plate. Fruiting is induced only after the original medium is washed off and replaced with a salt solution. Washing with water will do, but repeated washing tends to reduce the number of fruiting bodies. The requirement for ions seems apparent, for chlorides of Ca, Mg, Sr, K, and Na in the washing and soaking solution enhance fruiting. The divalent cations are more effective than the monovalent ones. $CaCO_3$ is preferred to $CaCl_2$, for removal of CO_2 in the air by KOH inhibits the fruiting process. For complete development, the cells must be continuously bathed in the calcium solution.

Cells are best induced to form fruiting bodies during the late exponential phase of growth.[244,1063] As the cells age, they tend to produce fewer and fewer fruiting bodies. The optimum temperature for fruiting is about 32°C. Although cells remain viable at 36°C, they are prevented from fruiting until the temperature is dropped to near the optimum. Thus, the developmental process is temperature sensitive, as has been found in other developmental systems.[141-148]

Not all aggregation centers are committed to become fruiting bodies. Some of them are doomed to perish before culmination. Indeed, some may disappear even as new ones appear.[1064] Presumably, aggregation centers that fail to become fruiting bodies disband, thereby releasing the component cells to join nearby centers. In addition, some cells may convert to myxospores before they become components of fruiting bodies, others even before they reach aggregation centers.[1063] The territorial extent of a center is not well defined. Consequently, the size of fruiting bodies is influenced by cell population density and the presence of other centers in the vicinity. The rate of fruiting-body formation in *M. xanthus* is dependent on cell population density.[971] At low density (a suspension of 2.5×10^8 cells $m\ell^{-1}$ spotted on an agar plate), development is suppressed, although there are enough cells to make up many fruiting bodies consisting of 2×10^4 to 4×10^4 cells each. High cell population density allows fruiting-body formation and further increasing the density results in an increase in the rate of fruiting. At very high cell population density (a suspension of 2×10^9 cells $m\ell^{-1}$ spotted on an agar plate), fruiting-body formation, without net vegetative growth, is rapid and practically synchronous.

Although the initial events in the aggregation process seem not to be light dependent, the subsequent morphogenetic movements have been demonstrated to be influenced by light.[1020,1065,1066] Cells of *Stigmatella aurantiaca* aggregate and develop into fruiting bodies, complete with stalk and sporangia, in the presence of incandescent light. In the dark, cells do aggregate but rarely develop into fruiting bodies. Aggregates formed in the presence of light, however, form fruiting bodies even in the dark. An early work on *M. fulvus* concludes that light does not influence the direction of migration of a swarm.[1067] In unidentified myxobacter species, however, it has been shown that light directs migrating swarms from points of high intensity toward lower intensity.[1068] A white variant exhibits the same response as a red-pigmented strain, ruling out the red pigment as a photoreceptor that directs the negative phototactic behavior.

Gene expression during the development of *M. xanthus* has been studied by analyzing patterns of protein synthesis.[1069,1070] Cells pulse-labeled with a mixture of ^{14}C amino acids at different developmental stages yield soluble and membrane proteins that show considerable changes demonstrable by autoradiography of SDS-polyacrylamide gel electrophoretograms.

At least a fourth of some 30 major soluble proteins, and several membrane proteins, change their rates of synthesis during development. Four types of developmental proteins are identifiable: those that show accelerated and continuous synthesis, those whose rate of synthesis peaks midway, those synthesized only near the end of development, and those that show a constant rate of synthesis throughout. When fruiting is suppressed by exogenous methionine, the changes during the early and middle periods of development still occur, but late proteins are no longer synthesized. Protein S, a developmental protein of the first type, is the most visible. Its synthesis occurs early, increases dramatically, and reaches a top rate that constitutes up to 15% of total protein synthesis. It is found in the soluble fraction during the early stages of development and in the insoluble fraction later. A component of the spore coat, it can be detached from mature spores with 1 *M* NaCl at 30°C, a treatment which does not hurt spore viability. Upon removal of the salt, purified protein S (molecular weight: 2.3 $\times 10^4$ daltons) reattaches spontaneously to spores that have been deprived of their protein S. Spores derived by glycerol induction do not contain protein S nor do they serve as sites for self-assembly. From its properties, protein S does not seem to be an aggregation protein.

A more interesting protein found in large quantities in the soluble fraction is myxobacterial hemagglutinin (MBHA).[1071] A development-specific lectin (apparent molecular weight: approximately 3×10^4 daltons), MBHA mediates the agglutination of erythrocytes from sheep and guinea pig, but not ox, chicken, or human erythrocytes. It is not detectable in extracts of vegetative cells, of cells starved in liquid culture, or of cells induced to sporulate by glycerol. Its rate of synthesis increases during development such that at its peak, it represents around 5% of total protein synthesis. MBHA is most abundant (up to 2% of total protein) in extracts derived from cells gathered in mounds. The hemagglutinating activity, not inhibited by simple sugars or amino sugars, is inhibited by the serum glycoprotein fetuin. The O-glycosidically linked trisaccharide glycopeptide (*N*-acetylneuraminic acid-galactose-*N*-acetylgalactosamine-serine polypeptide) of fetuin is itself inhibitory, but the trisaccharide alone is not. The penultimate galactose of the glycopeptide is implicated, because the inhibitory activity of asialofetuin is much reduced after periodate oxidation or treatment with β-galactosidase. It remains to be demonstrated that MBHA causes myxobacterial cells to stick together.

A cell-aggregation system that has more than an ephemeral existence must embody a mechanism for holding onto its component cells. Instead of being locomotor organelles, the proteinaceous fimbriae have been suggested to serve this purpose in myxobacterial aggregation, as has been suggested in other aggregation systems.[1072,1073] The presence of fimbriae correlates well with the formation of clumps in liquid cultures of *M. xanthus* and the motility of aggregates on agar. An inhibitor of swarming, EDTA also inhibits the formation of fimbriae. Antisera against fimbriae do inhibit swarming without disturbing gliding motility per se. Interestingly, fimbriae possess lectin activity. Red cells of guinea pig agglutinate in the presence of intact, but not depolymerized, fimbriae. The hemagglutinating activity is inhibited by mannose, *N*-acetyl-D-galactosamine, and to a lesser extent by fructose, raffinose, melibiose, and α-methyl-D-mannoside. Electron miscroscopy shows that the fimbriae of one cell adhere to the surfaces of its neighbor cells.

It is to be expected that during the developmental sequence, new proteins are synthesized while the synthesis of other proteins is shut off. For the synthesis of proteins, messengers must be synthesized by RNA polymerase. Conceivably, a defective RNA polymerase could lead to a disruption of fruiting-body formation. Rifampin-resistant mutants defective in development have been isolated.[1074] These mutants fall into two classes. Mutants of one class manifest very little inclination to form aggregates at high cell population density (8 $\times 10^9$ cells mℓ^{-1}) but may aggregate in flat mounds at low density (1.6×10^9 cells mℓ^{-1}). At either high or low cell population density, mutants of the other class form tight aggregates (translucent mounds). The mutants exhibit rifampin-resistant RNA-polymerase activity in vitro but no detoxifying activity. They are consequently concluded to possess an altered RNA polymerase. Because the fruiting defect and resistance to rifampin are co-transduced (>99% of the time), both phenotypes may be the consequence of a single mutation. The mutants show normal vegetative morphology and growth. The physical changes in the polymerase molecule must be subtle but effective enough to change the transcriptional specificities of the enzyme. The vegetative functions may remain unaffected, but the genes for development may not be efficiently recognized. Changes in transcriptional specificities have been proposed for other developmental systems.[1075-1077]

Other developmental mutants suggest the existence of intercellular signals for the coordination of the sequential steps of development.[1064,1078] The intercellular communication is necessary to guide the cells as regards the time and the place for aggregation. There are mutants which do not fruit among themselves, but they do fruit when combined pairwise with other mutants. Certain combinations with the wild type also allow sporulation. The synergistic effect is observed only with certain combinations, such that the numerous mutants fall into a very small number of synergistic groups (three, maybe four). Because these mutants can be induced to form myxospores in glycerol, their defects may be located in the aggregation steps. Although the synergistic effect may be ascribed to cross-feeding, the limited number of synergistic groups and the general health of the mutants make nutritional considerations less attractive. The synergistic effect may well be due to complementation of mutations in signal generation and reception.

Many nonmotile mutants have been isolated and studied.[1019,1025,1028,1031,1079] The ones that have proved most useful are those that become transiently motile upon contact with wild type cells or other mutants.[1025,1079] Pairwise combination analysis reveals that these mutants fall into six types (A, B, C, D, E, and F). Most of the mutants belong to type A. When paired with other types, type A remains nonmotile, whereas the other types mutually stimulate one another. Type A cells, however, are able to stimulate cells of the other types. The stimulation, which depends on material contact of interacting cells, is not due to genetic exchange, because no stably motile derivative is obtained. Linkage analysis by means of a generalized transducing phage allows the assignment of each of the stimulatable types to single distinct loci: *cglB*, *cglC*, *cglD*, *cglE*, and *cglF*. In contrast, type A mutants represent mutations in at least 17 genetic loci: 16 *agl* loci (characterized by slight motility at low temperature or on low-nutrient media) and the *mgl* locus (characterized by nonmotility under all circumstances).

Further genetic analysis has resolved gliding motility in *M. xanthus* into two multi-gene systems.[1031] System A (for adventurous, because single cells are preponderant) determines and regulates the motility of single cells, system S (for social, because groups are preponderant) the motility of groups of cells (Figure 21). System A comprises the 16 *agl* loci (*aglA* to *aglR*) and the five *cgl* loci (*cglB* to *cglF*). Nine genes make up system S: the *sgl* series (*sglA* to *sglF*) and the *tgl* locus, mutation of which confers transient motility stimulatable by contact. The *mgl* locus serves a common function, being required for both systems. Together, the two systems control gliding motility of myxobacteria as single cells and as a swarm. Single cells are motile if they bear an intact system A. Swarms of cells are motile if the cells bear an intact system S. At least one mutation in each system is required for the complete elimination of motility.

The fruiting process obviously denotes prior movement of cells or groups of cells. Would A^- and S^- mutations affect fruiting-body formation? A comparison between A^-S^+ and A^+S^- mutants shows that a majority of mutants of the first genotype (A^-S^+) are good fruiters, whereas a majority of those of the latter (A^+S^-) are nonfruiters or poor fruiters.[1031] Though not definitive, this evidence points to a possible role for system S in the fruiting process.

Aside from the morphological changes from vegetative cells to myxospores, there is the phenomenon of phase variation, which may have a bearing on development. In *M. xanthus*,

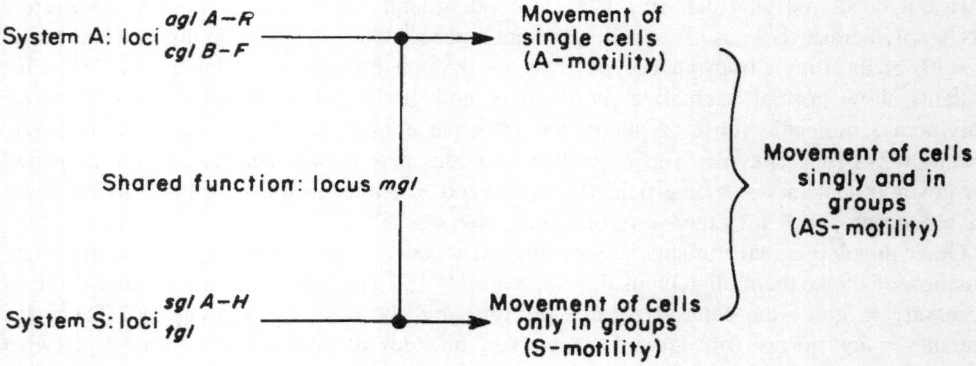

FIGURE 21. The two multi-gene systems that govern motility in *Myxococcus xanthus*. (From Hodgkin, J. and Kaiser, D., *Mol. Gen. Genet.*, 171, 177, 1979. With permission.)

four distinct colonial phenotypes have been identified: YS (yellow swarmer), TS (tan swarmer), YNS (yellow nonswarmer), and TNS (tan nonswarmer).[971,1062,1080,1081] Any one of them appears to be derivable from another. Although the interconversion is spontaneous, it can be enhanced by high temperature, UV irradiation, phenethyl alcohol, mitomycin C, and nalidixic acid. Variability in colony morphology has been observed also in *Archangium violaceum*.[1023,1082-1084]

One other observation, the developmentally induced autolysis of cells, merits discussion with regard to the fruiting process. Myxobacterial cells are known to lyse readily.[1080,1083,1085] What is more germane to the subject at hand is that there is extensive autolysis among the population of cells undergoing fruiting-body formation.[1086] In *M. xanthus,* only about 20% of the rod population differentiate into myxospores. The rest are fated to end up as cases of hara-kiri. The mass lytic event occurs past the aggregation stage, when the cells have gathered into mounds — to the many, sacrificial mounds. It is suggested that autolysis of the fraction of the population undergoing development provides nutrients for myxospore induction and, perhaps, nutrients for the construction of fruiting bodies. Autolysis has been encountered in other aggregation systems, including mating-aggregate formation during conjugation in *E. coli,* sex-directed flocculation in fission yeast, and the mating reaction in *Chlamydomonas*.

The diversity of myxobacterial species and the complexity of the developmental program demand morphological studies for laying the groundwork for the subsequent quantitative account of the aggregation process. Much effort has already been spent in describing pictorially by diverse techniques of microscopy the various aspects of aggregation and fruiting-body formation in different species. The generalized and simplified picture that emerges is derived from many reports that make up the rapidly expanding literature, including pictorial accounts that erstwhile have not been referred to but must be recommended to the reader for their sheer richness in visual detail.[1087-1095]

Given a few more years, the fruiting process in the myxobacteria could become a most rewarding model for development in higher organisms. There has already been much progress, thanks largely to the sustained vision and industry of Dworkin and his school in promoting research on myxobacteria as model developmental systems. Being considerably more complicated than endospore formation in *Bacillus* and *Clostridium,* the developmental sequence in the myxobacteria is nevertheless simpler than eukaryotic systems, if only because of its prokaryotic nature. The contribution of the initial event of cell aggregation can hardly be dismissed as minor in the developmental sequence. Whereas fruiting does not seem to be absolutely necessary for myxosporulation, there is little doubt that the aggregation process is a prerequisite to fruiting. Without cell aggregation, the fruiting body as a morphological entity would not be what it is, a multicellular construct of previously and potentially isolated cells. In the myxobacteria, prokaryotic aggregation reaches its greatest height, literally, exquisitely, in the macroscopic, colorful fruiting body.

IX. OTHER BACTERIAL AGGREGATION SYSTEMS

Cell clumping has been encountered in many a system, without being purposely looked for. One example is cell clumping in *Nocardia corallina*.[232] The random orientation of the cells in the clump rules out incomplete separation during cell division as the cause of clumping, which is observed over a wide range of hydrogen-ion concentrations (pH 2 to 9). The clumping has been attributed to a slimy secretion surrounding the cell. The slime can be washed off with distilled water.

Another bacterial species accidentally studied for its clumping behavior is *Corynebacteriam xerosis*.[97] Suspensions of cells grown at 30°C form clumps when rapidly cooled to 15°C with agitation. The extent of clumping increases as the temperature decreases from 30 to 15°C. It increases from mid-logarithmic to late-logarithmic phase of growth and decreases thereafter. It is observed over a broad range of pH values but is greatest around pH 3. Clumping in *C. xerosis* is a trifle different from most of the other bacterial systems in that a clump as observed with the electron microscope appears to consist of a single layer of cells collected on the surface of a gas bubble. Thus, instead of the clumps settling down to the bottom, they rise to the surface of the suspending medium. The clumping process is little affected by replacing the atmosphere with oxygen or nitrogen. It requires the presence of salt. The extent of clumping is reduced by uranyl nitrate and high concentrations of guanidinium chloride or urea. Isolated cell walls also form clumps. Proteases, papain especially, reduce clumping of both isolated walls and whole cells. But *p*-mercuribenzoate or mercaptoethanol has no effect, which would suggest that although proteins and phosphate groups may be involved, sulhydryl groups and disulfide links are probably not. The ability to form clumps is found in several, but not all, strains of *C. xerosis* and in other *Corynebacterium* species, including *C. aquaticum, C. mendiolanum,* and *C. ovis*. The association of the clumps with gas bubbles is reminiscent of pellicle formation in fimbriate bacteria (see Section IV of this chapter) and the interaction of bubbles with bacteria in the aquatic environment.[1096-1098]

Finally, without suggesting that the list of bacterial aggregation systems has been exhausted, we would like to mention in passing the well-known clumping of *Staphylococcus aureus* in the presence of fibrinogen.[1099-1103] An antigenic clumping factor associated with the cell surface is presumed to interact with fibrinogen to cause the fibrinogen-cell clumping reaction. Although much studied since its initial observation early in the century, we shall not discuss fibrinogen-induced clumping any further, because it is an example of an aggregation provoked by a serum component, a family of systems we have decided to exclude as a matter of convenience in this slim volume of microbial aggregation. A treatment of it at length would lead us inevitably to antibody-provoked cell agglutination, a subject that is most difficult to cover with justice.

[illegible]

[illegible]

[illegible]

[illegible]

[illegible]

Chapter 5

YEASTS

I. INTRODUCTION

Yeasts are fungi that are conspicuously unicellular for the greater part of their life cycle. The term is not precise. Many yeasts, if not all, may be grown as mycelial forms. On the other hand, many filamentous fungi not usually included in the term may be grown as predominantly unicellular cultures. The propriety of the term has been questioned and examined.[1104] With a long history from its Sanskrit origins but still currently useful, the term may be applied to microorganisms that look and behave like the one found in a labeled packet at the grocery store, *Saccharomyces*, the sugar fungus.

Taxonomically, the yeasts are a diverse collection. They may be divided into four groups.[1105] Most of them, including the better known ones, are found in the first group, which can form ascospores and therefore belong to the class Ascomycetes. The ascomycetous yeasts comprise 22 genera, among them the ones we shall be principally concerned with in this chapter, *Hansenula, Schizosaccharomyces*, and *Saccharomyces*. The second group consists of two genera in the order Ustilaginales of the class Basidiomycetes. The third group of three genera is characterized by the formation of ballistospores and is placed in Sporobolomycetaceae, a basidiomycetous family. The rest, some 12 genera, have no known sex cycle and are classified for the moment with the assorted group of *Fungi imperfecti* (Deuteromycetes).

Yeasts divide characteristically by budding, although certain ones divide by binary fission. They may be spherical, ovoid, or cylindrical. They are of course eukaryotes, cytologically plant-like but nonphotosynthetic. Some are quite useful for their biochemical activities. Where would Western civilization be, or who could possibly stand it, without bread and wine? Whereas the majority of yeasts are epiphytic or saprophytic, a few are pathogenic to plants, animals, and man. They have a colorful and distinguished history as experimental organisms. Their contributions to the foundations of biology, including biochemistry, genetics, and physiology, are too well known and too many to be enumerated. Some recent treatises, including the taxonomic work edited by Lodder, make a good introduction to the life of this morphological group.[1106-1110]

Unicellular though they may be, many yeasts form multicullular aggregates some time in their life cyle. Indeed, as experimental pawns for the study of cell-cell interaction, the yeasts antedate all other taxonomic groups. Some of them are studied quite intensively today as cell aggregation systems. In this chapter, we shall examine sexual agglutination in *Hansenula wingei*, sex-directed flocculation in fission yeast, sexual agglutination in *Saccharomyces cerevisiae*, flocculation of brewers' yeast, and cell aggregation in other yeast species.

II. SEXUAL AGGLUTINATION IN *HANSENULA WINGEI*

At least with respect to the recognition factors that mediate cell-cell interaction, the yeast *H. wingei* has been the most successfully studied. The success in unraveling the molecular basis of the mating reaction in this ascosporogenous species is to a large measure due to the ease with which the dramatic agglutination that precedes conjugation is demonstrated. On sexual agglutination, more than any other manifestation, is founded the reputation of this yeast as an experimental organism.

Sexual agglutination in fungi was discovered, together with *Hansenula wingei*, in 1956 by Wickerham.[62] The organism was isolated as a diplont from frass on a conifer. The isolate was allowed to sporulate and the resultant ascospores were recovered by differential heat treatment. Clones derived from the haploid ascospores were mixed in pairs. Certain combinations allowed cells to remain disperse, but others resulted in mass agglutination within a few seconds after mixing. The nonagglutinative mixture was concluded to be the consequence of combining cells of the same mating type. Agglutination was then due to complementary mating types. The mass agglutination reaction was so intense that the cellular mass rolled between the palms into a ball could be left to stand alone on a plane without flattening (Figure 1, Chapter 3). Agglutination was obviously sexual, for it was followed by conjugation. In the process, the haploid cells were converted to diploid cells, which were not agglutinative.

The sexual agglutination process promotes the conversion of the vegetative haploid cells by way of sex to diploid cells. The diploid cells are also vegetative. The consequence of fusion of two haploid cells in rich media is a zygote that subsequently undergoes mitosis instead of meiosis. The number of zygotes that convert to asci is very small.

Of the many strains isolated by Wickerham, the complementary strains 5 and 21 are most agglutinative. Subsequent work by other authors makes use of these two strains almost exclusively. Since the first report, much has been learned about the mechanism of agglutination, the physiology, the ultrastructure, and the molecular components of the interaction. Sexual agglutination in *H. wingei* has been reviewed by Crandall in several places.[16,98,515,516,1110] Wickerham has dealt with the taxonomy and the phylogeny of the genus.[36,601]

H. wingei Wickerham 1956 is one of 25 species and 6 varieties in the genus *Hansenula* H. et P. Sydow 1919.[601] It is a budding yeast. During logarithmic growth, cells may bear one to three buds. They may be spherical, ovoid, or oblong. Under certain conditions, cultures may produce true hyphae. The life cycle of the heterothallic yeast is shown in Figure 1. In the diagram, the complementary haploid strains are 5 and 21. When the haploid cells, germinated from their respective ascospores, are mixed together, they agglutinate immediately. Sexual agglutination is followed by conjugation of complementary pairs and the formation of diploid cells in the agglutinated mass. Diploid cells may reproduce by budding as such or, in appropriate media, may convert to asci with a maximum of four hat-shaped ascospores after meiosis. Two spores of strain 5 and two of strain 21 may be recovered from a single ascus. The haploid ascospores germinate in fresh media into vegetative haploid cells.

The life cycle has been examined in various ways for different reasons, but the agglutination reaction remains its most attractive feature. Many questions may be formulated with respect to sexual agglutination. Of first importance is the molecular basis for the interacting cells. What is particularly interesting to investigators concerned with receptors and recognition is that sexual agglutination in *H. wingei* is due to complementary molecules on the surfaces of cells of complementary mating types. The isolation and the identification of these molecules form the basis for further questions, such as their specificity, their genetics, their synthesis, their regulation and expression. By virtue of its molecular components having been identified first, the *Hansenula* system has become a sort of paradigm for other cell aggregation systems.

Being a mass event, agglutination is dependent upon cell population density. The higher the concentration of cells, the faster the reaction.[94] The coming together of the component cells is brought about by random collision when the mating types are mixed. For qualitative demonstration, liquid cultures are concentrated such that there are about 10^{10} cells $m\ell^{-1}$. The cell suspensions are placed as a drop on a glass slide. Agglutination is observed usually within 5 sec after mixing. The cells may be added to a drop of a test substance.

For rapid and quantitative assay of agglutination, cell suspensions are centrifuged down to a pellet, resuspended, and read in a colorimeter for turbidity. Pellets of nonagglutinative mixtures or of a single mating type are dispersable into free cells, agglutinative pellets not readily so. Resuspended cells that have agglutinated during pelleting settle out within 5 min,

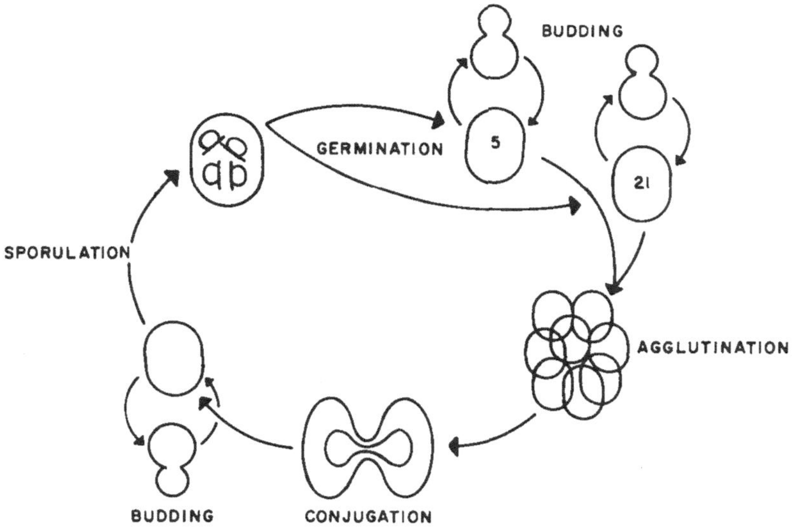

FIGURE 1. The life cycle of *Hansenula wingei*. Haploid strains 5 and 21 are complementary mating types. (From Crandall, M. and Caulton, J. H., in *Methods in Cell Biology*, Vol. 12, Prescott, D. M., Ed., Academic Press, New York, 1975, 185. With permission.)

leaving a clear suspending liquid above. The reduction in turbidity indicates the degree of agglutination.

Using the above quantitative procedure, Brock has shown that maximal agglutination, as indicated by maximal reduction in turbidity, occurs when equal numbers of strain 5 and strain 21 cells are combined (Figure 2).[94] Such an observation is expected if one mating type reacts exclusively with the other. When cells of one mating type are stained with a dye, so as to render them distinguishable, and then mixed with unstained cells of the other mating type, it can be observed with the microscope that the mating types are represented about equally in an agglutinated mass.

The combination between strain 5 and strain 21 is highly specific.[62,94] Cells of one mating type do not agglutinate among themselves. Under the conditions that the haploid cells are agglutinative, the diplonts are not. Nor does *H. wingei* interact with closely related *Hansenula* species and other yeasts. However, there is a cross-reaction between *H. wingei* and *Saccharomyces kluyveri*.[1111] Strain 3 of *S. kluyveri* agglutinates fairly well with *H. wingei* strain 21.

As is expected of a system that depends on random collision, sexual agglutination requires ions.[94] Washed cells do not agglutinate, save in the presence of salt. However, the chelator EDTA is ineffectual as an inhibitor, making a specific requirement for calcium ions unlikely. Heating cells of both strains at 100°C for 5 min does not eliminate agglutinability. On the contrary, heat-treated cells agglutinate more intensely. Maximal reaction is observed over a wide range of pH values: from 3 to 9. At the extremes of pH, it is severely inhibited.

It is clear that the reaction involves the cell surface. Cell-wall preparations isolated from mechanically disintegrated whole cells of complementary mating types agglutinate when brought together.[1112,1113] Thus, the reaction must depend on the properties of the interacting cell surfaces. The net electric charge on the cell surface could conceivably be involved, especially because it has been shown that dioxane, which lowers the dielectric constant of the reaction medium, inhibits agglutination. However, within the pH range of maximum agglutination, the electrophoretic mobilities of both strains are virtually the same, both strains being negatively charged to about the same degree. Too, between the strains, there is no

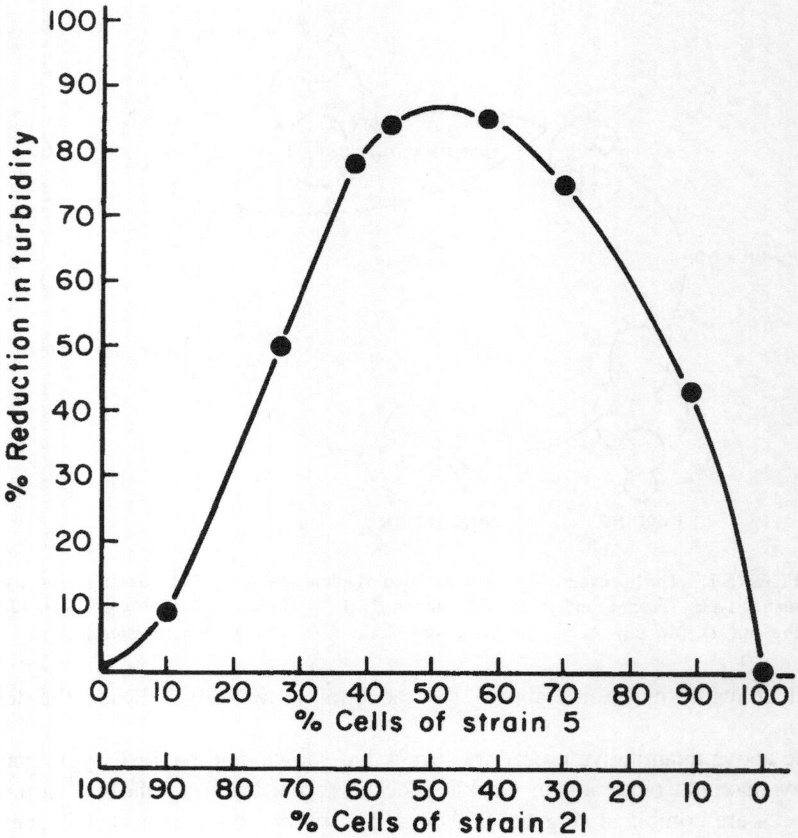

FIGURE 2. Ratio of strain 5 and strain 21 in the agglutinated mass. The degree of agglutination, indicated by percentage reduction in turbidity of the mixture, is highest when about equal numbers of the two strains are mixed. (From Brock, T. D., *J. Bacteriol.*, 75, 697, 1958. With permission.)

difference in the affinity for a dye. Hence, the interaction between mating types cannot be attributed to the attraction between cells of opposite or quantitatively different net charge. Moreover, it seems obvious that net surface charge can hardly confer biological specificity.

The first indications for the existence of specific mating components were derived from experiments designed to eliminate or alter the agglutinability of one mating type without altering that of the other.[1112,1113] Among extractants, 0.1 *N* NaOH destroys the agglutinability of both strains. Phenol (80%) destroys the agglutinability of strain 21 but not of strain 5. High concentrations of urea (8 *M*) inhibit the agglutination reaction reversibly, that is, without altering the agglutinability of either strain. Hot acid (0.01 *N* HCl) removes protein from both, but only strain 21 loses its agglutinative property. Mannan can be detected in the hot-acid extracts. Of various enzymes tested, proteases destroy the agglutinability of strain 21 without affecting that of strain 5, although much protein material is removed from both. Lipase also inactivates strain 21. Periodate oxidation inactivates both strains. Various sugars do not inhibit the agglutination reaction. A search for reactive groups on the agglutination components suggests that free amino groups are not involved. Free carboxyl groups, however, may be required for strain 21, which is inactivated by complexing with thorium or by esterification with methanol.

All these observations point to a highly specific interaction between two molecular structures found on opposite sides of the interacting cell surfaces.[1113] On the cell wall of strain

21 may be found a proteinaceous component that is complementary to a component, probably carbohydrate containing, on the cell wall of strain 5. The interaction between the complementary molecular structures may be envisaged to be mediated by hydrogen bonds between the carboxyl groups of the protein on one side and the hydroxyl groups of the carbohydrate on the other. A simplified picture indeed, but the tight fit of the molecular structures and their presence as integral components of the cell walls of the complementary mating types result in an agglutination that is quite specific and reminescent of an antigen-antibody reaction.

Immunological differences between complementary mating types have not been detected,[1114] but later work with chemical treatment of whole cells adds further evidence in favor of macromolecular interaction between walls of the two mating types. Nystatin-treated cells irreversibly lose their agglutinative property.[80] Washing heat-treated cells with a mixture of chloroform and ethanol reduces agglutinability. Strangely enough, when these washed cells of either mating type are combined with unwashed cells of the complementary mating type, normal agglutination occurs. Certain cell preparations that exhibit weak agglutinating activity may be made to agglutinate more strongly by extraction with high concentrations of urea, guanidinium chloride, or lithium bromide.[1111] The activation may be due to the removal of some macromolecular inhibitor from the cell surface. Neither *N*-maleimide nor acrylonitrile causes inactivation of either mating type, indicating that sulfhydryl groups are not necessary for sexual agglutination. The mating types differ in sensitivity to inactivation by various disulfide-cleaving agents. Mercaptoethanol completely destroys the agglutinability of strain 5 without affecting the agglutinability of strain 21. Other disulfide-cleaving agents, such as cysteine, sodium bisulfite, and periodate, confirm the greater sensitivity of strain 5. This observation suggests the presence of disulfide bonds in the strain 5 component. The time-course of inactivation by mercaptoethanol indicates that the number of this component on the cell wall is in excess of what is minimally necessary to hold the cells together.

Complementary factors specific to the mating type and involved in the sexual agglutination of *H. wingei* have now been isolated and characterized. They are compared in Table 1. In 1964, Taylor reported the isolation of 5-factor.[513] This report was followed the next year by an account of virtually the same factor, isolated in a different way, by Brock.[514] The isolation of 21-factor was reported by Crandall and Brock in 1968.[1115] The 5-factor (or 5f), derived from cells of strain 5 only, specifically agglutinates cells of strain 21. The 21-factor (or 21f), derived from cells of strain 21 only, inhibits the agglutinating activity of 5f. The 5-factor is multivalent, an agglutinin. The 21-factor is presumed to be univalent. Both are glycoproteins.

Taylor's original preparation was obtained by digesting whole cells of strain 5 with snail-gut enzymes.[513] It had a sedimentation constant of 31 $s_{20,w}$. It was isolated by adsorbing specifically the crude extract onto strain 21 cells and eluted off with K_2CO_3 (3×10^{-3} *M*). The possibility of its involvement in sexual agglutination and its specificity are indicated by the fact that it causes the agglutination of strain 21 cells only, specifically adsorbs to strain 21 cells, and is inactivated by mercaptoethanol, which destroys the agglutinability of strain 5.

Whereas Taylor's 5f preparations may be presumed to have been derived from cell walls, the 5f isolated by Brock came from cytoplasmic extracts of ruptured cells.[1115] Cytoplasmic extracts yield a heterogeneous collection of 5f particles with a sedimentation constant as low as 3.5 $s_{20,w}$. These smaller particles are found also in the culture medium of growing cells.[98] The excretion of 5f parallels the growth of the cell population. The mechanism of excretion is not known. Nonetheless, the smaller intracellular molecules may be the precursors of the larger agglutination structures found tightly bound to the cell wall.

No agglutinin activity can be detected in extracts of strain 21 cells.[513,514] Instead, there is in cell-free extracts of strain 21 an activity that satisfies the criteria required of a component

Table 1
COMPLEMENTARY AGGLUTINATION FACTORS OF *HANSENULA WINGEI* COMPARED

Property	5f	21f
Activity	Binds to and causes homotypic agglutination of strain-21 cells	Binds to strain-5 cells and inhibits agglutinating activity of 5f
Source	Cytoplasmic extracts or subtilisin digests of whole cells of strain 5	Cytoplasmic extracts or trypsin digests of whole cells of strain 21
$S_{20,w}$	Heterogeneous: 3.5, 6.5, 9.0, 15.4, 16.7, and 31	Homogeneous: 2.9
Mol wt	From 1.5×10^4 to 10^8 daltons	Ca. 4×10^4 daltons
Valency	Six or more	One
Chemical composition	Phosphomannan-protein	Mannan-protein
Carbohydrate content	From 50 to 96%	From 25 to 35%
Inactivation by	Disulfide-cleaving agents, Pronase, and mannanase	Heat, alkali, and other chemical denaturants

Table after Crandall, Egel, and MacKay.[515] Data summarized from the literature.[98,167,169,518]

that participates in the sexual agglutination reaction.[1115] The activity called 21-factor inhibits the 5f-induced agglutination of strain 21 cells. Any number of substances can possibly bring about such an inhibition. One inhibitory substance, called NSI (for nonspecific inhibitor), is found in all strains of *H. wingei* and adsorbs indiscriminately to cells of either strain 5 or strain 21. NSI comprises a heterogeneous population of large particles. The 21-factor differs from NSI in many ways, but it is fundamentally different in that it is found only in strain 21, specifically binds to strain 5 cells and not to strain 21 cells, and is removed from the cell walls by trypsin, which destroys the agglutinability of cells of strain 21. Although found in cytoplasmic extracts, 21f is associated with the cell wall, for trypsin digestion of whole cells yields much greater activity. It is isolated by adsorbing it to cells of strain 5 and eluting it with urea (8 *M*). The eluate, concentrated by precipitation with organic solvents, may be further purified by gel filtration in Sephadex® G-200. The contaminating RNA can be removed by starch-block electrophoresis. The purified 21f preparation is a fairly homogeneous population of small glycoprotein molecules with a sedimentation constant of 2.9 $s_{20,w}$, which corresponds to around 4×10^4 daltons.

The 21-factor is a glycoprotein composed of up to 35% carbohydrate.[98] The peak activity of a preparation eluted from a gel-filtration column is coincident with the protein peak. The ratio of carbohydrate to protein in the active fractions is invariant.There is some evidence that the carbohydrate is composed predominantly of mannose units. The protein nature of the factor is suggested by its inactivation by alkali, heat, high concentrations of urea, guanidinium chloride, lithium bromide, and other protein denaturants. It is stable to trypsin, understandably so, because it is isolated from the cell wall by trypsin digestion.

In contrast to 21f, purified preparations of 5f are heterogeneous with respect to molecular size. Phosphocellulose chromatography or gradient sedimentation of subtilisin digests of whole cells yields fractions with particle weights greater than 10^8 daltons.[1116] Highest activity is found among the larger particles, only a few of which are required per cell to agglutinate strain 21. The smallest 5f particle (about 1.5×10^4 daltons), obtained nonenzymically from cytoplasmic extracts, must be at least bivalent.[98] The larger ones, more biologically active, must contain more binding sites.[170,517]

The 5-factor is a phosphomannan-protein. It is inactivated by proteolytic enzymes (Pronase, trypsin, and chymotrypsin) and bacterial exo-α-mannanase.[168,169,1117] It is, however, resistant to subtilisin, which merely reduces it in size. It resists inactivation by boiling for 10 min, suggesting a lack of α-helicity in the secondary structure. Disulfide-cleaving agents, such as dithiothreitol or mercaptoethanol, inactivate it.[1118] Upon reoxidation, when the disulfide bonds cleaved by reduction are allowed to reform, 5f regains partial activity. Six disulfide bonds per particle have been estimated.[1119] The most highly purified preparations of 5f contain large amounts of carbohydrate, mostly mannose. Depending upon the preparations, the carbohydrate content can vary from 50 to 96%. Preparations of subtilisin digests purified by affinity chromatography and gel filtration yield homogeneous particles (9.6×10^5 daltons) consisting of 85% carbohydrate, 10% protein, and 5% phosphate.[168,169] Recent studies show that the phosphate, unlike the protein and carbohydrate components, is not required for the agglutination reaction.[518] Six active fragments, released by thiol reduction of 5f, have been isolated and estimated to be 1.2×10^4 daltons each.[517] The active fragment consists of 60 mannose units and 28 amino acids, all together making up a molecular weight of 1.25×10^4 daltons.[168,169] The remainder of the molecule, presumed to comprise the core of 5f, is made up of 90% mannose, in chains of eight mannose units attached to serine or threonine, and 10% protein, of which 62% consists of the two hydroxy amino acids.

The 5f molecule is a branched, randomly coiled polymer.[1117] Two plausible structures have been proposed (Figure 3).[168,169] In one model, the core consists of a single polypeptide chain. In the other, the core consists of six polypeptide chains, each with an active site. Either way, the six active sites are attached around the core by disulfide bonds.

The 5f agglutinin is assumed to bind to the 21f molecules on the surfaces of strain 21 cells, thus causing agglutination. Association constants have been determined for the specific adsorption of ^{35}S-labeled 5f to cells of strain 21.[166] Adsorption can be described by a single association constant that is pH dependent. The standard free energy of association, optimal at pH 4, is around -14.5 kcal mol^{-1}. For a reversible biological reaction, this free-energy value is high, seldom observed even in antibody-hapten reactions.[1120] The complementary binding sites might individually possess extraordinarily strong affinity for each other. Alternatively, the high value might be the result of cooperativity among the many binding sites.[167]

Together, 5f and 21f form macromolecular complexes in vitro.[170] The stable complexes are soluble because 21f, being univalent, cannot serve as a link. They have been detected by allowing the complementary factors to combine, destroying the 21f activity with alkali, and then recovering 5f agglutinating activity. As is expected from the heterogeneity of the 5f component, the complexes exist in diverse sizes. Gel filtration on Sepharose 6B reveals three activity peaks with estimated molecular weights of 0.5×10^6, 1.2×10^6, and 3.8×10^6 daltons, corresponding respectively to calculated combining sites of 6, 16, and 63.

With the demonstration of specificity and the existence of molecular complexes between the agglutination factors, the molecular basis for sexual agglutination in *H. wingei* appears to have been resolved. The only disturbing thing about 5f-provoked agglutination of strain 21 cells is that it is a homotypic interaction, whereas sexual agglutination between strain 5 and strain 21 is obviously a heterotypic system. Moreover, a cell-aggregation system need not be mediated by a multivalent agglutinin. All it requires is multivalent cells. Indeed, the multiple binding sites of a single molecule can hardly all be functional simultaneously *in situ*, for steric hindrance would not allow multiple docking of cells, unless the agglutination molecules reside in lengthy projections from the cell surface. Such a condition could be accommodated by fimbriae.

Fibrillar structures resembling fimbriae in bacteria have been observed in several fungal species.[125,541-544] They have been implicated in the sexual process in *Ustilago* and fission yeast and in the flocculation of brewers' yeast. Fimbria-like projections, distributed all over the cell surface, have been observed also in *H. wingei* (Figure 4). They are found in both strain 5 and strain 21. The agglutination factors, because of their location on the cell surface, could very well be associated with these surface filaments.

Because of their strong and specific affinity for each other, the agglutination glycoproteins determine recognition between mating types and enable cells to stick together for as long

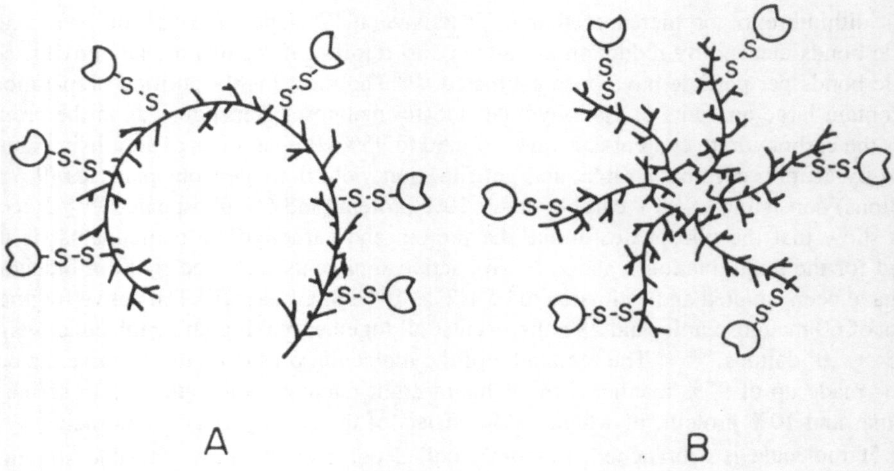

FIGURE 3. Alternative structures proposed for the 5-factor molecule. The core may consist of a single polypeptide chain (A) or of six polypeptide chains arranged radially (B). Branched carbohydrate chains, denoted by thin lines, stick out from the polypeptide chains, shown as thick lines. Tulip-form symbols found around the core represent combining sites. A glycopeptide structurally similar to the core, each combining site is attached to the core by a disulfide bond. (From Yen, P. H. and Ballou, C. E., *Biochemistry*, 13, 2428, 1974. With permission.)

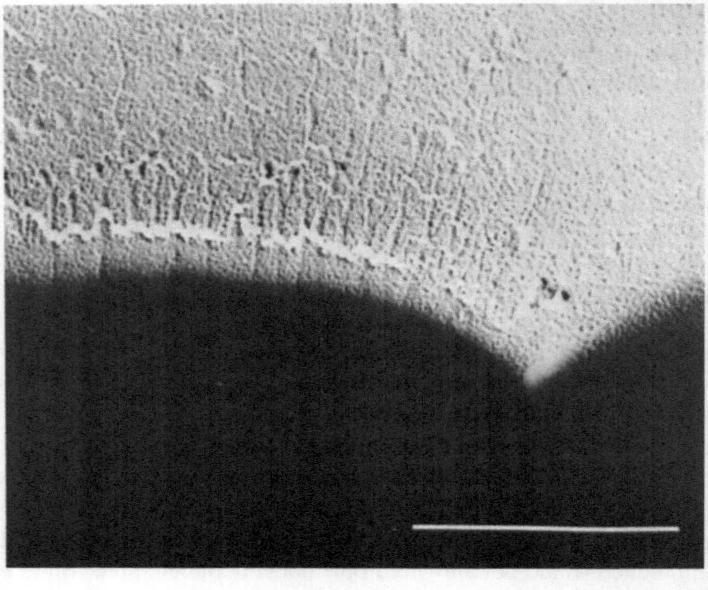

FIGURE 4. Fimbria-like filaments on the cell surface of *Hansenula wingei*. Bar represents 0.5 μm. Sample preparation for transmission electron microscopy as described by Poon and Day.[543] (Original photomicrograph kindly furnished by N. H. Poon).

as they remain functional. Once intimate contact is achieved, cells in the agglutinated mass fuse to become zygotes. It might be expected that when the zygotes are formed, or even earlier than that, the role of the agglutination glycoproteins *qua* recognitive and binding elements is over. Whereas the diploid progeny of the zygotes are not agglutinable, the zygotes themselves remain agglutinative, indicating that the glycoproteins are not destroyed, nor do they come off, and that they retain their activity.[98] Similarly, in sex-directed flocculation in fission yeast, postdevelopmental deflocculation occurs long after the sexual act has been consummated.[143]

Obviously, the ability to agglutinate is genetically controlled. Strains 5 and 21 are genetically identical except for functions related to mating type. Isolates from nature mate with either strain but not both.[35,36,62,1121] A microtiter plate assay for mating type has been described.[1122] It simply calls for challenging spore clones with heat-killed tester strains of 5 and 21. The mating types are allelic at the mating-type locus (*mat*). Asci derived from diplonts, which are of course the result of fusion of complementary haplonts, yield two spores of each mating type.[1121]

Many species of *Hansenula* are homothallic, but strains of *H. wingei*, by definition, are strictly heterothallic.[601,1123] The mating types are very stable. If they were not, a heterothallic clone would soon become homothallic. The frequency of spontaneous mutation from mating-type 5 to mating-type 21 is $<6 \times 10^{-8}$ per cell division, the reverse (from 21 to 5) $<3 \times 10^{-9}$ per cell division.[515] Thus, the haploid genome of *H. wingei*, unlike in *Schizosaccharomyces pombe* and *Saccharomyces cerevisiae* (see later sections in this chapter), may not contain the allelic information for both mating types.

Four chromosomes have been cytologically identified in haploid and diploid strains of *H. wingei* during mitosis.[1124] Four bivalents are observed during meiosis in the diplont. It is assumed that somatic pairing occurs in the diplont and, hence, the haploid number of chromosomes is four. Linkage analysis of six genes reveals two linkage groups, suggesting also a small number of chromosomes.[1125] In a preliminary genetic map, the mating-type locus is found closely linked to the centromere in chromosome I.[1367] Allelic though they may be, the genes for the agglutination factors must be markedly different from each other, for the primary structures of 5f and 21f appear to be quite dissimilar. It is, of course, assumed that the agglutination factors are the gene products of the mating-type locus.

The *mat* locus controls mating and associated events in the haploid cell. Mutants that do not mate are also nonagglutinative.[98] Sexual agglutination is a necessary condition for subsequent developmental events, including conjugation, cell fusion, and zygotic budding. In the resultant diplont, both allelic genes are present together. Nonetheless, the diploid cell is normally nonagglutinative. The synthesis of the agglutination factors is repressed. The information for their synthesis is there, because the zygote is derived from two cells of complementary mating types, both of which are recoverable as meiotic products in the ascus when the diploid cell is allowed to sporulate.

Diploid cultures do not synthesize the agglutination factors 5f and 21f and therefore do not agglutinate. Conditions for the induction of diplonts to become agglutinative have been found (Table 2, Chapter 3).[125,247] Because it contains the information for both mating types, a diploid cell can be induced to become one of either. The change of a diplont from its nonagglutinative state to a state in which it behaves as a cell of mating-type 5 (because now it synthesizes 5f) is called D →5 transition. The alternative transition is D →21. The D →5 transition is observed when diploid cells are grown to late-stationary phase in a medium containing yeast extract (0.67%) or in a synthetic minimal medium with vanadate (0.4 m*M* V^{4+}) plus molybdate. The latter condition is more effective when oxygen is limited. The D →5 transition may be induced also by growing the diploid cells to limiting cell population density in a synthetic minimal medium free of Ca^{2+} and Mg^{2+} or in a vitamin-limited medium and then aerating them in a phosphate buffer with glucose, Mg^{2+}, and pyridoxine. The D →21 transition may be induced by growing the diploid cells to late-log phase in yeast-extract medium to which EDTA (1 m*M*) has been added or to stationary phase in yeast-extract medium and then growing them in a synthetic minimal medium devoid of trace metals. Attempts to identify the inducing ingredient in yeast extracts have not succeeded.

In short, limiting Ca^{2+}, Mg^{2+}, oxygen, or vitamins or adding excess metal ions induces diploid cells to synthesize 5f; limiting the concentrations of trace metals induces them to synthesize 21f. These conditions by some as yet unexplained mechanism bring about the synthesis of the agglutination glycoproteins, which are constitutive in the haplonts but whose synthesis is repressed in the diplonts.

The induced diploid cell synthesizes either 5f or 21f and agglutinates with the haploid cell of either one of the two mating types. The subsequent events in mating — conjugation, fusion, and zygotic budding — remain repressed. Rare triplonts have been observed (at a frequency of about 10^{-4} to 10^{-3}).[125] The products of matings between a diploid and a haploid cell, triplonts do not agglutinate nor do they mate.

The inducing condition must influence the expression of the *mat* locus, presumably by way of regulatory genes.[98] Each of the component cells in an agglutinated mass bears a structural gene coding for the synthesis of its own agglutination glycoprotein as well as a regulatory gene that governs the repression of the synthesis of the complementary glycoprotein when the two genes are together in the diplont. Induction of the diploid cell to become agglutinative may consist in perturbing this mutual repression. There is some genetic evidence for genes regulating the expression of sexual agglutination in *H. wingei.*[1121]

No diffusible sex factors, analogous to the ones found in *Saccharomyces cerevisiae*,[90] have been found in *H. wingei*. Nonagglutinative diplonts or nonagglutinative haploid mutants cannot be induced by mixing them with the agglutinative mating types.[515] Conjugation-tube formation does not seem to be influenced at a distance, but rather occurs only upon intimate intercellular contact due to agglutination.[246] Mind you, there is evidence for sex-specific vegetative response. An enhancement of bud formation has been observed when cells of one mating type are mixed with UV-inactivated cells of the complementary mating type.[485] When placed approximately 20 μm apart on an agar surface, single cells of complementary mating types tend to bud in each other's direction.[1126] For sexual functions, diffusible factors appear to be redundant in the case of *H. wingei*, because the agglutination glycoproteins, upon which mutual recognition of mating pairs depends, are sufficiently effective. Furthermore, they are present at all times in the haploid condition.

Although haploid cells are constitutively agglutinative, there are certain growth conditions that make them agglutinate poorly or not at all.[98] If malt extract (1%), instead of yeast extract, is used as growth medium, cells of either mating type are less agglutinative. Sorbose at high concentrations inhibits growth, but at subinhibitory concentrations (1 to 2.5%), it renders cells of strain 5 nonagglutinative, without affecting the agglutinability of strain 21. The same results are obtained when cells of either strain are grown in yeast-extract medium containing glycerol, instead of glucose, as energy source. The synthesis of cell-surface inhibitors, such as NSI, may be involved in these phenomena.

The morphological events during the mating process have been monitored by light and electron microscopy.[246,485,1127] Sexual agglutination, presumed to be a hydrogen–bonded union among cells, is followed in the agglutinated mass by the irreversible pairings between cells of complementary mating types. A mating pair may be thought of as being joined by covalent bonds. At this stage, the pair is no loner susceptible to deagglutination by chymotrypsin or mercaptoethanol or by brief sonication. Conjugation tubes, in the form of protuberances at the region of contact, are then formed. A little later, the cross-wall between a pair is dissolved, thereby converting two cells into one. A transient heterokaryon is formed before the two nuclei move towards the middle to fuse into a diploid nucleus. The resultant zygote is a diploid cell that soon reproduces as a diplont by zygotic budding. The first diploid bud is found usually at the region between the two original cells. Although mating is normally a match between two cells, mating units consisting of triplets or quadruplets are also observed, indicating that mate selection is not under precise control. Multiple conjugation has been observed also in fission yeast.[121]

Conjugation is dependent on the physiological state of the cells.[246,485] Cells in stationary phase conjugate more efficiently than those in exponential phase. Although it can occur in media without a nitrogen source, conjugation requires a source of energy. No net synthesis of proteins is observed, but protein synthesis must be required, because cycloheximide, analogs of amino acids, and depletion of the amino acid pool inhibit conjugation. So does 8-azaguanine, suggesting the need for new messenger synthesis. Because the haploid cells may mate any time so long as they can agglutinate together with the cells of the complementary mating type, it seems reasonable to suppose that their coming together triggers a mutual exhortation to stop dividing and to start conjugating.

Sexual agglutination in *H. wingei* is one aggregation system which cannot be doubted as to its function to the individual cell. It is a prerequisite to sex, to conjugation, to the construction of a transient heterokaryon, thence to zygote formation and the creation of a vegetative diploid cell. The agglutination glycoproteins on the cell surface allow mutual recognition between two cells of complementary mating types. They assure the fusion of two dissimilar genomes, dissimilar at least as regards that segment which determines the mating process. There is the added possibility, because the complementary cells are not immediate descendants of a clone, to be genetically dissimilar elsewhere. With that possibility, the cell population can partake of the benefits of genetic mixing. Provisions for mass action, rather than the more precarious pairing in the open, help in enhancing the frequency of fruitful encounters and the chances for more protracted contacts. The species being heterothallic and the cells nonmotile, isolated haploid clones in nature rarely ever have the opportunity for sexual activity, until members of complementary clones are accidentally washed away together into, say, a dewdrop. Then, by sexual agglutination, the proper contacts between cells can be made. Individual encounters may be random, but the effective contacts are specific, specific in the sense that the effectively interacting cells and their interacting molecules are exclusively complementary.

To the species in the long haul, the Darwinian value of sexual agglutination lies in the consequent increase in ploidy: from vegetative haplonts to vegetative diplonts.[36] A haploid strain that evolves to acquire the ability to agglutinate can hardly remain haploid for long. The species as an evolutionary abstract reaps the advantages, as well as the disadvantages, that such an increase in ploidy entails. The diplont repetitively divides as such and, when a different set of environmental conditions prevails, undergoes meiosis and sporulates. The haploid spores simply wait for better times to germinate into vegetative haplonts. Thus, sexual agglutination helps in enlarging and diversifying the life cycle, not a bad strategem for the survival of the species. Similar events and functions are found also in the other yeast aggregation systems.

III. SEX-DIRECTED FLOCCULATION IN *SCHIZOSACCHAROMYCES POMBE*

Like *Hansenula*, *Schizosaccharomyces pombe* is an ascomycetous yeast.[1128] It is the type species of the genus, which comprises four recognized species and two varieties. It was first isolated by Lindner in 1893 from the Bantu millet beer ("pombe" in Swahili), hence the name of the species.[1129] Unlike most yeasts, which form buds, it multiplies by binary transverse fission, as the generic name implies. Also unlike most yeasts outside the genus, cells of *S. pombe* are cylindrical. During exponential phase in malt-extract broth, they measure (3 to 5) × (7 to 20) μm. The last generation prior to stationary phase produces cells that are almost as short as they are wide.

The pioneering work of Leupold in genetics and of Mitchison in physiology has made *S. pombe* a familiar eukaryotic model, especially for problems concerning the cell cycle, mating, and sexual development.[64,1130-1132] *Saccharomyces cerevisiae* aside, *Schizosaccharomyces pombe* is the best studied genetically among the yeasts. Three chromosomes have been cytologically identified.[1133-1135] A recent genetic map contains 72 markers in three linkage groups (Figure 5).[184] Reviews on the genetics, including the mating process, of fission yeast are available.[183,515,1130] The fission yeast fills a very prominent position in Mitchison's career-long preoccupation with the cell cycle.[1136,1137] Cell division in the organism has been reviewed recently.[1138] So, too, its sexual development, with emphasis on the cell wall as sex organelle.[483]

S. pombe is usually grown in the laboratory as stable haploid cultures, which may be either heterothallic or homothallic. The life cycle of a homothallic strain is shown in Figure 13, Chapter 3. In malt-extract broth at 32°C, strain NCYC 132 may be grown as a still culture with a generation time of about 120 min. The cell population density during stationary phase is 6×10^7 $m\ell^{-1}$. When aerated during stationary phase, cells form flocs, which become grossly visible 1 hr after the start of aeration. In the next 5 hr, flocs increase in size and number. Flocculation, the hydrogen-bonded union of cells, is followed by copulation, the covalent union of cells in the flocs. Thereafter follows conjugation, also called cytoplasmic fusion or dikaryon formation. When the two nuclei in the dikaryon fuse, a zygote is formed. The zygote then undergoes meiotic division, followed soon after by a second meiotic division. The sequential meiotic events yield four nuclear products, which are then packaged singly into four spores within the ascus. The first asci in the shaken culture may be observed at 8 hr after the start of aeration. Transferred to fresh medium, the ascospores, which may be liberated on further incubation or may remain confined within their asci, germinate to become vegetative haplonts once again.

There are further details to the cycle.[183,515] Heterothallic strains remain nonflocculent so long as they are grown as pure cultures. When complementary mating types are mixed together, they flocculate, but only after a protracted period of apparently mutual stimulation by contact.[67] Under certain conditions, the zygote of certain strains may remain a diplont and replicate as such. The diplont is not ordinarily stable. It may be prodded to haploidize mitotically or to undergo meiosis and sporulate.[1139] In *S. japonicus* or *S. octosporus*, mitotic replication of the four meiotic products occurs prior to their being packaged individually into eight ascospores.[1128,1140-1142]

Although the life cycle of *S. pombe* has been known for a long time, it is only in the past decade that the organism has been studied seriously as an aggregation system. Strain NCYC 132, earlier thought of as a sexually sterile strain and initially studied solely as a model for growth and cell division, was observed during stationary phase to form clumps when grown as shaken broth cultures.[65] As the phenomenon became more frequently encountered, it became a major source of annoyance as it required abandoning the errant culture and starting afresh. The observation even then was not new. In other laboratories, such an observation was often made and casual reference to it is found in the literature.[1128,1132] Persistent occurrence of the clumping phenomenon demanded that it be taken earnestly. Initial efforts were directed toward finding ways to eliminate it for the purpose of obtaining homogeneous culture samples for quantitative analysis. By the time ways had been found to eliminate it, much had been learned concerning its many consequences to the cell and to the population.

Clumping is best observed in malt-extract broth.[65] Aerated cultures grown to stationary phase form clumps, nonaerated cultures do not. Neither do actively dividing cells. The clumping phenomenon is termed flocculation, in deference to the well-known phenomenon in brewers' yeast. It occurs only during the stationary phase of a culture. The first flocs, microscopic in size, appear a few hours after the end of exponential growth. When maximal flocculation is attained, up to 70 to 75% of the cells may be found among the flocs.[1143]

The formation of flocs is not a passive process, that is, it is not due to the failure on the part of the sibs to separate materially after fission.[65] Cells in a developing floc have ceased dividing for some time. Only an hour or two before the appearance of the first flocs, groups of more than ten cells are not found in the culture. Besides, flocculation is not prevented from occurring after separation of pairs and larger aggregates by ultrasonic treatment at the

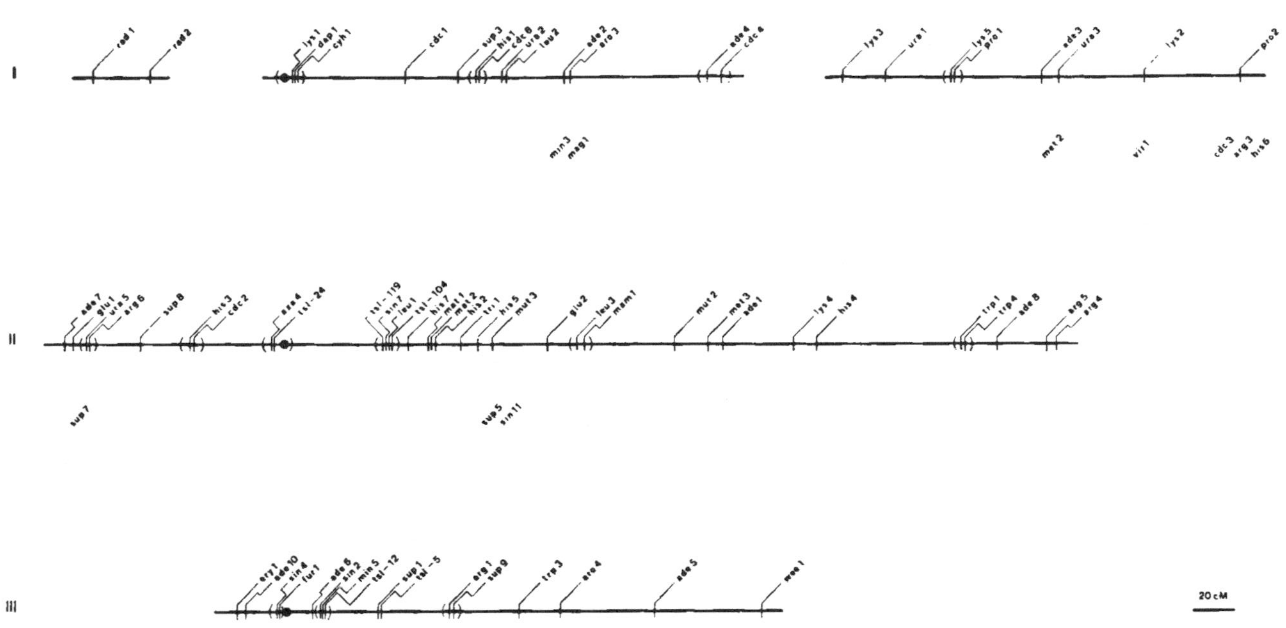

FIGURE 5. Genetic map of *Schizosaccharomyces pombe*. The dots are centromeres. The sequence of genes within parentheses remains to be resolved. Markers below chromosomes are placed to approximate their locations. Symbols: *ade*, adenine required; *arg*, arginine required; *aza*, azaguanine-resistant; *cdc*, cell division cycle deficient; *cyh*, cycloheximide-resistant; *dap*, diaminopurine-resistant; *ery*, erythromycin-resistant; *fur*, fluorouracil-resistant; *glu*, glutamate required; *his*, histidine required; *leu*, leucine required; *lys*, lysine required; *mag*, magdala-red positive; *mam*, mating-type auxiliary; *mat*, mating type; *met*, methionine required; *min*, methionine-inhibited; *mut*, mutator; *pro*, proline required; *rad*, radiation-sensitive; *sin*, antisuppressor; *sup*, suppressor; *tri*, trichodermine-resistant; *trp*, tryptophan required; *tsl*, temperature-sensitive; *ura*, uracil required; *vir*, vegetative iodine reaction; *wee*, small cells. The development-associated genes (*mat1*, *mat2*, *and mam1*) are found on chromosome II to the right of the centromere. (From Kohli, J., Hottinger, H., Munz, P., Strauss, A., and Thuriaux, P., *Genetics*, 87, 471, 1977. With permission.)

end of exponential growth. Obviously then, flocculation in fission yeast is an active process, a coming and sticking together of previously disperse cells.

Flocs do not disperse into free cells when the culture is diluted in distilled water. Some large flocs may break apart during the process of dilution, but the components subsequently reflocculate. Such a demonstration of stability indicates that there is in the spent medium no special ingredient that sustains floc formation. The stability of the flocs allows them to be separated from the free cells by gravity. A flocculated culture transferred to a tapered tube is left to stand undisturbed for 5 min. During this time, the flocs fall to the bottom while the free cells remain suspended in the supernatant, which is then replaced by water or phosphate buffer. The purification of flocs is accomplished by repeated sedimentation and repeated replacement of the suspending medium (Figure 11, Chapter 3). Free cells removed in the process do not flocculate among themselves. Purified flocs do not deflocculate in fresh medium until the component cells start growing and dividing again. The possibility that there is in the fresh medium a deflocculating agent that makes growing cells nonflocculent is thus excluded.

Purified flocs are stable at temperatures as high as 50°C or thereabouts (Figure 6).[65] More elevated temperatures bring about deflocculation, which is essentially complete at about 80°C. The temperature (T_m) at which half of the cell population in the flocs become free cells is about 61°C in distilled water. Thermal deflocculation, despite its lethal consequence, is reversible. Floc-derived cells freed by thermal deflocculation and therefore heat-killed at the same time, reflocculate when left to stand undisturbed at ambient temperature. The time (R_t) during which half of the cell population is observed to have reflocculated is 10 to 15 hr in distilled water. R_t values decrease to an hour or so upon continuous, gentle agitation of the suspension of thermally freed cells. In salt solution (0.1 *M*), R_t reduces to a few minutes.

Reflocculation of heat-killed and heat-dispersed cells indicates that flocculation itself, presumed to be distinct from the acquisition of competence to flocculate, is not a vital activity. Of course, heat-killed cells not derived from flocs do not form flocs among themselves. Free cells not found in flocs in the culture are considered to be incompetent. Purified cell walls from competent cells do flocculate, those from incompetent cells do not.[65] It is obvious that the site of cell-cell interaction is the cell surface: the wall or whatever appendages that may be attached to it. During the transition of a culture from exponential growth to stationary phase, differentiation among the cells occurs. The majority of cells become competent to form flocs. In these cells, it may be presumed, a material change in the cell wall has happened.

A floc of *S. pombe* has been described as an aggregate of more than ten cells (see Chapter 3, Section II).[76] Cells in groups of that number or less are identified as free cells. The cutoff aggregation number appears conveniently arbitrary. However, it is derived from the distribution of aggregation numbers of cells in broth cultures prior to flocculation (Figure 3, Chapter 3). The average aggregation number of such cells is less than 2.0. A microfloc, visible only in the microscope, is an aggregate of no more than 100 cells. A minifloc is an aggregate of more than 100, but unlike gross flocs, is not visible to the unaided eye. Gross flocs may consist of up to 10^5 cells each.

The development of competence to flocculate may be monitored hemocytometrically, turbidimetrically, and visually.[76] These methods have been described in Chapter 3 (Section II, Figures 7 to 10). They are also applicable to determining the rate and the extent of reflocculation as well as the purity and the stability of a purified-floc preparation. They have been used in studies such as determining the nature of the forces that hold the cells together and the physical properties of flocs (Table 2).

As noted above, cells freed from purified flocs by thermal deflocculation reflocculate upon cooling. The reflocculated mass may then be deflocculated again and subsequently

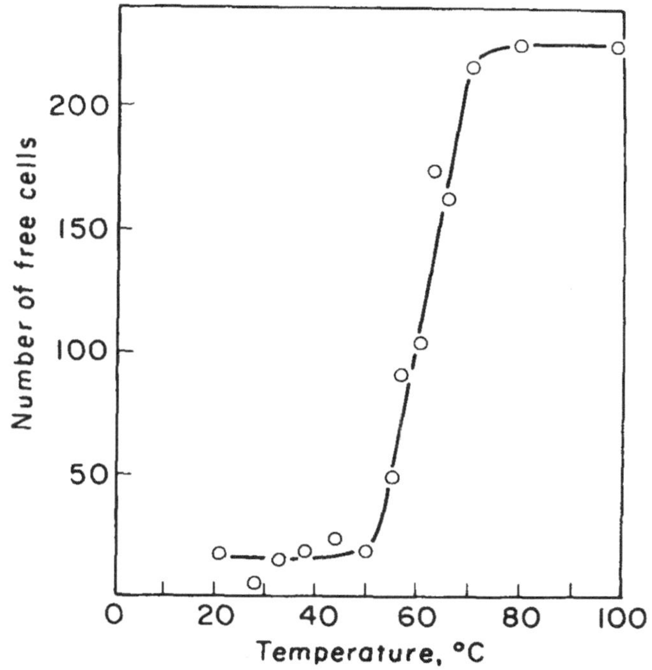

FIGURE 6. The effect of temperature on the stability of flocs of *Schizosaccharomyces pombe* in deionized water. The extent of deflocculation is indicated by the number of freed cells in the floc suspension. A 10-mℓ suspension of purified flocs in a tapered tube was incubated in a water bath, the temperature of which was raised 1°C every 5 min. At indicated points, the tube was withdrawn, inverted five times, and returned to the bath to remain undisturbed for awhile. At the end of 5 min, the supernatant was sampled with a Pasteur pipette introduced 10 mm below the liquid surface. (From Calleja, G. B., *J. Gen. Microbiol.*, 64, 247, 1970. With permission.)

allowed to reflocculate (Figure 14, Chapter 3). This sequential manipulation may be repeated over and over again.[99] The ability of the competent cells to reflocculate is not markedly altered when they are thermally manipulated. The presence of salt, NaCl or $CaCl_2$ for instance, increases floc stability as indicated by an increase in T_m. Salt, like gentle agitation, also increases the rate of reflocculation. The differences among cations in their effectiveness are not significant, various monovalent ions being just as effective as divalent ions. The apparent requirement for ions in reflocculation may be superseded by forceful contact. Whereas the extent of reflocculation within a short period of time increases with increasing concentrations of $CaCl_2$ to an optimal concentration of around 10^{-2} *M*, cell suspensions pelleted in a centrifugal field do not show such ion requirement. Incompetent cells, of course, when pelleted and resuspended do not flocculate. Calcium ions have been implicated in many aggregation systems. That they are not specifically required for flocculation in fission yeast is further indicated by the lack of effect of EDTA as a deflocculant. The role of ions in stabilizing flocs and in increasing the rate of reflocculation may be attributed to neutralization of the negatively charged cell surface by counter-ions, thus lowering the repulsive barrier between two cells approaching each other and increasing the probability of fruitful collision.

There are other ways of causing reversible deflocculation.[99] Purified flocs are stable over a wide range of pH values. Deflocculation occurs below pH 2 and above pH 11. It is

Table 2
PHYSICAL CHARACTERISTICS OF FLOCS AND FREE CELLS OF FISSION YEAST

Flocs (at maximal flocculation)	
Number of cells in flocs/mℓ of culture	5×10^7
Volume of flocs/mℓ of culture	5×10^{-2} mℓ
Number of flocs/mℓ of culture[a]	5×10^2
Number of cells in a single floc	1×10^5
Volume of a single floc[a]	1×10^{-4} mℓ
Diameter of a floc	0.6 mm
Time required for half of floc population to settle	60 sec
Free cells (just before the start of aeration)	
Number of free cells/mℓ of culture	5×10^7
Volume of free cells/mℓ of culture	5×10^{-3} mℓ
Volume of a free cell[a]	1×10^{-10} mℓ
Aggregation number of free-cell population	1.8

[a] Derived; the rest are average observed values.

From Calleja, G. B. and Johnson, B. F., *Can. J. Microbiol.*, 23, 68, 1977. With permission.

reversible, provided the disperse cells are not allowed to remain too long at those extremes. Blender treatment for up to 5 min brings about incomplete and reversible deflocculation. Deflocculation by ultrasonic treatment is complete, but reversible even if the treatment is followed by thorough washing of the dispersed cells. Guanidinium chloride (3 *M*) and urea (5 *M*), among various chemicals tested, are most effective as reversible deflocculants. Almost immediate reflocculation occurs upon the removal of either reagent. Whereas urea-provoked deflocculation is fairly rapid, reversible deflocculation caused by SDS is slow (Figure 7). The kinetics of defocculation suggests a cooperative effect.

Only proteinases, among the various commercial enzymes tested, cause deflocculation of purified flocs.[99,519] The more effective proteinases include Pronase, papain, trypsin, and clostripain. Nucleases, snail-gut enzymes, amylase, hyaluronidase, neuraminidase, denatured proteases, serum albumin, and peptones are ineffective. Proteolytic deflocculation is rapid and complete. It is irreversible in that it cannot be reversed by thorough washing of the dispersed cells. Irreversible deflocculation is brought about also by high concentrations of dithiothreitol or mercaptoethanol. Furthermore, the ability of heat-dispersed cells to reflocculate is abolished by other disulfide-cleaving reagents, such as cysteine, glutathione, cyanide, and bisulfite. Cystine, methionine, and oxidized glutathione are ineffective.

Sugars (mono-, di-, and trisaccharides), claimed in the literature to bring about deflocculation in brewers' yeast,[81,272] are ineffective deflocculants even at concentrations as high as 50%.[65,99] Likewise ineffective are common salts. High concentrations (up to 5 *M*) of NaCl or $CaCl_2$ serve to effect only marginal instability of purified flocs at ambient temperature. They do not prevent reflocculation.

In these experiments, it is assumed that reflocculation is a simple reverse of deflocculation and that a floc reconstituted from experimentally dispersed cells approximates, structurally and functionally, a floc in a broth culture. That assumption might not be entirely correct. Reflocculation, in whatever manner the precedent deflocculation is effected, is time dependent and incomplete. A reflocculation value of 80 to 90% is usually attained.[99,483] Part of the hysteretic loss is accountable to the presence of incompetent cells trapped accidentally in the native flocs. Mixing incompetent cells with purified flocs in a deflocculating medium enhances deflocculation and retards subsequent reflocculation. Although incapable of flocculating among themselves, incompetent cells (stained with basic fuchsin for purposes of

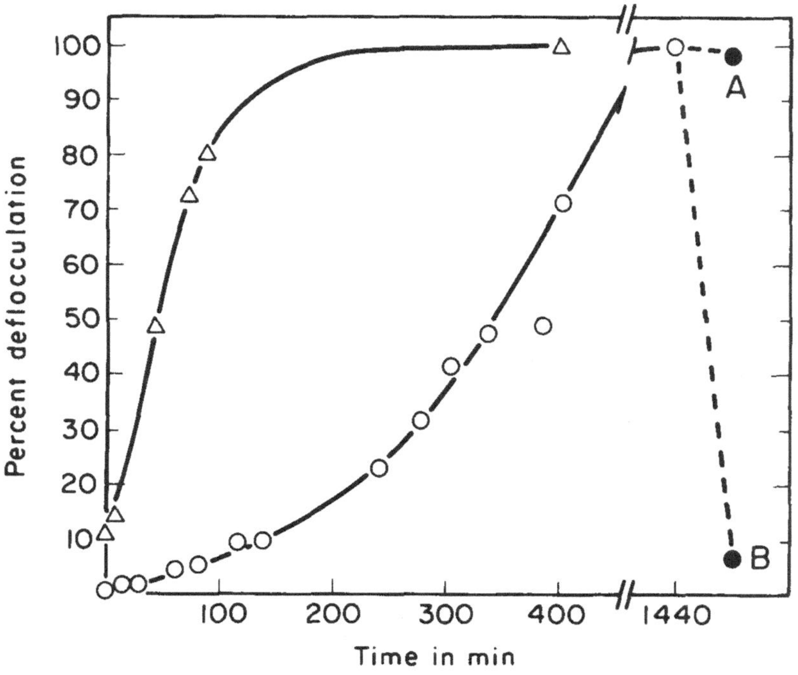

FIGURE 7. Reversible deflocculation of purified flocs of fission yeast by urea and sodium dodecylsulfate (SDS) at ambient temperature. Symbols: △, 5 *M* urea; ○, 0.5% SDS. (A) SDS-deflocculated cells after they were spun down and resuspended in the same SDS-containing dispersion medium. (B) Other half of SDS-dispersed cell suspension after spun down and resuspended in deionized water. (From Calleja, G. B., *Can. J. Microbiol.*, 20, 797, 1974. With permission.)

identification) are observed to bind weakly to competent ones. Thermal motion dislodges the incompetent cells when they are located at the periphery of a floc. Surrounded by competent cells and snugly ensconced in the native floc, they might have a difficult time competing favorably for positions in the reconstituted flocs during reflocculation.

Because common salts are ineffective deflocculants, the forces that bind cells together are not likely to be primarily ionic in character.[99,483] Instead, the various observations, especially that of reversible deflocculation by heat, mechanical shear, ultrasonication, guanidinium chloride, and urea, point to a major contribution of hydrogen bonds. A minor hydrophobic interaction is suggested by the slow, reversible deflocculation by SDS. By whatever means it is brought about, reversible deflocculation rules out the participation of covalent bonds between cell surfaces. It also excludes bifunctional ligands, because if they existed, they would be removed with the reversibly deflocculating reagents. The irreversible deflocculation by proteinases and disulfide-cleaving reagents and the lack of effect of non-proteolytic enzymes indicate a cell-cell interaction mediated by proteinaceous molecules containing disulfide linkages. These molecules are deemed to be covalently bonded to the cell wall even as their interaction is mediated by hydrogen bonds.

For many years, strain NCYC 132 was considered as sterile, sexless, and nonsporogenous. Microscopic examination of a flocculated culture does not readily reveal the presence of spores. The internal details of a floc are hard to discern in the compact arrangement of the many thousands of cells that make up a floc. The component cells, however, may be examined when a floc in SDS disperses slowly into free cells. As deflocculation progresses, the floc bares its components: single cells, copulant pairs, conjugant pairs, asci, and ascospores.[66] Pairs, asci, and spores appear only after flocculation of a culture. They are found only in flocs, not among the subpopulation of free cells. They are not found under cultural conditions that prevent flocculation or the development of competence to form flocs. Nonflocculent mutants do not form flocs nor spores.

There is no doubt now that NCYC 132 is a homothallic strain, but more important than that, flocculation is a prerequisite to sex. Hence, it is called sex-directed flocculation. It is analogous to sexual agglutination in other yeasts, such as *Hansenula wingei* and species of *Saccharomyces* and *Citeromyces*, and no longer, it seems, so analogous to flocculation of brewers' yeast. As if corroboration were necessary, at about the same time that sex-directed flocculation was initially reported to be followed by conjugation in a homothallic strain of fission yeast, conjugation in the heterothallic strains was first reported to be preceded by sexual agglutination.[66,67]

Sex-directed flocculation is just the most conspicuous beginning of a developmental sequence of morphogenetic events.[66,121] The time-course of flocculation in a culture and of the subsequent events identifiable with the light microscope is shown in Figure 8. The simplified sequence comprises flocculation, copulation, conjugation, and sporulation. Flocculation is the hydrogen-bonded union of cells. Copulation takes place between a pair of cells found in a floc. It is operationally defined as a covalent union that is no longer sensitive to disruption by proteolytic enzymes or by agents that break hydrogen bonds. It ends where conjugation begins. Conjugation is synonymous with cell fusion, whose instantaneous consequence is a dikaryon. The events after conjugation, zygote formation and meiosis, are not easily scorable with the light microscope until the appearance of asci and their ascospores. Notice that the culture remains flocculated to the very end of the sequence. Eventually, if the culture is allowed to incubate with agitation long enough, a slow, irreversible deflocculation, called postdevelopmental deflocculation, takes place.[143] After a week at 32°C, long after the end of development, all gross flocs are disbanded (Figure 9).

Hypoxic cultures do not form flocs. They can be made to do so when aerated in stationary phase. So long as cells are still actively growing and dividing, they do not flocculate even when aerated. Neither do they flocculate when grown in a defined medium containing Wickerham's nitrogen base (Difco) and a carbon source, nor in a medium containing yeast extract and glucose.[66] The defined medium can be rectified, so as to support flocculation, by commercial peptones or vitamin-free Casamino acids (Difco). It is sufficiently clear that the competence to form flocs is inducible in the minimal sense that it can be turned off and on.

The development of competence to form flocs requires aeration, as contrasted to mere agitation.[66] Whereas cells grown undisturbed in culture tubes do not flocculate, those grown undisturbed as a shallow culture in petri plates do. In addition, colonies on agar plates do sporulate, but not before cells become competent to form flocs. Competence may be ascertained by making an aqueous suspension of the colonies.

Induction of cells to competence involves the following procedure. Cells of the homothallic strain NCYC 132 are grown as a still culture in 10 mℓ of malt-extract broth (2%, w/v, Oxoid) in a tightly capped 20-mℓ bottle at 32°C. Incubated to stationary phase, the culture increases in cell number 100-fold (6 to 7 generations) from an inoculum of 6×10^6 cells derived from a 24-hr culture grown the same way. At 24 hr after inoculation, it is transferred to a 125-mℓ Erlenmeyer flask and shaken (150 rev min^{-1}) on a rotary shaker at 32°C. The first gross flocs appear within 1 hr after the start of aeration. Thus, induction is achieved by aerating stationary-phase cells grown as a still culture.

Grown as still cultures for varied lengths of time and subsequently aerated, cultures exhibit different induction lags and induction capacities.[84] Cultures in logarithmic phase of growth wait until stationary phase before they are ready for induction. However, prolonging the stationary phase before aerating the cultures results in the lengthening of the time before the onset of flocculation. Also, the capacity for induction to competence decays as a culture

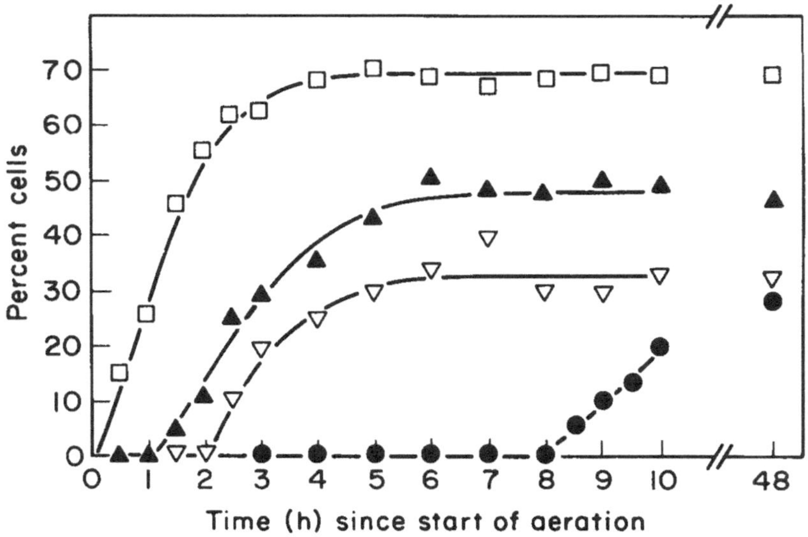

FIGURE 8. Time-course of sex-directed flocculation, copulation, conjugation, and sporulation in the homothallic strain NCYC 132. Symbols: □, flocculation; ▲, copulation; ▽, conjugation; ●, sporulation. Cultures were grown in malt-extract broth for 24 hr to stationary phase and then aerated starting at 0 hr. At each of the times indicated, a whole culture was sacrificed for cell counts. Incubation temperature was 32°C. (From Calleja, G. B. and Johnson, B. F., *Antonie van Leeuwenhoek J. Microbiol.*, 45, 391, 1979. With permission.)

ages, so much so that cells in stationary phase for 14 days, although retaining their viability, are no longer inducible. These observations indicate that cells are optimally poised for induction during a short segment of the stationary period. Efforts to bring about induction before that period is of no avail, whereas attempts later are met with increasing lags and diminishing capacities.

Heat-killed cells, of course, are not inducible, but do form flocs if induced before they are killed.[84] Metabolic inhibitors, such as cyanide, azide, and dinitrophenol, when added at the start of aeration, inhibit induction. They also inhibit growth, of course. Thus, whereas flocculation itself may be viewed as a mechanical colliding and sticking together of cells, the induction to competence to form flocs is a living process. UV irradiation, massive enough to cause failure of cells to grow into colonies on agar, does not prevent induction, thereby excluding a requirement for replication. Events subsequent to flocculation, however, are markedly inhibited by UV light. Other inhibitors of induction include membrane-specific antibiotics, such as polymyxin and nystatin, and inhibitors of cytoplasmic protein synthesis, such as cycloheximide, puromycin, and neomycin. Cycloheximide at concentrations as low as 50 μg $m\ell^{-1}$ is effective when added either at the time of inoculation or at the start of aeration. On the other hand, if added at the start of aeration, chloramphenicol (4.0 mg $m\ell^{-1}$), a specific inhibitor of mitochondrial protein synthesis, has no effect on induction. If added to cultures long before the last generation, it completely inhibits the development of competence to form flocs. Cells grown in chloramphenicol are not inducible, even after the drug is removed, but are viable when plated. All together, these observations strongly favor the conclusion that induction to competence to form flocs is a vital process that requires energy, mitochondrial function, cytoplasmic protein synthesis, but not DNA replication.

Subsequent experiments demonstrate that the increase in the extent of flocculation of a culture reflects the time-course of the induction process.[76] Because flocculation itself is not rate-limiting, the flocculation curve (seen in, for instance, Chapter 3, Figure 8 and Chapter

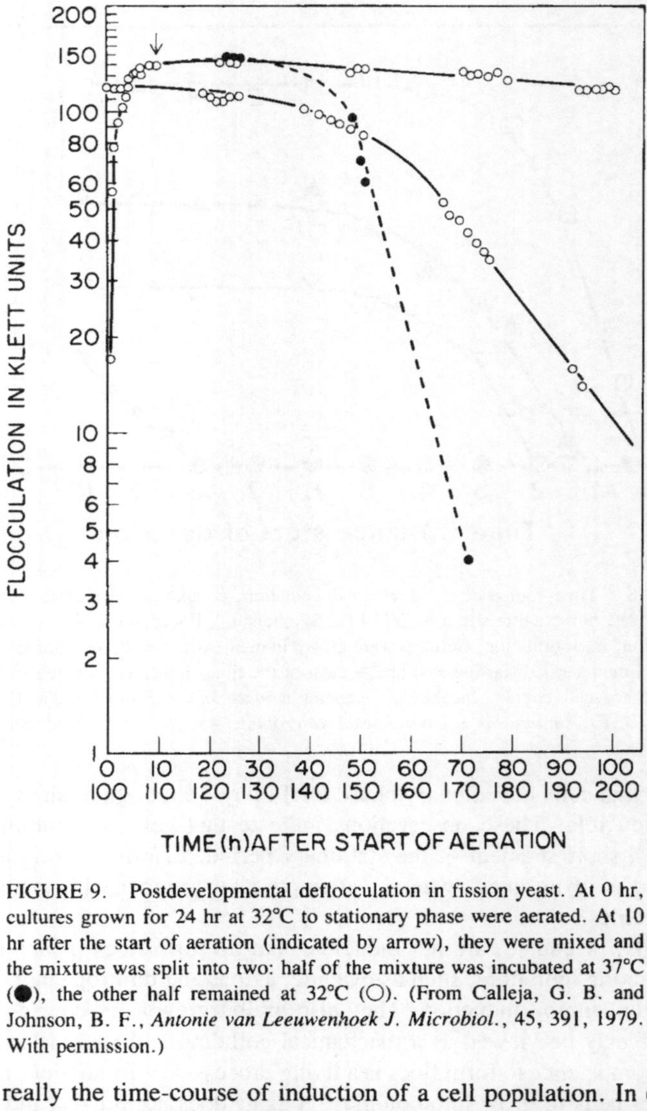

FIGURE 9. Postdevelopmental deflocculation in fission yeast. At 0 hr, cultures grown for 24 hr at 32°C to stationary phase were aerated. At 10 hr after the start of aeration (indicated by arrow), they were mixed and the mixture was split into two: half of the mixture was incubated at 37°C (●), the other half remained at 32°C (○). (From Calleja, G. B. and Johnson, B. F., *Antonie van Leeuwenhoek J. Microbiol.*, 45, 391, 1979. With permission.)

5, Figure 8) is really the time-course of induction of a cell population. In other words, the extent of flocculation of a culture is a direct measure of the extent of induction, which may be expressed in terms of how many cells in the population have been induced.

Sex-directed flocculation and its associated events are subject to catabolite repression.[134,1143,1144] Added at the start of aeration, high concentrations of glucose impair flocculation induction as reflected in the extent, and in the delay of onset, of flocculation. Copulation, conjugation, and sporulation, all are likewise inhibited. At very high concentrations of glucose (6% and above), induction is completely inhibited. The catabolite repression due to 1% glucose is not relieved by cyclic nucleotides. In fact, high concentrations of cyclic AMP ($>10^{-3}$ M), or cyclic GMP, also repress the developmental events, including flocculation induction. Results of experiments with actinomycin D are somewhat puzzling.[163] The antibiotic inhibits induction when added at the start of aeration. Added at the time of inoculation, it allows normal growth of the cell population, but nevertheless completely prevents induction. These experimental results appear to run counter to the report that cells of *S. pombe* are refractory to actinomycin D.[1145]

The fine structure of the developmental events following and including flocculation has been described.[121] Induction to competence to form flocs does not make induced cells look different from uninduced cells, except perhaps by the presence of sex hairs (Figure 20, Chapter 3). Comparable in appearance to fimbriae in bacteria[14,701] and in other yeast species,[543,544,516] sex hairs may serve as grappling hooks to hold the cells together in a floc. Presumably, the putative proteins involved in flocculation and found on the cell surface may reside in the sex hairs. Precisely because sex hairs appear to be distributed all over the cell surface, there seems no way of identifying on the cell wall a specific site for flocculation. During copulation, the hydrogen bonds that initially have held the cells together are supplemented and supplanted by covalent bonds between cell walls. About this time, conjugation tubes are formed. The mural areas, now fused to become a two-layer cross-wall, are then enzymically eroded. When the cross-wall is punctured, the cytoplasmic fusion of two cells is achieved. Further erosion removes the cross-wall completely and converts the two cells into a transient dikaryon. Soon after, karyogamy takes place. The zygote undergoes two sequential meiotic divisions. The final meiotic products are then packaged individually into spores.

The developmental sequence has been monitored with respect to respiration, macromolecular changes, catabolite repression, and commitment to sporulation.[1143] A prerequisite to flocculation induction is respiratory sufficiency. Respiratory activity increases about fivefold before the attainment of maximal flocculation (Figure 10). It remains high till the end of the sporulation process. When the increase in respiration is delayed, flocculation induction is similarly delayed. Prior to conjugation, total protein and total RNA first increase and slowly decrease thereafter. A round of premeiotic DNA synthesis takes place after copulation. There is no certainty as to when precisely it occurs, but there is no doubt that it occurs earlier than karyogamy.[130,1143] Before the burst of DNA synthesis, the DNA content of an induced cell is equivalent to that of a haploid spore. A condition for induction appears to be that the inducible cell be in the G1 stage of the cell cycle.[130-132] The same condition is required for conjugation in *Saccharomyces cerevisiae*.[128,129]

Commitment to development may be monitored by adding glucose at various times and noting when development becomes refractory to catabolite repression.[1143] As late as 4 hr after the start of aeration, exogenous glucose inhibits further flocculation induction (Figure 11). The inhibition at earlier periods is transient. After a period of gradual deflocculation, caused by the glucose-induced growth of cells, a gradual recovery of flocculation follows. The recovery seems due to the fact that most of the exogenous glucose has been consumed in the production of more cells. Thus, the developmental program is frustrated by glucose-induced reversion of the cells to the mitotic cycle. By the same approach, the period of commitment to sporulation has been determined. As early as 4 to 5 hr prior to the observation of the first spores in a culture, the majority of sex pairs are no longer repressible by glucose. Commitment to sporulation may come soon after copulation. It might well be concomitant with premeiotic DNA synthesis.

Conjugation is fraught with danger to a sex pair. Around 15% of sex pairs lyse spontaneously during conjugation at 32°C.[660,661] When a culture is transferred to distilled water, the percentage lysis may be as high as 40%. Vegetative cells, in contrast, are most difficult to lyse. The site of union of a conjugating pair is the site of lysis. The period in which cells are most susceptible to lysis corresponds to the beginning of conjugation, about 2 hr after the start of aeration. Conjugation-induced lysis in *S. pombe* is analogous to lethal zygosis in *Escherichia coli*[652-657] and to developmentally induced autolysis in myxobacteria.[971,1086] It is one evidence that the sexual process is subject to miscoordination, to a failure of controls over the glucanolytic activity during fusion, and erosion of the cell walls.

Strain NCYC 132 is routinely grown and induced to flocculate at 32°C. At 37°C, normal cell division is allowed, but not sex-directed flocculation.[143] Cultures grown to stationary phase at 32°C and then aerated at 37°C do not flocculate. Consequently, they do not sporulate. Grown to stationary phase at 37°C, cultures are not readily inducible when aerated at 32°C.

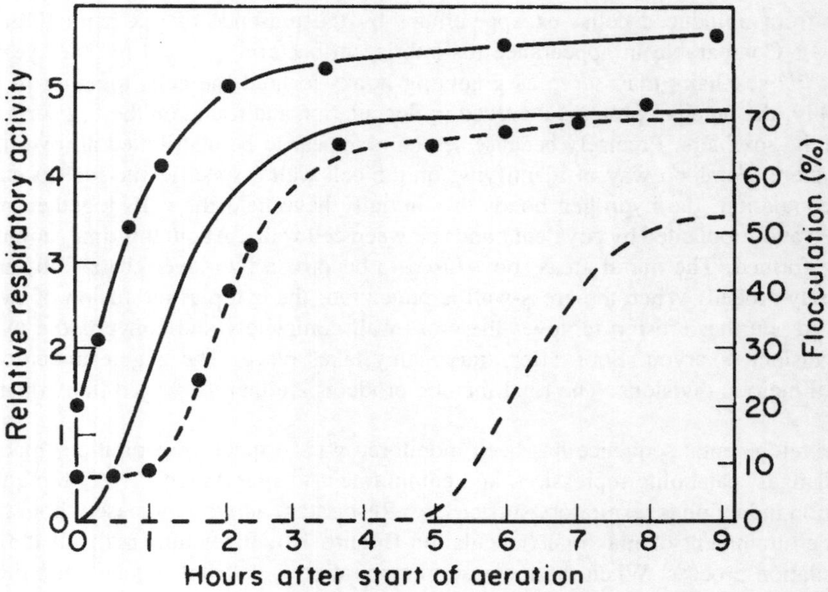

FIGURE 10. Respiratory activity during flocculation induction in fission yeast. Symbols: ●, respiratory activity; curves without dots, flocculation; solid lines, routine procedure for flocculation induction; broken lines, delayed induction. Flocculation induction was experimentally delayed by causing a greater degree of anaerobiosis before the start of aeration. (From Calleja, G. B., Johnson, B. F., and Yoo, B. Y., *Plant Cell Physiol.*, 21, 613, 1980. With permission.)

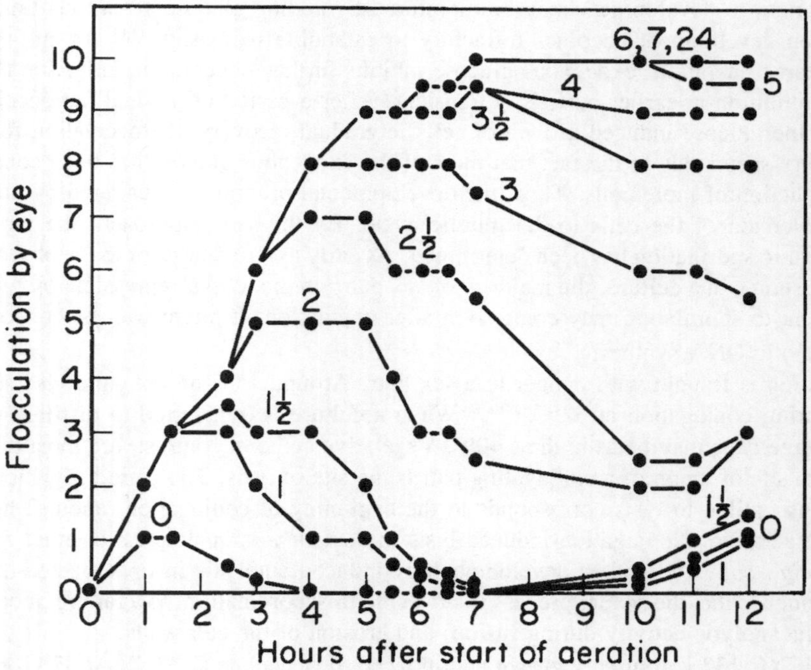

FIGURE 11. Glucose repression of flocculation induction in fission yeast. At 0 hr, a number of stationary-phase cultures were separately and simultaneously aerated. To each was added glucose (final concentration: 1%) at the hour indicated by the number above each curve. (From Calleja, G. B., Johnson, B. F., and Yoo, B. Y., *Plant Cell Physiol.*, 21, 613, 1980. With permission.)

Thermosensitve events in the developmental sequence are identified by growing and inducing cultures at the permissive temperature and then transferring them at different times to the restrictive temperature. These events include the following: development of respiratory sufficiency, readiness, induction to competence to form flocs, copulation, conjugation, and early sporulation (which should include meiosis). Readiness is operationally defined as the ability of a culture to be induced within 1 hr after the start of aeration at 32°C. In contrast, cell division (including mitosis), respiration, flocculation, and spore maturation are thermoresistant. At the restrictive temperature, conjugation-induced lysis and postdevelopmental deflocculation are enhanced (Figure 9). Conjugation and sporulation have been shown earlier in other strains of *S. pombe* to be thermosensitive.[67,515] The thermosensitivity of development is known in other fungal systems.[141,142,144-148]

Exposure of agar colonies of fission yeast to iodine fumes yields diagnostic colors. Homothallic colonies are black, heterothallic colonies are yellowish.[64,75,183,1130] Moreover, a black line, termed the iodine junction reaction, marks the boundary between contiguous heterothallic colonies of complementary mating types.The colonial color reaction points to the location of spore-bearing asci, which are the consequence of the zygotic union of cells. Under conditions that allow sporulation immediately after zygote formation, the color reaction is indicative of sexual interaction between complementary cells within a colony or along the boundary shared by contiguous, complementary colonies. Leupold's homothallic h^{90} strain yields homogeneously black colonies whereas the heterothallic h^{+}/h^{-} system yields homogeneously yellowish colonies and the iodine junction reaction between contiguous h^{+} and h^{-} colonies. A somewhat intermediate reaction is observed in the homothallic NCYC 132 strain. It consists of black streaks on a yellowish background, a mosaic.[1146]

Exposed to iodine, agar colonies of isolate 360-2, a direct derivative of the homothallic NCYC 132, may be sorted into two classes: P (with numerous radial, black streaks) and d (with scarcely any).[124] Neither class can be isolated as a pure culture. When replated, a P colony generates a P plate, which contains mostly P but also d colonies. In contrast, a d colony generates a d plate, which contains mostly d but also P colonies. The P/d colony ratio of a fresh isolate is high (if the isolate is derived as a P colony) or low (if the isolate is derived as a d colony). Upon repeated replatings of the isolate, the ratio falls, if originally high, or rises, if originally low. An equilibrium P/d colony ratio of about 0.5 is attained by isolates maintained for hundreds of generations. Furthermore, the two classes can be differentiated from each other by their interaction with their immediate neighbors. The iodine junction reaction is found between P and d colonies that are contiguous. It is not observed between two contiguous P colonies nor between two contiguous d colonies. Challenged with the heterothallic tester strains 975 h^{+} and 972 h^{-}, P colonies behaved like h^{+} colonies in giving the iodine junction reaction with contiguous h^{-} colonies, whereas d colonies behaved like h^{-} colonies in giving the iodine junction reaction with contiguous h^{+} colonies. A 2:2 segregation ratio of the classes indicates heterozygosity at a single locus.

Other than the iodine reaction, there are ways of telling a P colony from a d colony. The surface of a P colony exhibits more lines of depression, before and after exposure to iodine. On malt-extract-broth agar, a P colony is a shade darker than a d colony. Inasmuch as the iodine reaction is diagnostic of sporulation, a P colony contains more asci, more spores. Even more relevant to the subject matter, a subculture of a P colony in malt-extract broth shaken overnight is more flocculent than that of a d colony. All these differences between classes somehow relate to development.

The failure to purify the classes strongly indicates that a colony of either class is made up of two interconvertible cell types, P and d, identifiable after plating by the class of colony they generate. Thus, a P colony gives rise only to P plates (with both P and d colonies, but P in the majority), and a d colony only to d plates (with both d and P colonies, but d in the majority). The iodine junction reaction defines the complementarity of P and d at the colony level. Tetrad analysis confirms the complementarity of P and d at the level of the cell. An ascus is therefore the product of the union between P and d cells. The iodine junction reaction represents the line of encounter between P cells from a P colony and d cells from a d colony. The intracolonial iodine reaction consists of lines of encounter between subclones of one cell type and the clone of the other, which is the original and majority type in the colony. The observations and assumptions are summarized in Figure 12.

The disparity in number of P and d colonies from a culture after hundreds of generations is not due to selection by differential rates of replication of the two classes. Rather, it is indicative of differential rates of interconversion. The greater number of iodine streaks on a P colony and the lesser degree of mosaicness of a d colony suggest a higher frequency for the P → d interconversion compared to the reverse, the d → P interconversion. A comparison of the plates generated from P and d colonies confirms this suggestion. From an analysis of the P/d colony ratios of cultures, the rates of interconversion have been derived.[124,1147] The P→d interconversion occurs at the rate of 7.1×10^{-3} per cell division, the d → P interconversion at the rate of 3.7×10^{-3} per cell division.

The P/d sex interconversion system and the earlier work on other homothallic strains suggest that there might not be a homothallic fission-yeast cell at all, only a homothallic clone or a homothallic strain.[123,124,1148,1149] At any instant, a cell is in fact heterothallic in that it is capable of interacting with only one of two complementary and alternative mating types. Contrasted to a heterothallic clone of one mating type, a homothallic clone is a community of two cell types that are sexually complementary and interconvertible. The presumably complementary mating types that make up the homothallic h^{90} system are not separable nor readily identifiable, because they interconvert too often.[1149] When replated and exposed to iodine fumes, h^{90} gives rise to colonies all uniformly black. In contrast, the interconversion rates of the two complementary mating types that make up the heterothallic h^+/h^- system are so low that the colonies fail to give the intracolonial iodine reaction and are therefore uniformly yellowish.[572,1150,1151] The heterothallic mating types, identifiable by means of the iodine junction reaction, are maintainable as pure and fairly stable cultures. The interconversion rates of the homothallic P/d system are found somewhere between the extremes. Consequently, the complementary component mating types can be identified by the iodine reaction, intracolonial and junction, but are not isolatable as pure cultures. The iodine reaction of the homothallic NCYC 132 and its derivatives, which display the P/d sex interconversion system, is reminiscent of the iodine reaction given by isolates EF_4 and EF_5 of Gutz and Doe.[1151]

Much of what is known concerning the mating-type genetics of fission yeast derives from the early work of Leupold.[64,75,571,1130,1152] Three basic mating types are known: the homothallic h^{90} and the heterothallic h^+ and h^-, the latter two being self-sterile and complementary to each other. In addition, h^+ may be subdivided into h^{+N} (*N* for normal) and h^{+R} (*R* for recombination-derived).[571] The h^{+N} mating type mutates to h^{90}, h^{+R} to h^-. The homothallic h^{90} can mutate to h^{+N} or to h^-, both of which may be derived also from a mating between h^{90} and h^{+N}. Inasmuch as h^- does not seem to mutate to any of the mating types, it has been called h^{-S} (*S* for stable).[1150]

The instabilities and interconversions of mating types become more comprehensible when the complex mating-type locus, *mat*, is assumed to contain alternative structural genes that are closely linked.[515] The *mat* locus is located on chromosome II (Figure 5). It consists of the genes now called *mat1* and *mat2*, which are separated by 1.1 map units.[571] The following allelic combinations characterize the original mating types of Leupold:

$mat1^+\ mat2^+$	(h^{+N})
$mat1^-\ mat2^0$	(h^{-S})
$mat1^+\ mat2^0$	(h^{+R})
$mat1^-\ mat2^+$	(h^{90})

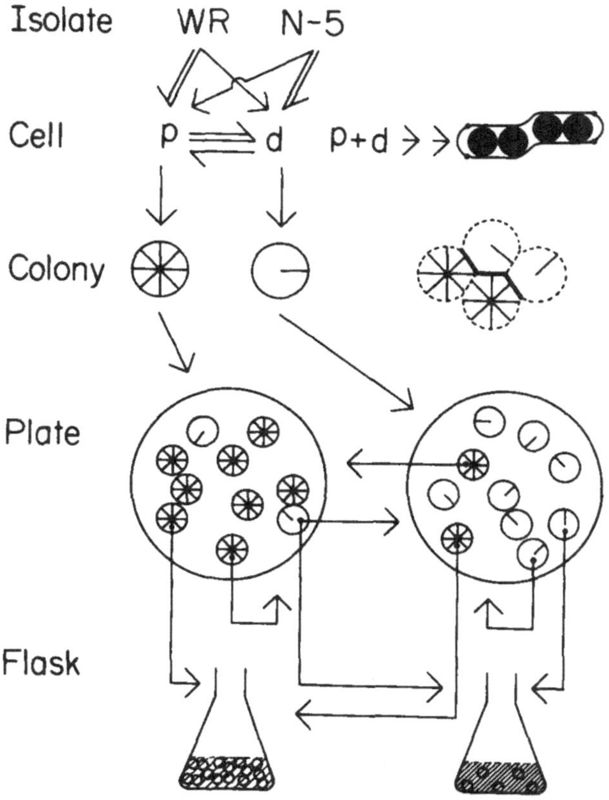

FIGURE 12. The homothallic P/d system: observations and assumptions. From strain 360-2 was isolated WR as a P colony, N-5 as a d colony. Both isolates are mixtures of cell types P and d. When first isolated, WR consisted of mostly P cells, N-5 mostly d cells. Whereas P and d cells are interconvertible, the P → d interconversion is more frequent than the reverse (d → P). P and d cells are complementary, an ascus being the product exclusively of the zygotic union of P and d cells. When plated, a P cell generates a P colony, which has numerous iodine streaks. A d cell generates a d colony, which has scarcely any. The iodine junction reaction is found between P and d colonies that are contiguous to each other, but not between contiguous P colonies nor contiguous d colonies. A P colony when replated gives rise to a P plate, which contains mostly P colonies; a d colony to a P plate with mostly d colonies. The resultant colonies from either plate give rise to plates corresponding to the class of the replated colonies. P colonies, when transferred to aerated malt-extract broth, give rise to aerated cultures that are more flocculent than those derived from d colonies. (From Calleja, G. B., Johnson, B. F., Zuker, M., and James, A. P., *Mol. Gen. Genet.*, 172, 1, 1979. With permission.)

The presence of *plus* and *minus* in one genome accounts for the homothallic behavior of h^{90}. The segment *mat1* may be either *plus* or *minus*, whereas *mat2* may be either *plus* or *null* (inactive). Sexual activity is the consequence of a complementary interaction between the *plus* allele of either *mat1* or *mat2* and the *minus* activity found only in *mat1*. Further genetic analysis by Egel allows the interpretation of mating-type instabilities in terms of gene transposition.[123,1148,1153] In the homothallic strains, the mating-type genes of complementary functions are found together and close to each other (*mat1*$^-$ *mat2*$^+$). The gene *mat2* may duplicate itself. When it does, the copy inserts itself in the position occupied by *mat1*$^-$ and becomes *mat1*$^+$. The *minus* function is thus lost, but is regained when the inserted copy is excised.

The duplication and the transposition of *mat* genes form the basis of Egel's flip-flop model for the control of mating-type interconversion in fission yeast (Figure 13).[123] The *mat2*$^+$ copy could be transposed to the *mat1* position in the normal direction or the reverse. Sequential repeats are for some reason unstable, but not if the repeat is in the reverse orientation. By displaced recombination, a sequential repeat may be lengthened or shortened. Inasmuch as there is no evidence for such an event taking place, the repeated elements are proposed to be of opposite polarities (P and P′, in the diagram). The *mat2*$^+$ segment, however, is itself flanked by repeated elements of the same polarity (c_1 and c_2). A regulatory element, R, with a promotor signal in one direction (a feature of insertion elements in bacteria) serves as a switch for the flip-flop. Presumably an enzyme system catalyzes the inversion of R, thus bringing about the expression of one *mat* gene segment at the expense of the other. P is read when the switch points to the right; M′ is read when the switch flips to the left. A secondary consequence of the presence of the invertible promotor, itself flanked by repeated elements of opposite polarities, is a hot-spot of interstrand recombination. The duplication of *mat2*$^+$ and the transposition of the copy affects both localized mitotic-recombination activity and mating-type functions. With the duplication-transposition of *mat2*$^+$ go the duplication and the transposition of R. Now that the structural gene for *minus* activity, M′, is disconnected from the promotor element, its expression is consequently inhibited. Although not identical to one another, the sequences a, b, and c share sufficient homology for occasional recombination. Site-specific recombination is most efficient when the segments of partial homology are arranged before transposition, in the configuration a′b. Known mating-type mutations are the consequence of deletions, which may be due in turn to the partially homologous repeats in one direction.

Homothallism in fission yeast has been proposed to be due to modifier action, without *mat2* being invoked as a separate gene.[1154] The more plausible model assigns the regulatory locus beween *mat1* and *mat2* as the site of the homothallic switch. Indeed, it has been shown recently that the *plus* segment is regulated in the middle of the *mat* region, between the *minus* and the *plus* segments.[1155] In the homothallic h^{90}, mutations termed "speckled" and "mottled" appear to diminish the rate of switching from one mating type to the other.[515,1151] These mutants are analogous to the P/d system. The "speckled" mutation maps somewhere in the *mat* region, whereas the "mottled" phenotypes may be the result of the switching activity of four genes called *mmo* (for mating-type modifiers), which are unlinked to *mat*.[515,1156] The promotor-containing locus in the flip-flop model may be the site of the "speckled" mutation, even as it may be the target site of the modifier genes and the still unmapped P/d switch. Sex interconversion in *Saccharomyces cerevisiae* has been explained in somewhat different terms (see following section).

The conversion from homothallism to heterothallism may be effected not by a switch but by more conventional mutations of additional genes other than, and not closely linked to, the *mat* region. The great majority of nonsporulating mutants are also sterile, i.e., unable to mate.[1157] Most of them are also unable to flocculate.[66] From the homothallic h^{90} may be derived mutants that have become heterothallic but with unimpaired *mat* functions.[1158] The defects of these mutants may be assigned to genes named (for mating-type auxiliary) *mam (minus)* and *map (plus)*. Mutations in *mam1* and *map1* result in noninducibility to flocculate. These genes may therefore control the synthesis of the flocculation molecules or the conditions for flocculation induction.The auxiliary gene *mam1* is located to the right of the centromere on chromosome II, but much farther to the right than *mat* (Figure 5). In contrast, *mam2* and *map2* mutants are able to flocculate, albeit the later stages of development are inhibited. Mating-type interactions, controlled by *mam2* and *map2*, as well as *mam1*, are necessary for conjugation, but not for meiosis, which nevertheless occurs in the diplont. However, *map1* is required for meiotic functions. Moreover, the structural gene *mat2*,

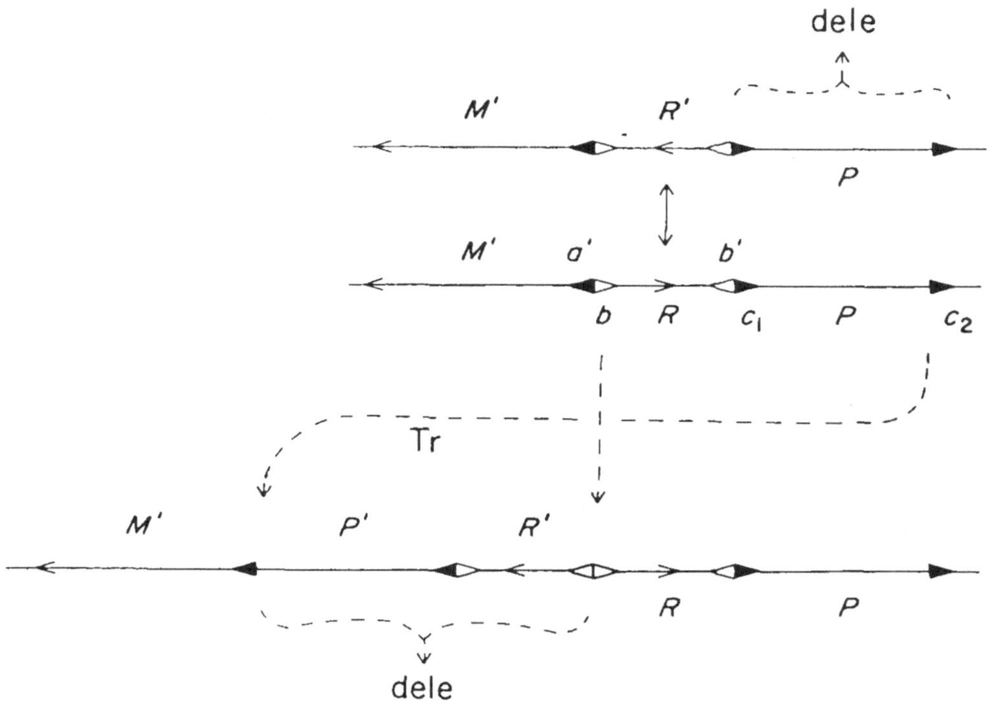

FIGURE 13. Flip-flop model for the control of sex interconversion in fission yeast. The *mat* locus is drawn to consist of two structural genes, *P* and *M'*, corresponding to *plus* and *minus* activity. Between these two DNA segments is a regulatory element, *R*, with a promotor. The arrows on the DNA segments indicate polarity. The prime tucked on to a symbol signifies polarity to the left. When the promotor points to the right, *P* is expressed, while *M'* is not. *M'* is expressed when the promotor flips over to the left. Responsible for site-specific recombination, the sequences *a*, *b*, and *c* are segments of partial homology. A transposition, Tr, consists of a duplication of *P* and *R* and the insertion of the copy (now *P'* and *R'* because of reverse polarity) between *M'* and *R* so that *M'* is cut off from the promotor. Known mating-type mutations are labeled *dele*, for deletion. Adapted from Egel.[123]

required for conjugation and meiosis, is required for the synthesis of a hypothetical *map1* gene product.[572] Other developmental mutations are called *fus* (for fusion), which prevents conjugation, *mei* (for meiosis), *mes* (for meiosis, second division), and *spo* (for sporulation).[1157]

Most of the genetic studies, it must be noted, have used Leupold's heterothallic h^+/h^- system and homothallic h^{90} and their derivatives. The sex-directed flocculation that initiates the developmental program in these strains differs somewhat from that of the homothallic P/d system, which has served as the mainstay for flocculation studies. For instance, Leupold's strains are not as glucose repressible.[67] Induction to the sexual process is achieved in a medium that is nitrogen-free, specifically, by withdrawing aspartate from the growth medium. Nevertheless, the sexual process in the heterothallic h^+/h^- system is also preceded by flocculation.

There is so far no evidence to suggest that flocculation induction involves diffusible factors from the complementary mating types.[67,121,132] When the heterothallic mating types h^+ and h^- are mixed together, they do not flocculate immediately. Instead, an extended period of time elapses before they become induced. Pre-incubation of the heterothallic cultures, separated by a filter through which molecules, but not cells, could pass, does not shorten the time of induction. Neither does a filtrate of induced homothallic cultures make them more inducible. These observations suggest that physical contact is necessary for mutual stimulation between the complementary mating types that make up the floc.

The recognitive and binding factors involved in sex-directed flocculation remain to be isolated and identified. Attempts along these lines have yielded meager results. From the filtrate of a culture undergoing postdevelopmental deflocculation has been isolated a fraction that reversibly inhibits the reflocculation of heat-killed, induced cells.[483,516] This fraction does not cause deflocculation. Resistant to boiling, the reflocculation-inhibiting activity is not found in uninoculated media nor in media that do not support the induction of flocculation. A molecular species smaller than 3×10^3 daltons is suggested by passage of the filtrate through Sephadex® G-25. It is possible that this fairly small molecule is but a fragment of one of the complementary flocculation structures. It may be released by postdevelopmental deflocculation into the culture medium from a much larger molecule that is covalently attached to the cell wall.

A sex pair in a floc consists of cells of complementary mating types. In the P/d interconversion system, it consists of a P cell and a d cell. Is the cell population of a floc made up of equal numbers of the complementary cell types? A d/P cell ratio of 1.27, fairly close to unity, has been obtained.[483] As observed also in *Hansenula wingei* and in *Saccharomyces cerevisiae*, a sexual aggregate of yeast cells comprises approximately equal numbers of the component cell types.[94,1159] Thus, it is not much more than an enlarged and extended interaction of complements. Recall, however, that in *Escherichia coli*, the mating-type ratio of mating aggregates may vary by a factor of four, depending upon the physiological state of the cells, the input ratio, and the cell population density.[28]

Whereas the complementary interaction between two cells is specific, the overall arrangement of the cells within a floc is less so. A floc is a random collection of many thousands of competent cells. The slight plurality of d cells in a floc is observed, because they constitute at equilibrium the majority in the interconverting population. It is assumed that the floc periphery is occupied mostly by d cells. That the d/P cell ratio, slightly but significantly, deviates from unity is further evidence that flocculation is not the consequence of committed pairs coming together: we know already that copulation takes place only between cells in flocs. Rather, cells in a floc may be viewed as if they were the denizens of a singles bar, where induced, swinging cells arrive unattached, uncommitted, and susceptible to the blandishments of willing and consenting neighbor cells. In the ambiance of a floc, where every cell is on the make, a pair of complementary cells, hydrogen-bonded by chance to each other and to a dozen or so other immediate-neighbor cells, commit themselves to each other by forming covalent bonds between them and thereafter committing themselves to an indissoluble union, for better or for worse and for good. If they escape conjugation-induced lysis, they may even become the posthumous progenitors of four little spores.

The antimetabolite 2-deoxyglucose is well known to cause lysis in growing yeast cells.[1160,1161] Its effect on flocculation and sexual development in fission yeast has been examined.[482,483] Although at high concentrations ($>2 \times 10^{-3}M$) 2-deoxyglucose inhibits flocculation induction and the subsequent developmental events, it is more effective as a lytic agent during vegetative growth. It also causes the lysis of conjugating cells. The site of lysis is that mural region involved in copulation, conjugation, and conjugation-tube formation. Moreover, at concentrations that permit flocculation induction, 2-deoxyglucose also brings about premature deflocculation. The various effects of the antimetabolite may be ascribed to faulty mural synthesis, which leads to the erection of defective cell walls with deranged activities. The derangement is manifested as a susceptibility of the cell to lysis, as a tendency toward formation of monster cells, as an impairment of copulation, and as an instability of the flocs. The mural defect and the consequent floc instability hinder copulation by interfering with the formation of covalent linkages that supplant the hydrogen bonds between flocculated cells.

Although the preferred sites of copulatory activity are the poles (with the old pole favored over the new), copulation may occur occasionally in the middle region of the cylindrical cell.[131,483] Thus, despite polar preference, the entire wall appears to be available for sexual activity. Incestuous matings (sexual interaction between sibs), quite common in the homothallic h^{90}, appear to be the exception in the homothallic P/d system.[64,483] They are, of course, extremely improbable in the heterothallic h^+/h^- system. Also, sex is not limited to pairs.[121,483] A small percentage (a little over 3%) of the copulating population may be found in *ménage à trois*, *quatre*, or even *cinq*. Supernumerary sexual arrangements are not unknown in other yeasts.[246,1162]

Though the *ménage* is intrinsically interesting, its rarity is the more enigmatic. A cell in the core of a floc is apt to be hydrogen-bonded to as many cells as can be sterically accommodated on its surface. That number is about 12. Because of the requirement for complementarity, it may be assumed that a P cell in the inner confines of a floc is surrounded by d cells, a d cell by P cells. Granted that there is polar preference, as well as polar discrimination, still there must exist a mechanism by which a third participant is excluded. When one pole of the cylindrical cell is already covalently linked with a pole of another, the remaining poles become unavailable for a third participant. Between the two cells, there is now a commitment to become one. An interpolar communication system, perhaps through mural modulation of some sort, has been invoked.[121,483] A failure in the communication system, such that the poles are unaware of each other's commitment, may lead to a *ménage*. If, however, a miscommunication in some manner makes one pole sense that the other is already spoken for, a cell may end up with no partner at all. In whatever way the exclusion of a third cell is effected, it does not lead to deflocculation. Indeed, deflocculation occurs only long after the developmental events are over (Figure 9).

The usefulness of sex-directed flocculation to the cell and to the species can hardly be questioned, but once sporulation is finished, perhaps even earlier than that, the value of the social arrangement rapidly vanishes. Postdevelopmental deflocculation, obviously a means to population dispersal, releases the components of the floc, which represents up to now a form of both security and bondage. The floc, now a very heterogeneous community of asci, liberated spores, abortive conjugants, copulation halves, unfulfilled cells unable to establish meaningful relationships, and dead cells, is gradually disbanded. Among the dead may be counted those that have perished by conjugation-induced lysis, those that have expired from exhaustion or from starvation in a depleted medium, and those that have passed on their genes to a new generation. Only the viable ones, mostly spores and a few hardy survivors of the orgiastic happening and a few straggling incompetent cells hanging around the flocs, will start vegetative life anew, as new colonies in a new environment.

IV. SEXUAL AGGLUTINATION IN *SACCHAROMYCES CEREVISIAE*

To researchers and nonresearchers alike, "yeast" means *Saccharomyces cerevisiae* Hansen, for good reason. The type species of the sugar fungus occupies a central position in the life of man and his history, not to mention his economics and his system of taxation. Its reputation rests on two metabolic products: ethyl alcohol and carbon dioxide. As the specific epithet indicates, *S. cerevisiae* is brewers' yeast, the yeast of the wine maker as well as the bootlegger. It is also bakers' yeast, the yeast of leavened bread. It helps feed man, helps him bear the woes of the world in good cheer, and possibly might supply his energy needs of the future. It is the yeast of the great biologists, of Cagniard de la Tour, Kützing, Schwann, Pasteur, the Buchners, Harden, Young, Neuberg, Warburg, Meyerhof, Hansen, Winge, Lindegren, and many more. No single eukaryotic species has been more useful as a biochemical tool. The very word "enzyme" is a Greek derivative meaning "in leaven". In genetics, *S. cerevisiae* ranks with the favored few: *Drosophila melanogaster*, *Zea mays*, *Neurospora crassa*, and *Escherichia coli*.

The species, as defined, is licentiously plastic. It includes a number of characters that are highly variable. The current taxonomic rules allow the inclusion of three varieties in the species: *S. cerevisiae* var. *ellipsoideus*, *S. cerevisiae* var. *tetrasporus*, and *S. willianus*.[1163] More than 50 synonyms are found in the extensive literature. The standard description of the organism allows the cells to be spheroidal, ovoid, ellipsoidal, and even elongate. This ascomycetous yeast may exist in various states of ploidy. It enjoys, aside from a favored status in the fermentation industry, a widespread distribution in nature.

Figure 14 depicts the life cycle of *S. cerevisiae*. Unlike the fission yeast, *S. cerevisiae* may be maintained in culture as a stable diplont. It reproduces by budding. Under the proper conditions, the diploid cell undergoes meiosis and becomes an ascus with a maximum of four haploid spores. Two of the spores are of the mating type **a**, the other two of the complementary mating type **α**. After the spores are germinated, the resultant vegetative haplonts may be maintained as stable haplonts. When the complementary mating types are brought together, sexual agglutination occurs. This is soon followed by cytoplasmic fusion and nuclear fusion. The zygote later forms a diploid bud where the two original cells have fused, as in *Hansenula*. The subsequent zygotic buds are also diploid. Occasionally, the transient heterokaryon may produce a haploid bud before nuclear fusion.

Although indications of sexuality in the species had been made much earlier, the first convincing demonstration of a haplophase and a diplophase linked by a sexual process must be attributed to Winge.[1164] He showed that a diploid cell forms haploid ascospores, which may germinate into vegetative haploid cells or may fuse in pairs. The vegetative haploid cells themselves may fuse. In either case, diploidy is regained after karyogamy. It remained for the Lindegrens in 1943 to demonstrate heterothallism, that the fusing spores or vegetative cells are of complementary mating types, **a** and **α**.[70,562] In an ascus, **a** and **α** are recovered in equal numbers, indicating that they are governed by alternative, allelic genes. However, because matings can occur within a clone, it is apparent that some strains are homothallic, even as others are heterothallic.[1165] Now we know that sex in a haploid clone is the consequence of mating-type interconversions. There have been a good number of reviews on the genetics and the physiology of the sexual process in yeast, meaning *S. cerevisiae* for the greater part because it is the best studied among the yeasts, but only the more recent ones are brought to the attention of the reader.[16,90,126,147,181,515,520,564,1166-1170]

It is surprising that while great strides were being made in the genetics of *Saccharomyces*, the cell-cell interaction that precedes genetic recombination attracted little attention. Doubtless, sexual agglutination in the species, although not as conspicuous as that in *Hansenula wingei*, must have been observed by earlier workers.[147] But whereas Wickerham's short list of sexual agglutinating yeast genera in 1958 included *Saccharomyces*, it did not include *S. cerevisiae*.[35] As far as could be ascertained, the first study on sexual agglutination in the species must be credited to Sakai and Yanagishima.[69] They showed in 1971 that sexual agglutination precedes conjugation. Since then, the subject has appeared in a number of reviews.[16,90,515,520,1169-1171]

Sakai and Yanagishima observed that agglutination occurred when the mating types **a** and **α** were mixed together.[69] Cells of one mating type did not agglutinate among themselves, nor did cells of diploid strains. Agglutination was sexual for it constituted the initial visible event of the mating reaction, which included cell fusion and zygote formation. Unlike that of *Hansenula wingei*, it was not instantaneous. It was not observed within the first 30 min after mixture of the mating types. Rather, it developed gradually over 4 hr. Cells prior to mixing did not agglutinate, nor did mixtures kept at 0°C. Moreover, cycloheximide, but not chloramphenicol, inhibited agglutination. Thus, it was concluded that a change in the property of the cells must have taken place when the mating types were brought together, a change that required cytoplasmic protein synthesis.

In certain combinations of strains of **a** and **α**, sexual agglutination may develop almost immediately after mixing.[1172] This rapid agglutination, contrasted to combinations that re-

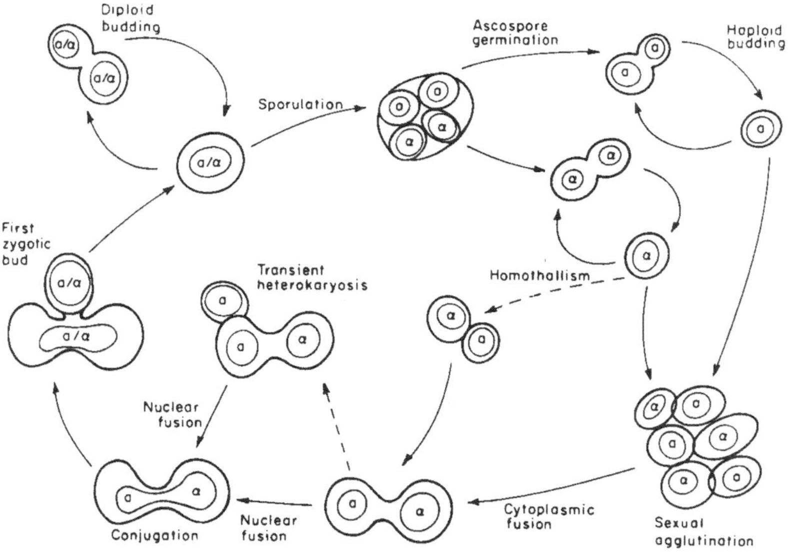

FIGURE 14. The life cycle of *Saccharomyces cerevisiae*. Dashed arrows represent pathways that are observed at low frequency or under special conditions. (From Crandall, M., Egel, R., and MacKay, V. L., *Adv. Microb. Physiol.*, 15, 307, 1977. With permission.)

quire several hours, is due to components of the mating reaction being constitutive for agglutination. Whereas **α** strains may appear to be invariably constitutive, there are two kinds of **a** strains with regard to agglutinability: constitutive and inducible, the latter being responsible for delayed agglutination. Even in the presence of cycloheximide, agglutination occurs within 30 min in constitutive strains, but not when an inducible **a** strain is involved. Further agglutination, however, does not develop beyond 30 min. Thus, the agglutination process may be divided into an early stage, which does not require protein synthesis, and a late stage, observed 2 hr after mixing, which requires protein synthesis and induction. Heat-killed cells, if constitutive, do agglutinate when mixed, but not if one component of the mating reaction is inducible. Protease-treatment of cells of either mating type inhibits agglutination, suggesting the involvement of proteinaceous materials on the cell surface.

Pre-incubation of cells of an inducible **a** strain in the cell-free culture filtrate of an **α** strain brings about induction.[1172] Such cells agglutinate with **α** cells even in the presence of cycloheximide. However, if the pre-incubation is performed in the presence of cycloheximide, induction does not take place. This observation suggests that there is in the filtrate of an **α** culture a substance that activates inducible **a** cells. The sex-specificity of this substance is shown by the fact that the activity, which is heat stable, is found only in the culture filtrates of **α** strains.

Diffusible sex substances in *S. cerevisiae* have been known for some time. Placed near each other, a pair of **a** and **α** cells tend to elongate toward each other until they touch and fuse.[71,1165] In liquid medium, both **a** and **α** cells may develop copulatory processes. On solid medium, only the **a** cells do so. Levi showed that **a** cells placed on an agar surface on which mating reactions have previously taken place undergo morphologically distinct alterations.[71] Such altered cells are called shmoos, after Al Capp's fabulous comic-strip characters (Figure 15, Chapter 3).[1173] Much later work by Duntze et al. demonstrates quite clearly that the pheromone that causes the formation of shmoos is excreted by **α** cells into the culture medium.[72] Called **α** factor, it specifically inhibits cell division and nuclear DNA replication in **a** cells.[128,129,496-499,1174-1178] It arrests **a** cells at the initiation of DNA synthesis, in late G1 of the cell cycle, but the syntheses of bulk RNA and protein and mitochondrial DNA remain unimpaired. The G1 arrest is reversible. When α factor is removed, fairly synchronous DNA replication among the arrested cells is observed.

It is obvious that α factor allows cell-wall synthesis. However, the mural material that is synthesized in response to the pheromone appears to differ in certain ways. Compared to the walls of untreated cells, it contains less mannan and more glucan.[1179] The mannan of **a** cells treated with α factor comprises a greater amount of unsubstituted backbone mannose and of shorter side chains. A grosser difference is that shmoos are more readily lysed by Glusulase. Also, the wall at the growing tip of a shmoo is considerably thinner. The new wall material is also different in that it tends to bind more of the lectin concanavalin A.[90,1180]

But more relevant to our subject matter, the culture filtrate that causes increased agglutinability in **a** cells has turned out to be α factor.[173,174,252-255,1172,1181-1183] There has been much controversy as to whether or not the same factor that causes shmooing and G1 arrest also brings about the induction of agglutinability of inducible **a** cells. The main objection has been that there is the possibility of a number of pheromone species in the culture filtrate of α cells. With the advances in the purification and characterization of α factor, the reservation as regards α factor being an inducer of agglutinability in inducible **a** cells may now be withdrawn.

The sex factor that is excreted into the culture medium by α cells and to which **a** cells respond has been purified independently in three laboratories.[173,174,252,255,496-498,1184-1186] It is a dodeca- or tridecapeptide with the following sequence of amino acids:

H_2N-(Trp)-His-Trp-Leu-Gln-Leu-Lys-Pro-Gly-Gln-Pro-Met-Tyr-COOH

In some preparations, the N-terminal tryptophan may be absent, the methionine may be oxidized into its sulfoxide form. Although cupric ions may be complexed with purified α factor, it is doubtful whether the metal contributes at all to the activity of the pheromone. The preparation of Tanaka et al. does not contain copper.[1186]

The primary structure of α factor has been confirmed by chemical synthesis of both the dodecapeptide and the tridecapeptide with biological activity.[1185-1188] Synthetic analogs (substitution with D isomers of leu 6 or lys 7, norleucine for methionine) are less biologically active than the native tridecapeptide, but less subject to proteolytic degradation.[1189] The binding of α factor and its subsequent metabolism by **a** cells may be deduced from the spontaneous recovery of the treated cells from G1 arrest and the loss of α-factor activity from the culture fluid.[1174,1190,1191] The upake of the pheromone and its consequent concentration in the nucleus of **a** cells, as well as its degradation, have been demonstrated.[1192,1193] Neither α cells nor diploid cells take up α factor. The degraded peptides are not taken up by **a** cells either. Synthesis of α factor is sensitive to inhibition by cycloheximide and by conditional mutations that inhibit the synthesis of RNA and protein.[1194] The pheromone is found in the culture medium at the end of logarithmic growth, accumulates during early stationary phase, and is degraded into hexa- and heptapeptides (the split occurring between leu 6 and lys 7) during late stationary phase.[1186,1194,1195] Thus, the pheromonal activity of α factor is secure in the literature. What was controversial for awhile is whether α factor does bring about the induction of agglutinability of **a** cells.

Although the purified α factor of Tanaka et al. was at first reported not to cause enhanced agglutination,[1186] more recent reports leave little doubt that α factor, native or organically synthesized, stimulates the agglutinability of **a** cells.[1183,1196,1197] Moreover, earlier work in Yanagishima's laboratory convincingly showed that α substance-I_A, which has turned out to be one of the α factor peptides, induces agglutinability of inducible **a** cells and enhances that of constitutive **a** cells, indeed, at concentrations several orders of magnitude lower than concentrations that cause shmooing or inhibition of DNA synthesis.[173,174,252,253,1182,1198] The optimal concentration of **α** substance-I_A (synonymous to **α**2, which is the dodecapeptide without the N-terminal tryptophan[498]) is 1 ng $m\ell^{-1}$ (Figure 15). That concentration is about a 1000-fold lower than that required for 50% inhibition of DNA synthesis in an **a**-cell population. Observed within 20 min, induction attains a maximum in 1 hr. The primary action of **α** factor appears to be the induction of agglutination. Inhibition of DNA synthesis and shmooing may be secondary consequences. Apparently, what **α** factor does is to elicit from **a** cells the induction of **a** factor, which acts in turn on **α** cells.

The sex factor from **a** cells has not been as well studied. However, its existence is implied from the early work on diffusible sex factors.[71,129,1165,1173,1175,1176,1199,1200] When **α** cells are positioned near an overnight growth of **a** cells on agar, a few of the **α** cells are converted to shmoos.[1173] The morphological response, however, is less evident than that of **a** cells treated with **α** factor. Incubated in the culture filtrate of **a** cells, **α** cells do not form shmoos, but within 2 hr, they are transiently arrested in G1.[1199] Mitochondrial DNA synthesis, however, remains unabated as the treated cells increase in volume.[1201] Thus, sex factors from both components of the mating reaction may lead to mutual inhibition of nuclear DNA synthesis and to a synchronization of the participating cells.[128,1176] The less-than-dramatic response of **α** cells to the putative **a** factor may be due to an inherent difference in the response of **α** cells or due to an **a** factor that is not constitutive, not easily diffusible, or even more labile.[1170] An induction mechanism for **a** factor is favored by the finding that **a**-factor activity is detectable only when the **a** cells are grown in the presence of **α** cells.[1170] A low level of uninduced activity, however, has been reported.[1202] Whether induction means *de novo* synthesis or the removal of an inhibitor remains unresolved.

The pheromone from **a** cells has been isolated and partially characterized.[1202] It is a large glycoprotein (greater than 10^6 daltons), heat resistant, and active between pH 4 and pH 6, although stable over a wider range of hydrogen-ion concentration. The molecule, if it is made up of one molecular species, possesses both G1-arrest and shmooing activities on **α** cells. There is also good evidence that agglutinability of **α** cells is enhanced by **a** factor.[254,1171]

In several ways, the structurally different pheromones from the complementary mating types are functionally similar, better still, reciprocal. Above all, each appears to be synthesized exclusively by one mating type and to bring about specifically the enhancement of agglutinability of the other. The pheromone synthesized by **α** cells, **α** factor, stimulates **a** cells, whereas the **a** factor from **a** cells stimulates **α** cells. They do not themselves mediate the agglutination reaction. Rather, they bring about, in as yet unknown way, the synthesis or the unmasking of agglutination substances on the surfaces of the responding cells.

Like all microbial aggregation systems studied intensively, sexual agglutination in *S. cerevisiae* is presumed to be mediated by macromolecules on the surfaces of the interacting cells. In constitutive cells, these agglutination substances are constitutively found on the walls. They are inducible in inducible strains. Both constitutive and inducible **a** strains are known. Although early reports concluded that all **α** strains are constitutive, more recent ones report the characterization of inducible **α** strains.[520,1203] When inducible strains are mixed together, a time lag, usually about 30 min or so, is observed before agglutination, suggesting mutual induction or the induction of at least one component.[1172,1204] When constitutive strains are the components, agglutination begins immediately (Figure 16). Nevertheless, it takes 2 hr before maximal agglutination is achieved, which would also suggest further induction, even as some of the cells are already competent to agglutinate prior to mixing. One would suspect that the mechanical coming together of competent cells in a shaken liquid medium does not take that long to attain a maximum. Agglutination is followed, about 1 hr after mixing, by the appearance of conjugating cells.

The increase in aggregation number of a mating mixture is shown pictorially in Figure 17. Initially, small aggregates of less than 50 cells are formed.[1170] Larger aggregates are

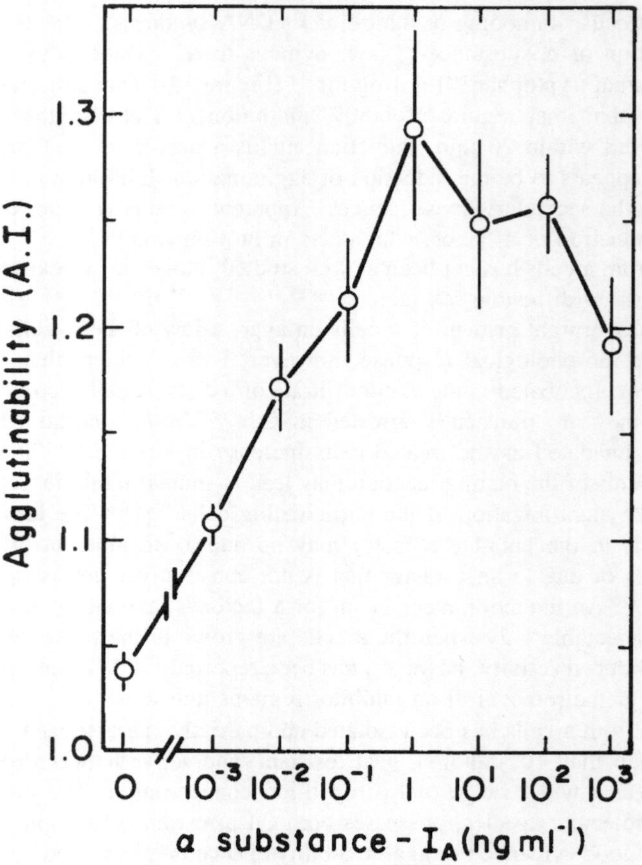

FIGURE 15. The effect of **α** substance-I_A (**α** factor) on the induction of agglutinability of inducible **a** cells. Cells were incubated with the pheromone for 2 hr before assayed for agglutinability (expressed as A.I., agglutination index). Vertical lines represent standard errors. (From Shimoda, C., Yanagishima, N., Sakurai, A., and Tamura, S., *Plant Cell Physiol.*, 19, 513, 1978. With permission.)

later formed by the coalescence of primary aggregates. In liquid suspension, the coming together of cells to form mating-specific aggregates is due to random collision.[1205] Although it might be expected that only unbudded cells be found in aggregates since G1 arrest is a prerequisite to mating, it has been shown that the agglutinated mass contains cells at various stages of the cell cycle.[90,520] The cell-cell interaction being sex specific, and subsequent bilateral conjugation consummated with high fidelity,[1206] it may be presumed that the ratio of the complementary mating types in the aggregate is close to unity, despite the random process of coming together. Indeed, it is claimed that regardless of input ratio, the resulting agglutinated mass is made up of about equal numbers of **a** and **α** cells.[1159] Ratios of close to unity have been obtained for other yeast aggregation systems.[94,483] However, the invariability of the sex ratio reported in *S. cerevisiae* could suggest that pair formation precedes

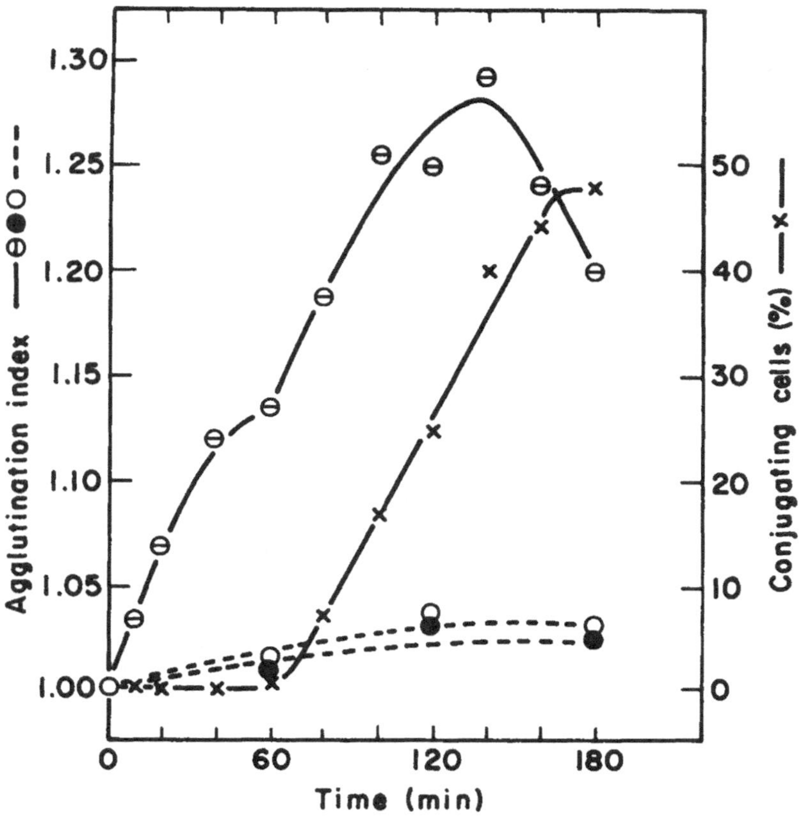

FIGURE 16. Time-course of sexual aggutination and conjugation of constitutive **a** and **α** cells. Symbols: ⊖, agglutination index of a mixture of **a** and **α** cells shaken in a growth medium at 26°C; X, percentage of conjugating cells in the mixture; ○ and ●, agglutination indices of separate cultures of **a** and **α** cells, respectively. No conjugants were observed in the separate cultures. (From Osumi, M., Shimoda, C., and Yanagishima, N., *Arch. Microbiol.*, 97, 27, 1974. With permission.)

mass agglutination, rather than occurring only among agglutinated cells, a surprising implication to which we shall come back shortly.

Yanagishima et al. have succeeded in releasing the substances responsible for sexual agglutination by treatment of isolated cell walls with snail-gut enzymes (Glusulase).[162,171,172] After Glusulase digestion, the walls of one mating type are boiled for 5 min and the supernatant is adsorbed onto cells of the complementary mating type. The eluate at elevated pH is further purified by gel filtration. The agglutination factors of either mating type do not cause agglutination: they are not agglutinins. Instead, they are assayed by their ability to inhibit the agglutination reaction by specifically inhibiting the agglutinability of the cells of the complementary mating type. They are thus presumed to be univalent. Both factors prepared in this manner are large glycoproteins (about 10^6 daltons), which are sensitive to proteolytic degradation, isolated or *in situ*.

Smaller agglutination molecules have been derived by a simpler method of releasing the activity from the cells. Both **a** and **α** agglutination factors can be solubilized by autoclaving for a few minutes cells in Tris buffer (pH 7.0) containing EDTA.[521,522] The extracts of

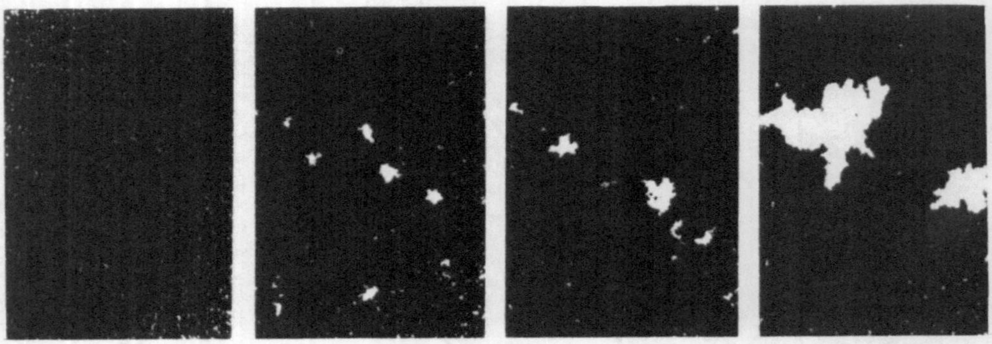

FIGURE 17. Microscopic visualization of the increase in aggregation number of a mixture of constitutive **a** and **α** cells. Mixture was shaken in growth medium at 26°C. (From Yanagishima, N., Yoshida, K., Hagiya, M., Kawanabe, Y., Shimoda, C., Sakurai, A., Tamura, S., and Osumi, M., in *Growth and Differentiation in Microorganisms*, Ishikawa, T., Maruyama, Y., and Matsumiya, H., Eds., University of Tokyo Press, Tokyo, 1977, 193. With permission.)

autoclaved cells are purified by ion-exchange chromatography on DEAE-cellulose and by affinity chromatography on Con-A Sepharose. The **a** agglutination substance is recoverable from **a** cells only and causes the loss of agglutinability of **α** cells, whereas the **α** agglutination substance is recoverable from **α** cells only and causes the loss of agglutinability of **a** cells. Diploid strains yield neither agglutination substance. The purified **a** agglutination substance (2.3×10^4 daltons) is much smaller than the **α** agglutination substance (1.3×10^5 daltons). The **a** molecule contains about 61% carbohydrate, the **α** molecule about 47%.

The isolated univalent substances are deduced to have been responsible for agglutination while *in situ*, because each is synthesized exclusively by one mating type and specifically adsorbed and inactivated by the other. They are presumed to be complementary molecules, a presumption that finds ample confirmation in the formation of a molecular complex in vitro.[521] The isolated substances are mixed together and the unreacted molecules are removed by adsorption to constitutive **a** and constitutive **α** cells. The added cells having been removed and the agglutination-inhibition activities of the mixture having been determined to have become nil, the complexes are broken apart into their complementary components by adjusting the pH of the mixture to 9.5. The respective activities may then be recovered and separated by column chromatography, by which it may be shown that they are found in those fractions where the free molecules were found in the first place during purification.

The functional site of the complementary agglutination substances is the cell surface, from which they are detached by the autoclaving of whole cells or by the Glusulase digestion of isolated cell walls. The complementary cells recognize each other by the specific interaction of their wall-bound agglutination substances. The site of synthesis is the cytoplasm, for these same autoclaved cells upon rupture serve as additional source of agglutination-inhibition activities.[1203]

As the time-course of agglutination shows, the extent of agglutination falls off shortly after the maximum is reached (Figure 16). The time of the decline corresponds to about the time when the maximal number of conjugants is first observed. As in *Hansenula wingei*, diploid cells of *S. cerevisiae* are not ordinarily agglutinable. Recently, it has been demonstrated that zygotes are self-agglutinative.[1203] Both **a** and **α** agglutination substances can be recovered from isolated zygotes. The first diploid bud from the zygote is also self-agglutinable, but subsequent diploid buds are less and less self-agglutinable.

Of what function is sexual agglutination in *S. cerevisiae*? The term, of course, already embodies a presumed function: sex. Still, it must be demonstrated that agglutination is a prerequisite, not merely a precedent, to the conjugation of **a** and **α** cells and the formation of zygotes. That agglutination precedes conjugation, there is little doubt. However, in the light of the finding that, irrespective of input ratio of complementary mating types, their ratio in the agglutinated mass is unity, the question must be asked whether pairing does not precede the formation of primary aggregates of 50 to 100 cells and the coalescence of primary aggregates. Or does pairing occur exclusively in the agglutinated mass? There are claims that specific pairing is an earlier event.[250,1205] On the other hand, it has been reported that pairs are rarely observed outside the aggregates.[1170] Furthermore, when stained **α** cells are added to a mixture of the mating types at a point halfway in the time-course of agglutination, the added cells are observed to be directly incorporated into the agglutinate. Zygotes are formed preferentially *en masse*: the small number of pairs found outside the agglutinated mass might have been detached by mechanical shear later in the mating reaction

Nonagglutinating mutants form zygotes at markedly decreased frequencies and mutants isolated as nonmaters are nonagglutinative.[520] In a very large collection of nonmater strains, more than half of the **α** isolates have defects in the production of **α** factor and all **a** mutants have lost the ability to respond to **α** factor.[1173,1199] Genetic analysis of some of these mutants reveal single mutational events in one of a number of loci, linked or unlinked to the mating-type locus. Upon further analysis, **a** mutants isolated for their lack of **α**-factor response in the form of shmooing or G1 arrest have turned out to be sterile or nonmaters.[1207] Four nonmating **α** mutants that do not show signs of agglutinability have been isolated.[1170] No **α** agglutination substance could be obtained from these mutants. Thus, it appears that the agglutination substances mediate the agglutination reaction. The issue regarding the function of agglutination, however, is by no means settled, but the enumerated pieces of evidence, plus the fact that physiological conditions restricting agglutination also restrict sexual activity, all these point to agglutination enhancing, if not a prerequisite to, the mating reaction.

Among proteases tested, alkaline and neutral proteases from *Bacillus subtilis*, Pronase, and chymotrypsin abolish the agglutinability of both constitutive **a** and **α** cells.[171,253,1170] Acid protease from *Rhizopus chinensis* and trypsin inactivate **α** cells but not **a** cells, whereas mercaptoethanol or cysteine destroys the agglutinability of **a** cells but not of **α** cells. The effect of proteases and sulfhydryl compounds on the isolated agglutination substances is the same as on the agglutination of whole cells, that is, both **a** and **α** agglutination substances (**a**-AS and **α**-AS) are inactivated by neutral protease from *B. subtilis*, but only **α**-AS is inactivated by trypsin, whereas only **a**-AS is sensitive to mercaptoethanol or cysteine. These results suggest that, whereas both **a**-AS and **α**-AS are proteinaceous, only the agglutination substance from **a** cells contains reducible disulfide linkages.

How do the agglutination glycoproteins bring about agglutination? Anchored at one end to the cell surface, they may mediate cell-cell interaction by serving as recognitive and binding sites (Figure 18). These substances are already present on the walls of constitutive cells prior to mixing. When cells of complementary mating types are mixed, the complementary agglutination substances on their respective surfaces recognize, and bind to, each other. Although the agglutination substances themselves are univalent, the cells are presumed to be multivalent, hence the three-dimensional cell-cell interaction (shown as two-dimensional in the diagram). When the isolated **α**-AS is added to a suspension of constitutive **a** cells, the agglutinability of the cells is masked, for the molecules of **α**-AS cover the reactive sites of the **a**-AS structures attached to the cell surface. The agglutinability of **α** cells is likewise masked, when isolated **a**-AS is added to constitutive **α** cells. The complementarity of the agglutination substances is shown by the formation of a specific molecular complex *sans* cells, by agglutination in the presence of complementary constitutive cells, and by the masking of agglutinability by the isolated substances. In a diploid cell, the agglutination substances somehow are not synthesized or not active.

In inducible cells (Figure 18), the agglutination substances on the walls are initially absent. When **a** cells come in contact with **α** factor (which is excreted by **α** cells), they are induced to synthesize **a**-AS molecules, which are then assembled on the walls, thus rendering **a** cells competent to agglutinate with **α** cells. In turn, **α** cells coming in contact with **a** factor are induced to synthesize **α**-AS molecules and assemble them on their walls, thus bringing about their competence to agglutinate with **a** cells. When these induced cells are mixed together or with constitutive cells of the complementary mating type, agglutination takes place.

That the effect of the pheromone of one mating type is the induced synthesis of the agglutination substance of the other mating type has been confirmed.[253] When inducible **a** cells are incubated with **α**-substance-I (**α** factor), **a** agglutination substance is recoverable from Glusulase digests of their cell walls within 2 hr. No detectable activity is observed when **α** factor is not added or when cycloheximide is added with the pheromone. Incubated with **α** factor, constitutive **a** cells are observed to increase their amount of recoverable **a**-AS activity. Such an increase is not observed in the presence of cycloheximide. It then appears that the pheromone, aside from causing G1 arrest and shmooing, induces the synthesis of the complementary agglutination substance in inducible cells and enhances further synthesis in constitutive cells.

The induction of the synthesis of agglutination substances requires energy, a fermentable sugar such as glucose, and a nitrogen source.[253] Because induction is inhibited by cycloheximide, cytoplasmic protein synthesis appears to be required. However, it is not yet possible to tell whether the required protein synthesis is the synthesis of the proteinaceous agglutination substances. It is possible that the assembly of preformed structures on the cell surface requires protein synthesis or merely that newly synthesized enzymes are necessary to activate preassembled structures.

Cultural conditions influence the sexual agglutinability of constitutive cells. Extreme aeration of a culture impairs the development of agglutinability.[254] Cells in stationary phase are less agglutinative than those in exponential phase. Those grown in glucose as carbon source are more agglutinative than those grown in glycerol. Indeed, constitutivity with regard to agglutinability of a strain does not denote that the agglutinative activity is always present, that its development is not repressible. It simply means that the activity may be demonstrated to be present even before contact with the cells or the factor of the complementary mating type.

Elevated temperatures during growth do not allow the development of agglutinability of constitutive cells.[141,142,254,1203] A survey of strains from various sources reveals that the thermosensitivity of the development of agglutinability is the rule rather than the exception. Of 31 strains constitutively agglutinative at 25°C, only two were found agglutinative when grown at 37°C. Loss in agglutinability at the restrictive temperature could be due to a loss in the capability of the cells to synthesize the agglutinative substances, because the machinery

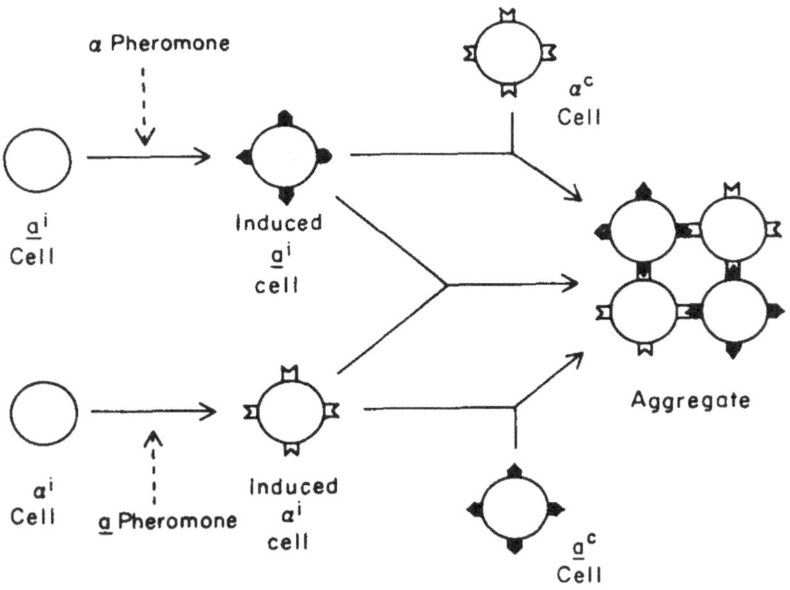

FIGURE 18. Proposed mechanism for sexual agglutination in *Saccharomyces cerevisiae*. Symbols: $\mathbf{a}^c$ and α^c, constitutively agglutinative **a** and α cells, respectively; $\mathbf{a}^i$ and α^i, inducible **a** and α cells, respectively. Complementary agglutination substances are shown as wall appendages: **a** agglutination substance is a black wedge; α agglutination substance is a V-shaped socket. Further explanation is found in the text. (From Yanagishima, N., *Bot. Mag. Spec. Issue*, 1, 61, 1978. With permission.)

for synthesis is thermosensitive or because the cells have simply lost their constitutive capacity; in other words, they might have become inducible. The loss is reversible. When the cells rendered nonagglutinative at restrictive temperature are incubated with live cells of the complementary mating type, they recover their agglutinability even at the restrictive temperature. Recovery is not achieved in the presence of heat-killed cells. Hence, constitutive cells, at the restrictive temperature, lose their constitutive agglutinability and become inducible.

The time-course of the loss in agglutinability during growth at the restrictive temperature indicates a fairly rapid conversion from constitutivity to inducibility, suggesting that the conversion is not entirely ascribable to dilution by population growth.[142] The reverse conversion at the permissive temperature, from inducibility to constitutivity, is more gradual. There is genetic evidence to suggest that the conversion is governed by a gene closely linked to the mating-type locus. More recent evidence points to a concomitant loss of agglutination substances during loss of agglutinability at the restrictive temperature.[1203] In neither mural nor cytoplasmic fractions could agglutination substances be recovered when cells are grown at a temperature greater than 35°C. In contrast, the sex pheromones and the substance that binds α factor are synthesized even at 36°C. The repression of the development of agglutinability and the synthesis of the agglutination substances are reversible even at 36°C by the addition of complementary pheromone to the culture. Thus, the thermosensitive event, perhaps regulated by mating-type genes, is not necessarily the synthesis of agglutination substances, but the maintenance of the state of being constitutive. The induction of sex-directed flocculation in fission yeast, as well as subsequent events in that developmental system, has been observed to be thermosensitive.[143]

Other components of sexual agglutination in *S. cerevisiae* have been identified. A factor which irreversibly destroys sexual agglutinability of constitutive **a** cells specifically has been reported.[1208] Whereas its target is sex specific, its source is not. The heat-sensitive cell-free activity is found not only in both **a** and **α** haploid cells but in diploid cells as well. It is thought to be a protein, a proteinase, perhaps.[1204] A substance that binds **α** factor has been extracted from **a** cells.[1190] The binding activity is recovered from both constitutive and inducible **a** cells boiled for 5 min. Purified by elution with a linear gradient of NaCl on DEAE-cellulose, the isolated fraction binds **α** factor and inactivates it. The proteinaceous binding substance could well serve as the receptor for the pheromone. The mating signal from an **α** cell, **α** factor is received on the cell surface of an **a** cell by the binding substance. After recognition and processing of the signal, the receiving cell initiates the process of becoming competent by synthesizing its agglutination substance. However, because the binding substance is claimed to be excreted by **a** cells into the culture medium, it may in effect be similar to what is called the barrier factor.[1209]

As already mentioned, when **a** cells are placed near a colony of **α** cells, they form shmoos. A barrier of **a** cells placed between inhibits the shmooing reaction.[1209] The barrier phenomenon is explained by assuming that **a** cells synthesize constitutively a diffusible substance that binds or inactivates **α** factor or, perhaps, renders **a** cells refractory to the pheromone. The barrier factor is not likely to be the same molecular species as **a** factor, for mutants that lack barrier-factor activity nonetheless possess **a** factor, even as mutants that exhibit the barrier phenomenon may be deficient in **a** factor. The putative binding substance of **a** cells could act directly by removing **α** factor from the medium.[1191] The recovery of **a** cells from G1 arrest due to **α** factor indicates that a mechanism exists for inactivating the mating signal. Recovered cells resume growth and cell division. They become refractory to **α** factor for some time until further division has taken place. Cells sensitive to **α** factor, such as **a** haploid cells, can inactivate the pheromone, but nonmating **a** mutants and the **a/α** diplonts cannot. The inactivation mechanism appears to depend on cellular contact, for the cell-free culture medium of **a** cells fails to inactive **α** factor.

It is but expected that cells capable of receiving the mating signal be capable of processing that signal in such a way as to terminate it when it is no longer needed. The molecular signal may be bound, taken up, inactivated, and metabolized. There are convincing examples of signal reception in other microbial sexual systems. The male gametes of the watermolds, *Achlya* and *Allomyces*, do inactivate the sex hormones generated by the female gametes.[494,495,1210-1212]

Whereas the agglutination substances are assumed to reside on the cell surface, there is no way of telling at the moment where exactly they are located. On the walls of conjugating cells, surface fuzz has been observed.[101] It is discernible as an electron-dense material between two walls in the process of fusion. Because fimbria-like structures have been implicated in the apparently nonsexual flocculation of brewers' yeast, in sexual agglutination in *Hanseula wingei*, and in sex-directed flocculation in fission yeast,[121,247,516,541-544] it might well be that in *S. cerevisiae*, the agglutination substances are located in similar structures. Because the cell walls contain a major amount of mannan,[1213] and the isolated agglutination substances are presumed to be wall-bound glycoproteins,[172,521,522] it is interesting that the lectin concanavalin A, at concentrations that hardly affect zygote formation, inhibits sexual agglutination.[1214] Contrasted to the agglutination substances, which apparently inhibit the initial formation of aggregates consisting of no more than 50 cells each, univalent concanavalin A seems to interfere with the subsequent coalescence of these primary aggregates. This observation might be taken to mean that the rest of the carbohydrate surface, although not the site of the agglutination substances, may contribute to the efficiency of sexual agglutination. Mere steric hindrance, however, cannot be ruled out. At concentrations that completely inhibit conjugation, 2-deoxyglucose hardly affects the development of agglutinability.[100] In a general way, the glucose analog may interfere with mural synthesis. A structurally defective wall may serve as a poor site for cell fusion, but may not disturb the assembly of the agglutination molecules.

The morphological changes that accompany the mating reaction in *S. cerevisiae* are qualitatively similar to those that occur in *Hansenula wingei* and the fission yeast.[121,1127,1141] There are variations, of course, but they are slight. For instance, the shmooing response seems to put gross mural extension temporally ahead of sexual agglutination, at least in certain situations, such as when complementary cells are not allowed immediate cellular contact. The mural activity following the mating signals leads to an alteration of the cell wall. Morphological studies show that cells in contact, not only entrap the electron-dense material between them, but also extend themselves toward each other in the region of contact.[101,1214] The result is rudimentary conjugation tubes on both sides of what has now become a cross-wall between the conjugating cells. Thinning of the walls at the region of contact starts at about this time. Mural fusion also must have occurred before further erosion of the two-layer cross-wall. The eventual dissolution of the cross-wall results in cytoplasmic fusion and the formation of a dikaryon. With the cross-wall barrier between them gone, the nuclei are free to do their own thing: nuclear fusion, the formation of a zygote. The transient dikaryon is now a monokaryon, in fact, a diplont. Mitosis follows soon after. The first diploid bud arises, after mitosis, from the neck of what has been once a pair of haploid, complementary cells. Subsequent diploid buds may arise from either pole.

Although effective procedures for mass mating have been around for some time now,[126,147,249-251,1206,1216] the biochemical aspect of the mating process, beyond signal generation and reception and sexual agglutination, remains relatively unexplored. The synthesis of nuclear DNA, of course, is arrested, but a complex and protracted sequence of events as the mating process, doubtless requires RNA and protein synthesis. In temperature-sensitive mutants defective in RNA or protein synthesis, but not in mutants defective in DNA synthesis, conjugation is impaired at the nonpermissive temperature.[1177] There is evidence that mitochondrial protein synthesis is not required.[69] Conjugation does not appear to be subject to catabolite repression.[147,251,1217] In fact, exogenous glucose (up to an optimum of 10%) enhances zygote formation. So do inorganic salts, peptone, yeast extract, and aeration. Aerobic growth or mitochondrial function, however, may be dispensed with, for respiratory-deficient strains are capable of mating.[1218] An optimum hydrogen-ion concentration of pH 4.5 has been reported.[1204]

There is evidence indicating an increase in autolytic activity during conjugation.[1219] A greater amount of protein and sugar can be recovered from the cell-free medium of a mixture of **a** and **α** cells than from the media of separate cultures. Conjugation-induced lysis, also reported in fission yeast,[660] is further suggested in *S. cerevisiae* by the difference between zygote counts determined microscopically and zygote-colony counts of an auxotrophic mating mixture plated on minimal medium.[251] The synthesis of glucanases, however, has recently been shown to remain steady during conjugation, only to surge when the first diploid buds appear, long past the sexual process.[1220] A study of the mating reaction in protoplasts concludes that mating and the morphological changes caused by pheromones are dependent on the integrity of the cell wall.[1221] The mural lytic event during cell fusion takes place only after intimate intercellular contact.

Sexual interaction in *S. cerevisiae* is genetically governed by the mating-type locus *MAT*. The locus is found on chromosome III, some 20 map units to the right of the centromere.[181,182,562,563] It consists of two alleles: the genotype *MAT***a** determines the phenotypic behavior of a haploid **a** cell, the genotype *MAT***α** the phenotype of a haploid **α** cell. The diplont resulting from the fusion of the sexually complementary **a** and **α** cells is represented by the genotypic notation *MAT***a**/*MAT***α**.

Heterothallic (*ho*) strains are stable in the laboratory; they stay haploid and they retain their mating-type designation. If and when they change their mating type, it is at a frequency compatible with that expected of common mutations (about 10^{-6}).[1222,1223] In contrast, there are the homothallic strains (*HO*), which cannot be maintained as haplonts for long. They tend to diploidize with little or no provocation. Their instability as haplonts is due to frequent interconversion of one mating type to the other.[565-570,1223] A haploid spore of mating-type **a** or **α** gives rise after two generations to four progeny, two of which may be of the complementary mating type. If permitted to fuse, the progeny will produce stable diplonts, whose descendants will remain diplonts until they are allowed to sporulate into haploid spores. Diploid cells are nonmaters. They are ordinarily not sexually agglutinative. Hence, homothallic strains, unlike heterothallic ones, are not sexually agglutinative because they tend to diploidize readily. The resultant diplonts and their diploid descendants soon outgrow and replace the haplonts in the population. Provisions for mass mating would be gratuitous when incestuous mating is freely available.

In addition to the *HO* gene that determines whether a strain is homothallic or heterothallic, there are the genes *HML* and *HMR*. Together, they promote the interconversion of mating types.[565-570,1223-1228] The gene *HO* (its recessive allele is designated as *ho*) controls the switch that brings about alterations of the *MAT* locus from *MAT***a** to *MAT***α** and vice versa. For the maintenance of the altered mating type, *HO* (which is not linked to *MAT* but remains unmapped to date) is not required. The loci *HML* (formerly called *HM***a**)and *HMR* (formerly called *MH***α**) are loosely linked to *MAT* on chromosome III, approximately 64 cM and 65 cM to the left and to the right of the centromere, respectively.[1226] The *HML* locus may exist in one of two co-dominant alternative states: *HML***a** and *HML***α**; in like manner, *HMR* may be represented either as *HMR***a** or as *HMR***α**.[1128] The activation of *MAT***a** requires the presence of either *HML***a** or *HMR***a**. For the activation of *MAT***α**, either *HML***α** or *HMR***α** must be present. Most heterothallic strains in the laboratory have the genetic constitution *HML***α** *HMR***a** *ho*.[1229]

Sex interconversion in *S. cerevisiae* and related isolates is generating much interest lately but, alas, also much confusion. The plasticity of the taxon has not helped any in making the literature easy reading, nor have the continual changes in the genetic nomenclature, not to mention the inconvenience of the **a**-and-**α** notation for mating types. The rapid progress in the field, however, appears to shed more light now so as to dispel the confusion that usually precedes simplification.

The homothallic nature of some strains of *Saccharomyces* was evident to early workers.[1164,1165] Homothallic strains grown from a single haploid spore generate diploid cells that can sporulate. In contrast, heterothallic strains, of one mating type, do not sporulate, unless first allowed to mate with the cells of the other mating type. Various genes responsible for the state of homothallism in various strains and species of *Saccharomyces* have been described in the literature.[565-569,1224,1230-1235] These homothallism genes are found in *S. cerevisiae* by way of crosses with other species in the genus or by way of mutations from heterothallic strains.[1224,1130] The more intensively studied of these are the dominant genes *HO, HML***α**, and *HMR***a**.[567-569,1234-1236] They bring about sex interconversion (from **a** to **α** and from **α** to **a**) by causing a heritable change in the mating-type locus, perhaps, as proposed by Oshima and Takano, via a controlling element that can reversibly interact with the mating-type locus.[568,1235] The genes *HM***α** (now called *HMR***a**) and *HM***a** (now called *HML***α**) could produce the sex-specific controlling element. The mating-locus *MAT* on chromosome III could serve as the site of affinity for the controlling element arising from either side of the same chromosome. When the **a** controlling element from the right (from *HMR***a**) binds to *MAT*, the **a** mating-type allele is expressed. Similarly, the **α** mating-type allele is expressed, when the **α** controlling element from the left (from *HML***α**) is bound. The gene product of *HO* probably catalyzes the reversible binding. This model, as we shall soon see, has led to the more current concepts of sex interconversion.

Using the shmooing reaction of **a** cells to **α** factor, Hicks and Herskowitz have been able to demonstrate the action of the *HO* gene by single-cell pedigree analysis, without losing the analyzed cells to the mating process.[1223] Descendants of homothallic spores are individually tested for their mating type by their sensitivity to **α** factor. By this elegant technique, they have shown that sex interconversion may occur as often as every cell division. Thus, from a single haploid spore may be derived a clone of four cells, two of which are of the mating type different from that of the original spore cell (Figure 19). Because a daughter is easily distinguishable from its parent, it can be deduced that the original spore cell (*S*) in the diagram switched at the two-cell stage of the growing clone, while the daughter (*D1*) maintained its mating type. At the four-cell stage, *S* may switch again to recover the original mating type, even as *D1* switches for the first time and *D2* (the second daughter) and *D1-1* (the first daughter of the first daughter) stay put.

A number of switching rules have been derived.[1225] Experienced cells (those that have budded at least once) are capable of switching sometime during the next cell cycle, though they need not switch every cell cycle. Naive or inexperienced cells (spores and buds, which have not ever budded) rarely undergo sex conversion. The frequency of switching from **a** to **α** and that from **α** to **a** are approximately the same (0.71 vs. 0.73 per cell division, respectively). The frequency of switching being sufficiently greater, in either case, than 50%, the switch must be directed, rather than the result of a random choice between alternative states. In addition, it is noteworthy that cells which have undergone sex conversion come in pairs, that is, mother and daughter are invariably of the same type. This observation suggests that the change in mating type occurs prior to DNA replication in the cell cycle.

The fact that cells in a clone can switch with apparent ease from one mating type to the other and back again indicates that cells have the capacity, and the information, to be **a** or **α**. At any time, however, a particular cell is either **a** or **α**, but not both. A defective *MAT***α** in a sterile strain (*mat***α**$^-$) can convert to a functional *MAT***a**, which later gets converted to a functional *MAT***α**.[1238] The recovery of *MAT***α** function from the defective *mat***α**$^-$ within as short a time as three generations suggests the prior existence of *MAT***α** information that is silent (not expressed) but nonetheless present somewhere in the genome and just waiting to be activated.

These observations on homothallic behavior have been generalized by Hicks et al. into what is called the cassette model of mating-type interconversion.[570] Obviously inspired by the controlling-element model of Oshima and Takano,[568] they propose that the mating type of a given cell is the expression of a silent copy (a cassette) of the mating-type information transposed and inserted into the mating-type locus *MAT* by the action of the *HO* gene (Figure 20). The **α** information resides in *HML***α**, the **a** information in *HMR***a**. The cassettes, which may come in the form of transposable elements or jumping genes, may be thought of as no more than prerecorded tapes informative only when inserted into the playing machine, the *MAT* locus.

Recent findings offer abundant support to the cassette model. The healing of mating-type mutations have been found to be a general property of sex interconversion.[1239] The process of interconversion allows the recovery of either *MAT***a** or *MAT***α** functions from a variety of *mat***α**$^-$ and *mat***a**$^-$ mutations. Conversely, mutations at one of the loci that furnish the silent copies generate defective mating-type alleles, suggesting the activation of *MAT* by transposition.[1228,1240,1241] There is genetic evidence also for the activation of the silent information at *HML* and *HMR* by chromosomal rearrangements.[1241] The model predicts that the *MAT***a** and *MAT***α** alleles are nonhomologous blocks of DNA and that copies of each should be found to the left and to the right, respectively, of *MAT*, where *HML* and *HMR* are located. A functional copy of the **α** gene has been cloned by transformation in yeast.[1228] DNA sequences homologous to the cloned fragment used as probe have been located at chromosomal sites corresponding to *MAT*, *HML*, and *HMR*. Although the complementary alleles at the three loci show partial homology, restriction fragments of the **α** allele at any of the loci are characteristically shorter than those of the **a** allele. This difference implicates

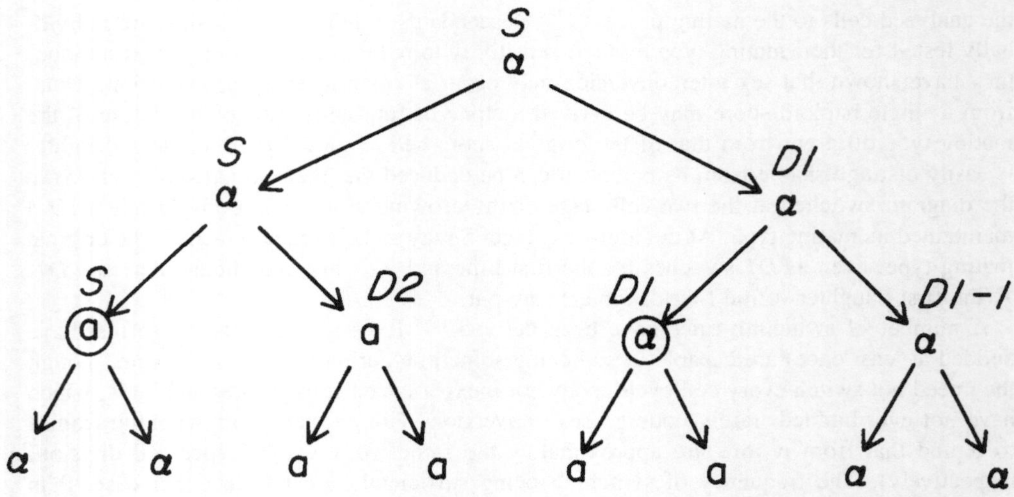

FIGURE 19. Sex interconversion in a homothallic strain of *Saccharomyces cerevisiae*. Symbols: *S*, the original spore cell; *D1*, first daughter; *D2*, second daughter; *D1-1*, the first daughter of the first daughter. Circled cells are capable of switching. (Adapted from Herskowitz et al.[1237])

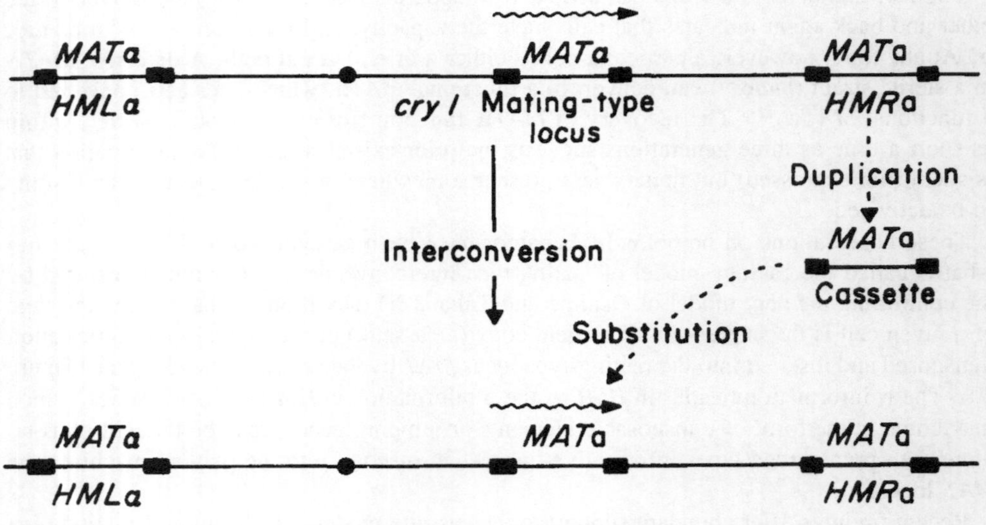

FIGURE 20. The cassette model of mating-type interconversion. Adapted from Strathern et al.[1239] Top line is chromosome III of an **α** cell, bottom line that of an **a** cell. The centromere is represented as a dot on the line. *MAT***α** and *MAT***a** are cassettes containing the information that determines mating types **α** and **a**, respectively. They are located at *HML*, *HMR*, and the mating-type locus. Expression of the cassette information at the mating-type locus is represented by a wavy arrow. The information at *HML* and *HMR* is not expressed. Interconversion from **α** to **a** occurs when a duplication of *MAT***a** from *HMR* is inserted into the mating-type locus. The marker *cryl* (about 4 cM from the mating-type locus) stands for cryptopleurine sensitivity.

DNA rearrangement at *MAT* during sex interconversion, as implied in the cassette model. A mutation influencing the efficiency of sex interconversion has been identified.[1242] Called *swi*1-1, the centromere-linked mutation appears to be nonlinked to *MAT* or to other known homothallic genes. It reduces the switching frequency of strains bearing *HO* to less than 10% of the wild-type switching frequency. The phenotype is detected as colonies comprising nonmating diploid cells and equal numbers of **a** and **α** haploid cells.

Heterothallic strains may be viewed as homothallic strains defective with respect to sex interconversion. They rarely give up their mating type, maintaining it for many, many generations. As a consequence, a heterothallic clone of either mating type may remain sexless, until it comes in contact with another clone of the complementary mating type. When that happens, sexual agglutination occurs. Like that in *H. wingei*, sexual agglutination in heterothallic strains of *S. cerevisiae* is a helpful adjunct to the sexuality of haploid cells. It effectively leads to a higher ploidy, since the diploid cells are stable, unlike those of *Schizosaccharomyces pombe*. The diploid cells may elect to sporulate when the conditions are right.

Because of sex interconversion, homothallic strains tend to diploidize readily. In the resultant diploid, heterozygous at *MAT* (**a**/**α**), the mechanism for sex interconversion is turned off, although it remains operative in homozygous diplonts (**a**/**a** and **α**/**α**).[1243-1245] Unlike the homozygous strains, however, **a**/**α** diploid cells can undergo meiosis and sporulate.[515,1246] They exhibit increased survival to X-irradiation and increased frequency of mitotic recombination after UV-irradiation. But with rare exceptions, **a**/**α** diplonts are nonmaters. In contrast, **a**/**a** and **α**/**α** diplonts display the same mating reaction as their corresponding haplonts.[1246,1247] The **a**/**a** diplonts, for instance, respond to **α** factor produced by **α** haplonts or **α**/**α** diplonts.[72] Except for the original zygote formed by the union of **a** and **α** haploid cells in the agglutinated mass, as well as the first zygotic buds, **a**/**α** diplonts normally are not agglutinative among themselves or with haploid cells.[126,1203] Nor do they need to be. But nonmaters, such as certain strains used by the brewers, may be flocculent, as the brewery people describe yeasts that tend to engage in apparently nonsexual agglutination. The overall effect of sex interconversion in homothallic haplonts and of sexual agglutination in heterothallic haplonts is the facilitation of conjugation, the creation, from two haploid cells, of a diploid cell capable of reproducing as a diplont and, if its immediate environment demands, of converting itself into four haploid spores.

V. FLOCCULATION OF BREWERS' YEAST

By brewers' yeast, we mean principally two taxonomic groups of primary importance to the ancient tradition that is beer making. Strains of the first taxon, *Saccharomyces cerevisiae*, are known as top-fermenters. They make the stout or the ale that goes with cakes. Strains of the second, *S. uvarum (carlsbergensis)*, are bottom-fermenters. They make the lager that goes with pretzels and oompah-pah. Their taxonomic propriety as a separate entity is a matter of contention.

Toward the end of fermentation, the yeast that makes the beer tends to aggregate in masses of many thousands of cells. A good brewers' yeast stays homogeneously disperse during much of the period of fermentation, monitored as a continuous decline in the specific gravity of the wort (Figure 21). When its metabolic role is about finished, it forms cell aggregates and, as if taking a cue from the brewmaster, separates out of suspension. To the brewmaster, that means racking time is just around the corner. Lager yeasts fall to the bottom, ale yeasts rise with the sparkle to the top. Brewers call this aggregation phenomenon flocculation (from the Latin "floccus", a tuft of wool) and the aggregates flocs. The terms are quite apt. In some flocculent strains, flocs in fact look like tufts of wool. Nonflocculent yeasts are described as powdery.

Flocculation is obviously of great practical concern to the brewer. The course of fermentation depends on the concentration of active cells in suspension.[92,93,103,1248-1250] Through the cell surface pass the sugars that are fermented intracellularly into alcohol that eventually diffuses out. A floc has an effective surface that is considerably less than the collective surface of monodisperse cells. Moreover, it is effectively removed from the bulk of nutrients in the wort. As a consequence, early flocculation leads to incomplete fermentation or, in

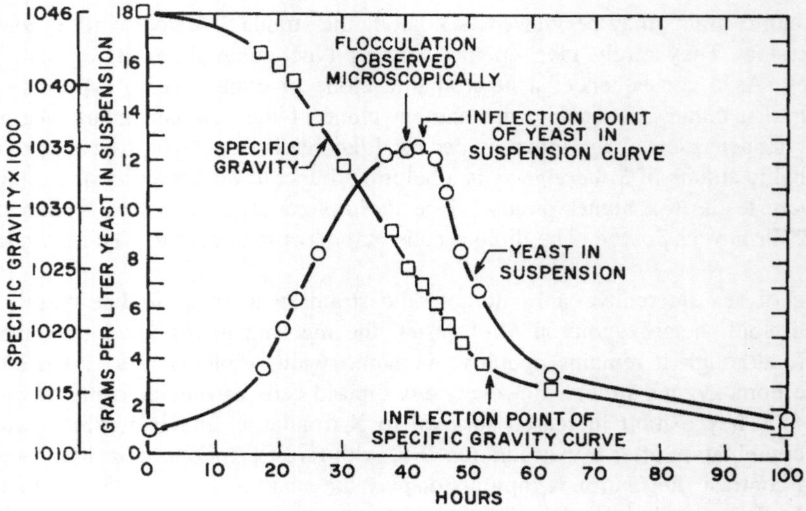

FIGURE 21. The onset of flocculation during fermentation of wort under simulated-brewery conditions. Note that the appearance of the first flocs, by microscopic inspection, precedes the fall of concentration of yeast in suspension. Class II yeast was used to ferment brewery worts of original gravity 1046. (From Gilliland, R. B., *Wallerstein Lab. Commun.*, 20, 41, 1957. With permission.)

the language of the brewery, reduced attenuation. That means the brew has a specific gravity higher than desired, contains less alcohol, and is sweeter.

On the other hand, the property of a strain to separate cleanly from the fermented wort, once fermentation is over, is of enormous value. There would be fewer yeast cells in the green beer at the time of racking, which may then be started earlier. Secondary fermentation would be less vigorous, but sustained by a higher concentration of left-over substrate. The beer would be freer from off-flavors, brighter. A bigger crop of yeast can be recovered, more conveniently. A minor disadvantage is that the richer brew that contains fewer cells could promote unwanted contaminants. Another is that there may not be a sufficient number of cells to carry on secondary fermentation. Clearly, the perfect yeast must be monodisperse during the course of fermentation until such time predetermined by the brewmaster. It does not exist, nor does the perfect beer.

Much work has gone into trying to manage and, only incidentally, understand flocculation in the brewery. Of course, the perfect beer is perfect profit. Although not the first microbial aggregation system to be discovered, flocculation of brewers' yeast was first to be studied with great vigor, and with great naiveté, from which other systems have abundantly benefited. Pasteur described it in his classic *Etudes sur la Bière* in 1876, though he did not pursue the matter any further.[46] The earliest investigations were reported from the breweries of Germany. Seyffert, in 1896, ascribed the loss of flocculence of a pure culture to the softness of the water in the brew.[47] When lime was added to the brewing liquor, the culture regained its flocculence. In 1897, Kusserow reported that peptones added to wort brought about flocculation and that asparagine inhibited it.[48] As is wont to happen, a dissenting opinion was raised 2 years later by Lange, who found no such difference.[49] He maintained that peptones, which precipitate in alcohol, merely drag the yeast down to the bottom. And thus, before the end of the last century, the lively and conflicting literature on the flocculation of brewers' yeast had begun to accumulate.

A survey of the early literature has been made by Jansen, from whose review the brief historical account here is derived.[1251] Many of the first efforts were devoted to describing the influence of wort components, including salts and growth-promoting substances, on flocculation.[47-53,103,104,1252-1257] Another early preoccupation was the treatment of yeast flocculation as a colloidal phenomenon, obviously an aftermath of the success of colloid science, then in style.[149-160] Very little can be gained from reading the early literature. Much of it is only of historical interest. However, no less important than pointing out what is right, the early papers collectively show, without even trying, why things then thought to be right are wrong. Glaring examples are the fruitless approach through colloidal theory and the attempt to explain flocculation in the brewery as the consequence of contamination by bacteria or by wild yeasts.[54,158,1258] These outmoded theories are by no means dead, but they have very few living champions. Knowing what is fruitless can be fruitful — no small beer, that.

After more than half a century of flocculation studies that seemed to lose strength, a meeting of the European Brewery Convention in Brighton in 1951 brought a new surge in interest. This meeting, over beer and flocs, is a convenient watershed. Papers before Brighton may now be condescendingly labeled early, papers after may be presumed to make up the modern view. The published proceedings of this meeting have been reviewed by Comrie.[1259] Many reviews since then have served to monitor the progress of research on the flocculation of brewers' yeast.[9,87,1248,1260-1270]

Several points of clarification concerning terminology are in order. Flocculation of brewers' yeast, as we define it here and as defined by the Congress of the European Brewery Convention of 1958, must be limited to the reversible aggregation of cells. Chain formation, the consequence of incomplete separation of parent and bud after cell division, is out. Sedimentation is also out, both for reasons already discussed in Chapter 1. Too, all yeasts, flocculent or not, or for that matter all particles with dimensions similar to a yeast cell, form sediments sooner or later. Rapid separation, on the other hand, is the consequence of cell aggregation. Whether it is observed as a head (in ale strains) or as a deposit (also called sludge or sediment, in lager strains) does not matter much.[1268] Either way, the effective separation of cells from wort is about the same: the aggregation of cells has already occurred prior to gross separation. A head is formed when flocs entrap gas bubbles and rise to the surface of the brew. To some extent, even single cells may be carried by bubbles to the top, but only when flocculation is allowed to occur may the head remain stable.[1259] Flocs of highly flocculent ale yeasts, however, do coalesce and sink to the bottom of the fermenting vat.[91,1248,1262,1271] Head formation by ale strains is also called barming.[1272] It must be distinguished from the formation of a surface film; neither may be called flocculation. In the literature, in older papers especially, flocculation may be referred to as clumping, agglutination, agglomeration, settling, sedimentation, flocculence, and by various other names, including flotation and coagulation. Flocculence must be restricted to the property of a culture or a strain to form flocs. What constitutes flocculation is the intercellular process of coming and sticking together. It must be reversible in that a floc may be dispersed into free cells and the freed cells may then be made to flocculate again.

Most of the early knowledge about flocculation of brewers' yeast derives from work with bottom-fermenters, but top-fermenters have caught up lately. Understandably, much of the work has been carried out under brewery or simulated-brewery conditions. The early hopes for an easy management of flocculation, for it was then believed to be a simple phenomenon, gradually faded when it became apparent that flocculation is after all a very complicated system. Wort itself is complex enough, much more so the yeast cell.

In the brewery, flocculation normally occurs when fermentation is about complete (Figure 22A). At or near the end of fermentation, just short of a week from pitching, the bulk of the sugar has been used up, the temperature of the brew has gone down a few degrees, the hydrogen-ion concentration has reached a pH minimum, and the specific gravity has settled to about the desired value. When a flocculent yeast is used, the turbidity of the brew suddenly drops, marking the start of flocculation. Microscopic observation of samples from the middle

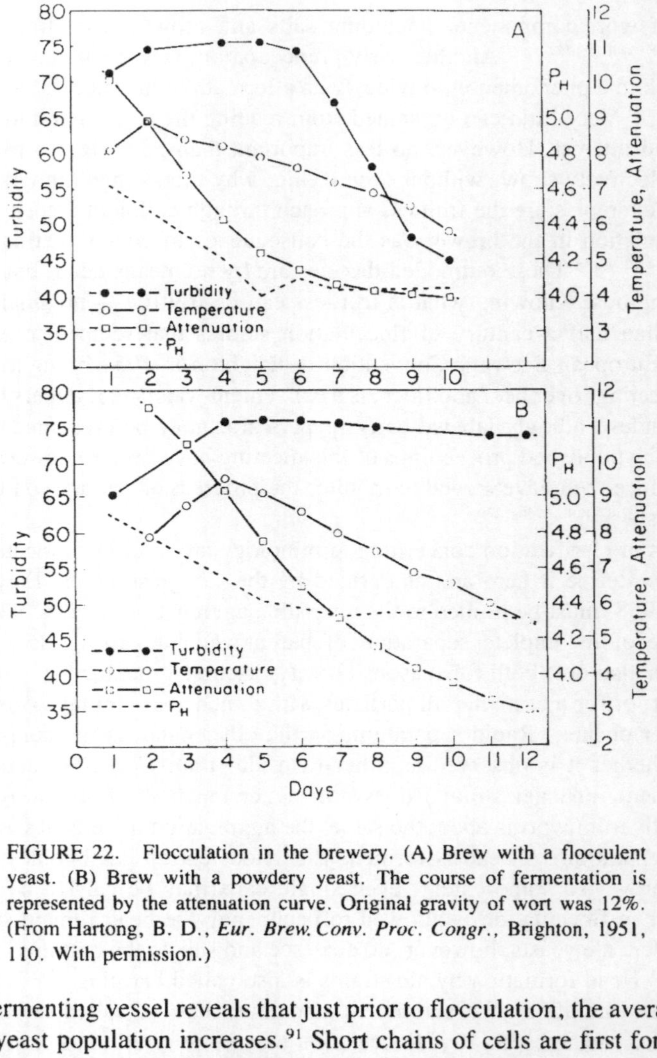

FIGURE 22. Flocculation in the brewery. (A) Brew with a flocculent yeast. (B) Brew with a powdery yeast. The course of fermentation is represented by the attenuation curve. Original gravity of wort was 12%. (From Hartong, B. D., *Eur. Brew. Conv. Proc. Congr.*, Brighton, 1951, 110. With permission.)

portion of the fermenting vessel reveals that just prior to flocculation, the average aggregation number of the yeast population increases.[91] Short chains of cells are first formed and maintained, later coalescing into loose clumps of 100 or more cells. When the clumps sink to the bottom or rise to the top, the amount of yeast in suspension (monitored as turbidity or mass) drops. In a brew with powdery yeast, such a drop in turbidity does not happen (Figure 22B).

At the onset of flocculation, cells have stopped growing and dividing. In effect, they have reached stationary phase of growth.[1273] It is well known that growth-promoting substances in the wort inhibit flocculation.[103,104,262] Indeed, flocculated cells disperse when introduced into fresh medium. Conditions that promote growth do not, as a rule, permit flocculation to take place.

There are those who argue that the end of fermentation brings about flocculation. Balderdash, it is more likely, others say, that the converse is true: flocculation ushers in the end of fermentation. Redispersion of flocs in the fermented wort allows further attenuation.[155] There is no real controversy, however, for both dogmatic views are neither completely right nor completely wrong, as most dogmatic views are. Neither flocculation nor the end of fermentation needs the other, although either can bring about the other. They are separable events. Flocculation can occur even before the end of fermentation; the end of fermentation may come, as it has to come somehow, without flocculation. Exactly when flocculation occurs depends on the strain and the brewing conditions.

Strains of brewers' yeast have been classified a number of ways according to their flocculating properties.[79,91,278,1248,1262] Gilliland's old system of classification, still most useful, recognizes four classes.[1262] Class I gathers all nonflocculent strains together. Their cells remain homogeneously disperse (or in clumps of less than ten cells) through all the stages of fermentation. In general, nonflocculent strains fail to meet the standards of the brewery. They make a poor beer that is hazy and yeasty, yield a poor crop of cells, and leave insufficient nutrients in the brew for secondary fermentation. Class II comprises the standard ale strains. Completely disperse initially, they form flocs of up to a thousand cells toward the end of fermentation (Figure 21). They are good and rapid fermenters, separate well from the brew, attain proper and consistent attenuation, and allow for good secondary fermentation. Highly flocculent yeasts make up class III. Like those of class II, they are completely disperse initially, but their extra flocculence, manifested toward the end of fermentation, allows them to separate out rapidly, making for a beer that lacks attenuation. In class IV are contained the chain formers. They separate out before fermentation is over, leaving much unfermented wort for the growth of unwanted contaminants. The chain formers are themselves unwanted in the brewery.

The powdery yeasts of class I throw homogeneously and readily dispersible sediments. The granular flocs of class II are dispersible by agitation, whereas those of class III are not. The flocs of class IV are fluffy and considerably larger. Gilliland's classification scheme was originally devised for ale strains. Yeasts of the first three classes are found also among lager strains, but no lager yeasts have so far been identified as chain formers.[257,258]

Outside the brewery, the flocculation characteristics of yeasts hardly ever bother the wine maker, the distiller, the yeast manufacturer, or the baker. Traditional wine making relies on the yeasts on the bloom of the grape. The distiller used to prefer powdery yeasts, whereas the yeast manufacturer would rather use slightly flocculent strains.[1251] But with modern equipment for cell separation and collection, the flocculence of a strain now assumes minor importance so long as it does not impair the fermenting capacity of the yeast. The baker, however, understandably still favors the homogeneous dispersibility of powdery yeast for easy blending with the dough. Recent developments in the design of continuous fermentation require flocculent strains.[13,116,1274] Alcohol or beer production in tower-fermentation systems depends on flocculence to maintain the cell population density in the continuous fermentor. A powdery yeast, because it would be homogeneously dispersed in the fermentation soup, rapidly washes out. The more flocculent a yeast, the better it is retained, the better it is able to bear an increase in the flow rate of the system.

Yeast flocs are usually large enough to be visible to the unaided eye. In most instances, flocculent and nonflocculent strains are visually distinguishable from each other. Compared to those of bottom-fermenters, flocs of top-fermenters are less conspicuous. Indeed, what are flocculent ale yeasts to the British brewers may not be flocculent enough by German standards.[1251] Subjective estimation by eye, while adequate for some purposes, such as instant nondestructive comparison of many strains or of stages of one strain in the course of fermentation, is not good enough.[77,79,85,87,262,266,271,282]

Many of the apparent contradictions in the literature doubtless arise from strain variability, but many are due to differences in the quantitative procedures for the measurement of flocculation. It is not that there is a dearth of facile methods. Rather, there are too many. Every laboratory, it seems, has its own peculiar method adequate to its objective. A quantitative all-purpose procedure acceptable to all remains a major illusion. Many a serious worker in the field seems to be aware of this situation and, in an effort to remedy it, merely makes matters worse by adding to the clutter his own version of one of the many methods already available.

Just about every workable quantitative procedure for flocculation assay appears to be a variation of gravity sedimentation. According to Stokes' law, the rate of fall, V, of a spherical particle acted on by gravity g in a viscous medium is a function of the particle radius r:

$$V = \frac{2gr^2(d_1 - d_2)}{9\eta} \tag{1}$$

where d_1 is the density of the particle, d_2 the density of the medium, and η the coefficient of viscosity. As a yeast floc, being made up of thousands of cells, is considerably larger than a single cell, it is clear that flocs sediment faster than those cells with aggregation numbers of ten or less. Highly flocculent yeasts settle within a few minutes, powdery yeasts may stay suspended for hours (Figure 23). The rate of sedimentation or the amount of sediment at the end of a predetermined period of time may be measured. The sedimentation rate, a function of floc size, tells us merely how large the flocs are, nothing about how many. It is assumed quite reasonably that floc size is a measure of the intensity of flocculation. A function of the number of cells competent to flocculate, the amount of sediment may be expressed as volume, weight, or number of cells, in absolute terms or as a fraction of the total cell population. It tells us how much biomass is in flocs. Both kinetic and thermodynamic approaches are valid. Both have been used, ofttimes indiscriminately, to measure purportedly either intensity or extent of flocculation. It is often difficult to tell from the literature if a given strain is flocculent because it throws more sediment or because it sediments faster. The problems regarding quantitative estimation of yeast flocculation have been discussed in Chapter 3, Section II, as well as in the literature.[76,87,1270]

In general, a sample of yeast in suspension (at least a few milliliters in volume) is allowed to stand undisturbed in a conical tube for a period of time (5 to 10 min for fairly flocculent strains), during which flocs settle to the bottom. At the end of the period, the volume of the flocs may be instantly estimated by simply noting the height of the sediment, if the tube is graduated.[87,103,105,106,1270] Alternatively, the supernatant fluid may be removed and the flocs at the bottom determined for dry weight or, after resuspension and preferably deflocculation, for turbidity, number of cells, or colony-forming units.[76,95,104,278] As well, the top portion of the yeast suspension may be sampled for turbidity measurements or cell counts, before and after standing.[76,91,283] The difference in turbidity or cell count between the whole sample (before standing) and the free-cell population (after standing) is due to cells in flocs. Other investigators prefer to monitor the clearing of the supernatant or the formation of sediment.[92,95,96,107-114,277] The sedimentation rate, expressed as unit over time or the time at which a certain sedimentation value is reached, is readily determined by photometric devices.

In addition to quantitative methods based on gravity sedimentation, other methods are available. One method, for example, makes use of the deflocculating ability of sugars (mannose or maltose).[272] A suspension of flocs may be titrated with mannose and the added amount that completely brings about deflocculation determined. Another procedure calls for thermal deflocculation.[102] The turbidity of a stirred suspension of flocs is monitored as the temperature is raised step-wise (Figure 24). As the flocs disperse into free cells, turbidity increases up to a certain temperature, T_f, above which no further increase is observed. T_f is an estimate of the stability of the flocs. The difference between the initial and the final turbidity values or their ratio may be equated to a measure of the flocculated fraction in the sample. In the brewery, the flocculation point, the time at which a brew begins to form flocs, may be determined by sampling the middle portion of the vat content and observing microscopically for flocs (Figure 21).[91]

These objective procedures have been used, together with subjective visual estimation, to determine the flocculence of strains, the effect of various inducers, potentiators, and inhibitors, and the development of flocculation in the brewery and in the laboratory, all in the hope of unraveling the mechanism of cell-cell interaction. When the development of flocculation is monitored, many authors have not been careful to distinguish the flocculation of cells already competent to flocculate from the development of competence.

Because flocculation occurs in wort, it seems logical enough to expect that there is something in the wort that makes the phenomenon happen. Much effort has been spent in

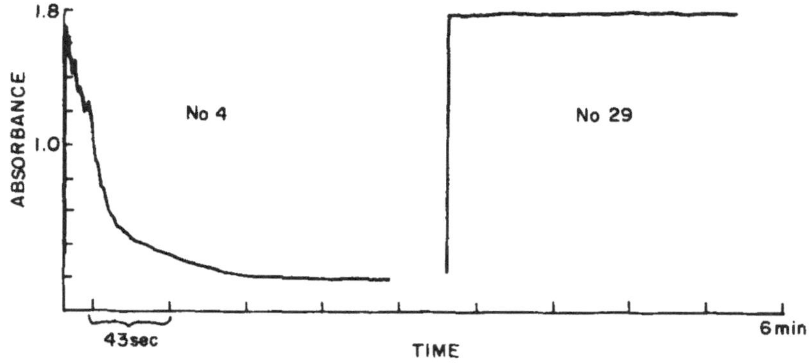

FIGURE 23. Sedimentation curves of flocculent (No 4) and nonflocculent (No 29) yeasts. Cells grown on agar slopes were resuspended in acetate buffer. Absorbance was read at 670 nm. (From Greenshields, R. N., et al., *J. Inst. Brew. London*, 78, 236, 1972. With permission.)

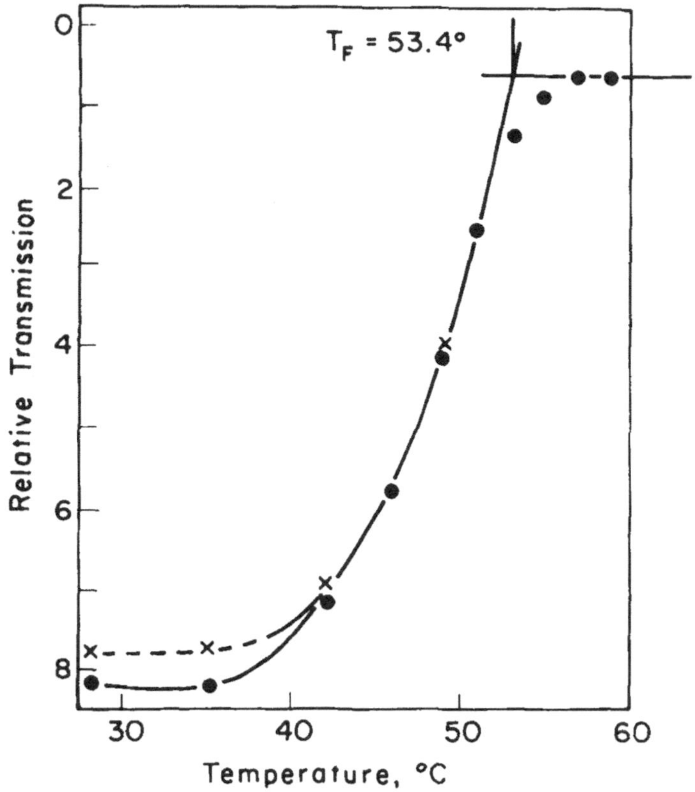

FIGURE 24. Determination of flocculation and floc-dissociation temperature. Cells of *Saccharomyces cerevisiae* NRRL Y-265 were suspended with stirring in a medium containing 0.01 μ Tris, pH 7.6, + 10^{-6} *M* Ca. After equilibration at 28°C, temperature was raised step-wise. Turbidity values were recorded at indicated temperatures after equilibration. Symbols: ● ascending; x, descending. Compare with Figure 6. (From Taylor, N. W. and Orton, W. L., *J. Inst. Brew. London*, 81, 53, 1975. With permission.)

dissecting wort for the purpose of identifying components that promote flocculation. Unfortunately, just about everything has been claimed, one time or another, by somebody or other, to promote flocculation, only to be dismissed later as of no consequence, or even to be inhibitory (Table 2, Chapter 3).

One early and popular view was that components of the brew bind to the cells and render them precipitable. Various substances have been implicated, including proteins, peptones, tannins, and calcium oxalate.[49,91,104,1257,1275-1277] Later investigators claimed to have discovered in wort other flocculation-promoting substances, such as humic acids and a polysaccharide called treberine.[258-261,263-266,268,269] Needless to say, the bulk of this approach has not been very instructive. For instance, it was believed that powdery yeasts are rich in proteinases, which digest away protein precipitates on the cell surface, whereas flocculent yeasts, deficient in such enzymes, form flocs because they cannot brush themselves away clean.[1278] Deliberate addition of protein, such as casein, to wort does not enhance or hasten the onset of flocculation in the brewery.[91,108] Besides, flocculation can occur in synthetic media.

Instead of the floc-promoting capacity of wort, its floc-dispersing activity has been looked into by others. As cells during the early stages of fermentation remain disperse, only to flocculate when fermentation is about complete, it is presumed that there is in the wort something that prevents flocculation. It is a common observation that flocculated cells derived at the end of fermentation become disperse when inoculated into wort.[81,103,114,262,1257,1262,1279] Although disperse in wort, the cells remain flocculent when tested during the early period of culture. The original cells grow and divide, then lose their apparent flocculence, before the culture eventually regains it during stationary phase. A few hours after inoculation, the cells when tested are found to have lost their ability to form flocs.[81,103,114] Cells dispersed in wort or in distilled water, however, when resuspended in beer or in buffer, readily reflocculate. Something in the wort then may prevent flocculation or there is lacking in the wort some component that sustains flocculation.

Prolonging exponential growth of a culture leads to a delay in the onset of flocculation.[262] In wort with a high concentration of assimilable nitrogen, flocculation onset is delayed; a high concentration of lysine completely prevents its happening at all. A lower nitrogen content of wort, on the other hand, enhances floc formation and leads to a better head.[267,1280] If growth of cells prevents their flocculation, then conditions that promote growth could promote deflocculation.

Wort proteins, it has been theorized, adsorb to cells, protect them, and render them nonflocculent.[267,1280] Coating cells with gelatin or other proteins has been reported to prevent flocculation.[263] However, wort materials, be they proteins or humic acids adsorbed onto the cells, have been convincingly shown to be of no consequence to floc formation.[77,272,1279] Although the tendency of flocs to disperse on washing or dilution in distilled water suggests that there is in the wort something that holds the cells together and that is being washed away, repeatedly washed cells derived from flocs will reflocculate in a solution of calcium salt. In contrast to mature cells, young cells are not competent to form flocs. When incubated in glucose in the absence of a source of nitrogen, they do not grow, but they eventually develop flocculence.

Without dismissing the possibility that there is in the wort some component that brings about in one way or another the formation of flocs, it must be concluded that there is no wort component that is absolutely required and that is adsorbed onto the cell surface to make the cells competent to form flocs. Quite the contrary may be plausible: that there is in the wort something that brings about the exposure of some wall material that somehow enables the cells to stick together. Alternatively, there may be inhibitory wort materials adsorbed onto the wall and finally dislodged at the end of fermentation, thereby allowing cell-cell interactions to build up into flocs. For the moment, it is instructive to look at flocculation as an inducible system, in that, under the best of brewery conditions, it happens only during stationary phase.

There are dissenting reports. For instance, one study shows that if inositol is left out from the medium, flocculation may occur even during early logarithmic growth.[278] Although flocculation may appear only during the end of logarithmic growth, the ability to form flocs, as tested by resuspending washed cells in a buffer with calcium, is never lost in an aerated culture.[1281] The failure of the cells to flocculate before stationary phase may simply be due to the lower cell population density during the early phases of a batch culture. It is possible, however, that growth under anaerobic or semi-anaerobic conditions leads to the loss of flocculence. More about the effect of aeration later.

Because the pH of the brew decreases before flocculation becomes apparent (Figure 22A), it might well be that the hydrogen-ion concentration is a factor. Indeed, pH used to be considered an important part of the problem, especially during the period when its determination was a novelty and a tedium.[154-160] Flocculation tends to be favored in one way or another by acidification of the starting wort. It was earlier thought that the increase in the hydrogen-ion concentration of the wort in the course of fermentation causes the precipitation of proteins and other substances on the cell surface, thereby causing floc formation. Although a particular strain may flocculate in the beer when a distinctly predictable pH value is reached, deliberate acidification of the brew to the characteristic pH value does not bring about flocculation.[153] It is now quite obvious that whereas pH influences yeast flocculation, it is of minor importance under brewery conditions.[91,93,103,111,114,153,267,278] Flocculation occurs anywhere between pH 2 and pH 8 (Figure 25). The range can be narrower or broader, depending upon the strain and other experimental conditions. The optimum is variable from strain to strain, anywhere from pH 3 to pH 6, well within the range of brews of commerce (pH 3.5 to 5.5).

The enhancement of flocculation by inorganic salts was known to early workers.[50,1252–1257] Cells in flocs are dispersible in distilled water. They reflocculate when resuspended in a solution of salts, such as chlorides and sulfates of calcium, magnesium, potassium, and sodium, but not barium chloride nor copper sulfate.[91,103,108] An ionic strength in the vicinity of 0.01 μ is reported to be about optimum.[102]

Calcium is believed by many to be a specific requirement for the formation of flocs.[50,81,102,108,267,271,277,279-281,1252,1253,1257,1282-1285] Of the various ions reported to sustain flocculation, calcium is the most effective. The addition of calcium ions to cells dispersed from flocs in distilled water results in reflocculation. The use of the chelating agent EDTA, which disperses flocs into free cells, results in reflocculation inhibition, which is reversed by the addition of calcium. Floc cells bind more calcium than free cells do and more calcium than other ions.[271,280] Isolated walls of flocculent cells bind more than those of nonflocculent cells.

It has been proposed that the apparent effectiveness, if not the specificity, of the calcium ion is due to its serving as a bridge between adjacent cell walls.[281] The proposal, of which there are several variants, has gained many adherents.[81,271,277,282,1283-1285] One model calls for calcium complexes with the carboxyl groups of proteins.[271,281,282] Another points to the phosphodiester linkages in the mural mannan serving as anionic sites.[1283-1285] In addition to the ionic bridges, hydrogen bonds involving the hydroxyl groups on mural polysaccharides may help stabilize the intercellular connections.[81]

Taylor and Orton contend that calcium is specifically required, since its floc-promoting activity is antagonized by the calcium-complexing compound EDTA and by the alkaline-earth metals strontium and barium.[102,279] The sodium ion is a weak antagonist. The inhibition of calcium-provoked flocculation can be reversed by the addition of calcium ions only and not by magnesium. Furthermore, the required concentration of calcium salt is low enough, about 10^{-8} M, to exclude the cation merely acting as a counter-ion.

The absolute requirement for calcium ions does not remain unchallenged.[114,278] The ability of magnesium and manganese to partially substitute for calcium has been reported.[282] There are claims that magnesium is just as effective as calcium;[108,1282] and that other cations are

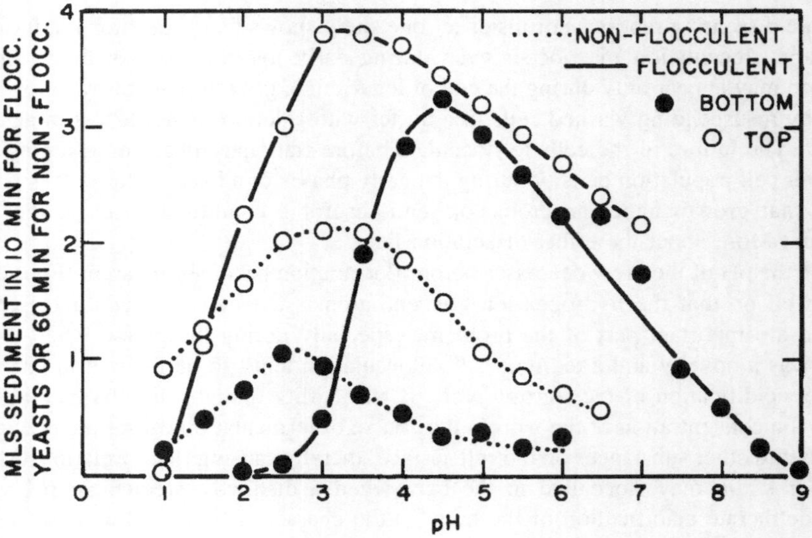

FIGURE 25. Effect of pH on flocculation as determined by sedimentation of cells. The pH of the cell suspension was adjusted with HCl or NaOH. A point corresponds to the mean of several experiments with flocculent top and bottom yeasts and nonflocculent top and bottom yeasts. (From Helm, E., Nøhr, B., and Thorne, R. S. W., *Wallerstein Lab. Commun.*, 16, 315, 1953. With permission.)

just as good.[103,111,267] Different strains differ as to the specific requirement for calcium.[77,111,271,282,285] Moreover, the ion effect is pH dependent. At high pH values, low concentrations of such divalent cations as calcium, magnesium, zinc, and cadmium are equally effective.[111,267]

Obviously, the contradictions are due to differences in strains, cultural conditions, and experimental techniques of quantitation. It cannot be doubted that ions are necessary for flocculation. If for no other reason, they are required as counter-ions to a system that involves the collision of similarly charged particles. Under certain conditions, calcium is definitely more effective than other cations and is effective at very low concentrations. What is questionable is not its role as a counter-ion, but its proposed role as a bridge, for such a bridge connotes simple ionic interactions among cells in flocs.

Heat-killed cells do flocculate, provided they are flocculent to begin with.[81] Hence, flocculation per se is not a process that requires vital activity. Isolated walls from flocculent cells do flocculate among themselves.[284] Walls from nonflocculent cells do not. It is obvious that flocculation is a wall-wall interaction.

The cell wall of brewers' yeast is made up of roughly equal amounts of glucan and mannan (about 40% each), proteins (12%), lipids (6%), and small amounts of chitin and phosphates.[1286-1288] The mannan content of flocculent walls has been claimed to be greater than that of nonflocculent walls.[1284,1285,1289-1291] The glucan content appears to be invariant. Coincident with the onset of flocculation is an elevated synthesis of mannan.[280] During the period of flocculation development, there is a subsequent decrease in mannan synthesis. Despite a contradictory report that flocculent walls contain less mannan than do nonflocculent walls,[1292] it is commonly believed that the phosphomannan-protein complex is the active site of the cell-cell interaction.

It is claimed that cells of flocculent strains contain more phosphorus and bind more calcium than do cells of nonflocculent strains.[1283-1285] Esterification of cells with 1,2-epoxy propane reduces calcium-binding only slightly, whereas treatment with trypsin releases much of the phosphorus, accompanied by a drop in flocculence and in the capacity of the cells to fix calcium ions. Moreover, it is maintained that there is no correlation between flocculence and the content of acidic amino-acid residues. Therefore, it is concluded, flocculence is a function of the phosphate groups in the phosphomannan-protein complex.

The possible role of mural phosphates has been further examined.[277] Removal of phosphorus with HF also removes small quantities of mannan, glucan, and proteins from flocculent as well as nonflocculent strains, without apparently touching the carboxyl content of the outer wall. The surprise is that HF treatment enhances the flocculence of flocculent strains and renders powdery strains flocculent. Esterification, in contrast, diminishes flocculence. Thus, it is concluded that the carboxyl group in the protein, and not the phosphodiester linkage in the mannan, is involved in flocculation. Other workers have not found significant differences in phosphate content between flocculent and powdery strains.[1289,1290,1293]

Early comparative studies concerning the gross composition of cell walls of flocculent and nonflocculent strains have not revealed very meaningful differences.[87,258,1289,1293] Treatment of whole cells with trypsin, followed by mercaptoethanol and another trypsin digestion, solubilizes about 15% of the dry weight of either flocculent or nonflocculent cells.[1294] Components of high molecular weight from flocculent cells have been found to contain more carbohydrate and less protein than do the corresponding materials from nonflocculent cells. They contain significantly greater amounts of proline, lysine, and arginine. Materials released by Pronase show only minor differences between flocculent and nonflocculent strains. SDS-polyacrylamide gel electrophoretograms of deflocculation extracts from flocculent and nonflocculent strains look pretty much the same.[87] A peptide material (about 3.7×10^4 daltons) that is prominent in flocculent digests and minimally detectable in nonflocculent digests remains to be verified.

Shadow-cast with tungsten oxide and examined with the electron microscope, cells of flocculent and nonflocculent strains show little variance in cell surface topography.[87,1270] But when the cells are washed with ether prior to shadowing, there is an outstanding difference (Figure 26). The surface of flocculent cells appears covered by hair-like extensions or fimbria-like material. All flocculent strains examined display this surface characteristic, whereas nonflocculent cells are smooth. Brief blender treatment causes deflocculation, so does treatment with NaOH at pH 12. Both treatments result in the removal of fibriae. As noted a number of times now, hair-like extensions from the cell surface seem to be a common feature of many cell-aggregating systems (Table 6, Chapter 3).

Other attempts to find differences between flocculent and nonflocculent strains with regard to mural characteristics include the selective binding of dyes. Alcian Blue binds specifically to the mannan component of the yeast cell wall.[1295] In a thorough study, all strains tested adsorb the dye, the amount adsorbed depending on the strain, but there seems to be no obvious correlation between amount adsorbed and flocculence.[87] However, either spheroplasting or protease treatment causes a marked decrease in the adsorbing capacity of the cells for the dye. Cultures deflocculated by esterification with 1,2-epoxy propane show diminished dye adsorption, whereas cultures insensitive to deflocculation by esterification do not suffer reduction of binding of Alcian Blue. Treatment with dithiothreitol or mercaptoethanol releases mural material of both low and high molecular weights, including glycoproteins, but little phosphorus.[1296] As with the esterification treatment, strains flocculent in both wort and defined media are resistant to deflocculation by sulfhydryl reagents; they retain their capacity to bind Alcian Blue. Those strains flocculent only in wort readily deflocculate and concomitantly show reduced capacity to bind the dye.[87]

Despite the difficulty in showing the differences in wall composition of flocculent and nonflocculent cells, it is clear that cell-cell interaction resides in the phosphomannan-protein complex, which is the outer layer of the cell wall, rather than the inner glucan layer. Protease treatment of whole cells causes the release of large amounts of mural protein and mannan, without significantly altering the glucan content.[81,1285,1297] As a consequence of protease

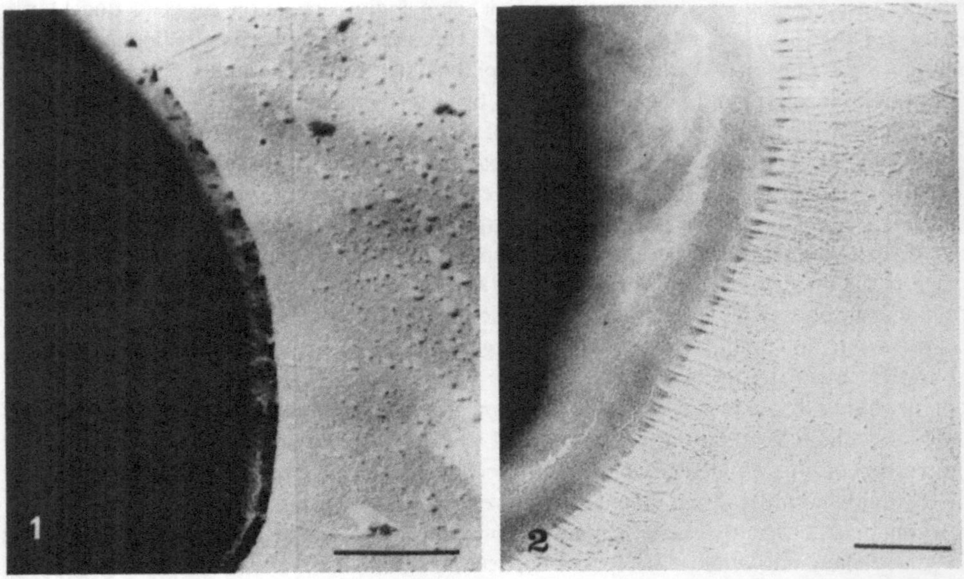

FIGURE 26. Electron micrographs of brewers' yeast (*Saccharomyces cerevisiae*).[87,544] (1) Nonflocculent; (2) flocculent. Bar represents 0.5 μm. Samples were shadow-cast with tungsten oxide after ether washing. (Original photomicrographs kindly furnished by N. H. Poon.)

treatment, flocculent cells become irreversibly nonflocculent. Among the proteolytic enzymes found to cause irreversible deflocculation are Pronase, papain, trypsin, chymotrypsin, pepsin, and proteinase K.[81,488,1281,1285,1297,1298]

The proteinaceous nature of the macromolecules involved in cell-cell interaction is further substantiated by chemical modification of the cell surface. The flocculence of whole cells is destroyed by SH-alkylation, disulfide reduction, esterification, acylation, deamination, iodination, and photooxidation.[1298] Sulfhydryl oxidation is ineffective as a procedure for deflocculation, suggesting that sulfhydryl groups in the protein are not meaningfully involved. Inhibition of flocculence by disulfide reduction, unlike the other chemical treatments, is reversible. Upon reoxidation of the treated cells, flocculence is restored; flocculation reappears, but not in the presence of high concentrations of urea.

It was once in vogue to explain flocculation of brewers' yeast exclusively by the dictates of colloidal theory.[150,152,1299] Flocculation was thought of as nothing more than the consequence of cells falling upon one another at the isoelectric point. That simplistic explanation has now been made obsolete by data from many laboratories.[108,159,160,284,286,1282,1300,1301] At no time during fermentation do yeast cells lose their negative charge. Indeed, if there is any quantitative change at all in electric charge, it is negligible and does not suffice to account for the sudden propensity of the cells to form flocs.

Ionic charges, of course, are initially important when cells are in a collision course. Other forces assume primacy as soon as intercellular contacts become more intimate. Because deflocculation and reflocculation can be caused to occur rather freely, the contribution of covalent bonds need not be considered at all. Instead, much weaker interactions are more likely to be the main forces that keep cells together. Reversible deflocculation caused by heat, urea, or an increase in the dielectric constant of the suspending medium suggests the primary participation of hydrogen bonds.[81,1302] Flocs may be further propped up by hydrophobic interactions.

The involvement of proteins is indicated by irreversible deflocculation provoked by proteases and by diverse chemical treatments that modify the reactive groups in polypeptides.

Mannan is also implicated by virtue of its being a large component of the outer wall and by the selective activity of mannose as deflocculant.[272] Furthermore, flocculation is reversibly inhibited by concanavalin A, a mannose-specific lectin.[1303] In addition, the apparent specificity of the calcium ion must be somehow accommodated in an overall molecular mechanism.

A central role for the calcium ion, acting as ionic bridge between negatively charged walls, is the gist of the one model that has attracted many champions.[81,281] However, the model seems inadequate, for it fails to make room for the specificity of the cell-cell interaction. To account for specificity, macromolecular connections may consist of complementary carbohydrate-carbohydrate, carbohydrate-protein, or protein-protein interactions. Models of carbohydrate-protein interaction between walls have been advanced.[1302,1303] Two kinds of receptor sites are assumed to be located on the cell surface (Figure 27). The mannan site is found on the walls of all cells, competent or incompetent. Found on the walls of competent cells only, the disulfide-containing protein site binds a calcium ion in some way. Wall-wall interactions between two competent cells are bilateral connections between mannan and protein sites on the surface of one cell and the same kinds of protein and mannan sites on the surface of the other cell. It may be added that the lectin-like complementation is apt to be mediated by hydrogen bonds.

No microbial aggregation system has generated more controversies with regard to cultural conditions influencing aggregation than does flocculation of brewers' yeast (Table 2, Chapter 3). For instance, various components of wort have been reported by certain workers to be stimulatory, even as others, including oft times the very same workers, have reported them to be inhibitory or simply innocuous. Amino acids and sugars also belong to this contentious group of nutrients. It is unfortunate that in a great majority of studies with inducers and inhibitors, flocculation per se of already competent cells was not isolated from the development of competence nor even from population growth, thereby rendering such studies hard to evaluate.

Depending on the ionic species and the concentration, inorganic salts in the medium have been claimed to promote or inhibit flocculation (Table 2, Chapter 3). Most reports on salt effects must be interpreted as merely confirming the requirement for gegenions in effecting fruitful collisions of cells after they have become competent to form flocs. Calcium ions have been implicated for the most part, but one particular study favors magnesium ions in the culture medium as a specific nutritional requirement.[284]

Exogenous undefined substances appear not to be required for the development of competence, because flocculation can develop in synthetic minimal media. A need for specific inducers not found in minimal salts media, notwithstanding claims in the literature,[85,87] may also be excluded. The reports that peptones, wort components, or yeast extracts promote flocculation may simply mean that initial good growth of the population contributes somehow to the subsequent development of competence. As long as the growth of the population remains an integral part of the test condition, nutrients such as vitamins, amino acids, and fermentable sugars may seem helpful, even required.

The fact that flocculation of a culture occurs after the end of exponential growth implies that the development of flocculation is inhibited by compounds that permit growth, if they are added to the culture when the cells are supposed to become flocculent. A condition short of complete depletion of nutrients, as a rule, favors flocculation, but because flocculation is not inducible in a solution of inorganic salt when cells have not yet become competent, its development to competence must require some source of exogenous energy.

Alcohols and solvents, ethanol in particular, have been observed to stimulate the development of flocculence.[114,275] This observation is only topically related to the report that lipids as nutrients enhance the development of flocculence of a culture.[115] The flocculation of a culture is inversely related to the concentration of fatty acids in the medium.[1304] Among many other compounds believed to influence flocculation is glycogen.[273,1305-1309] Also studied

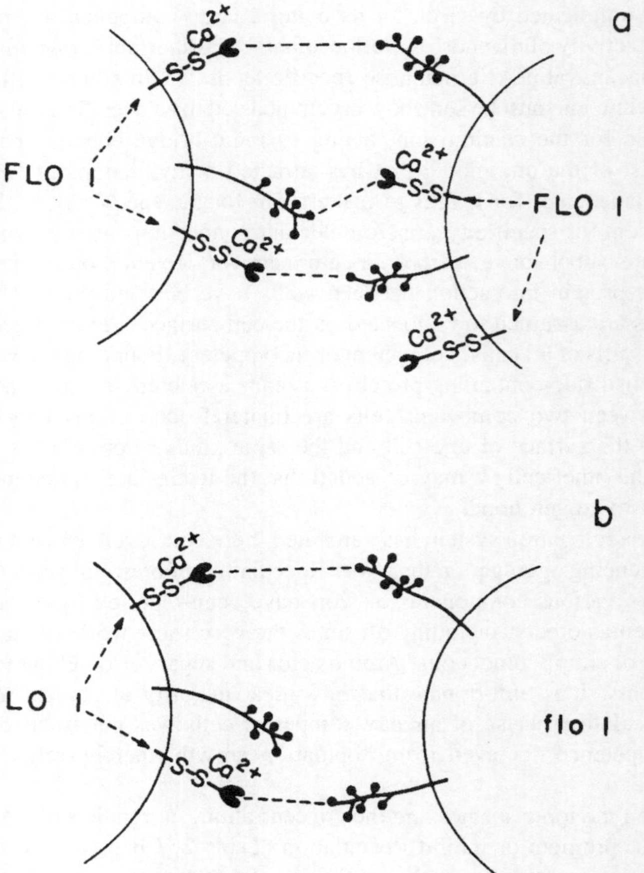

FIGURE 27. A model for cell-cell interaction in brewers' yeast. (a) Interaction between competent cells (FLO 1). (b) Interaction between a competent cell (FLO 1) and an incompetent cell (flo 1). Mannan sites are represented by multi-knobbed appendages. A protein site, found in FLO 1 only, is represented by a claw attached to the cell wall by a disulfide bond. (From Miki, B. L. A., Poon, N. H., James, A. P., and Seligy, V. L., *J. Bacteriol.*, 150, 878, 1982. With permission.)

are the changes in the activities of enzymes as correlative to flocculence development.[1310-1312] Needless to say, these are somewhat roundabout approaches that have proved to be minimally instructive.

Elevated temperatures, in general, cause the suppression of flocculence development.[91,104,105,267,1305,1313] A thermosensitive mutant derived from a flocculent strain has been reported to be nonflocculent when grown at 30°C, but flocculent when cultured anaerobically at 22°C.[115] This mutant is nonflocculent when aerated or cultured in a medium containing ergosterol and unsaturated fatty acids. Whereas some workers observe that aeration inhibits flocculation,[280,1271,1313] others report just the contrary.[259,1314,1315] Insufficient aeration of the brew may cause the onset of flocculation to occur earlier.[91] In one particular strain, flocculence is maintained by an aerobic culture.[1316] It is lost during growth under anaerobic conditions and then regained during stationary phase or when the culture is shifted to conditions of continuous aeration.

The development of flocculation appears to require mitochondrial functions of some sort. Respiratory inhibitors and uncouplers, such as cyanide, azide, and dinitrophenol, inhibit the development of competence, suggesting the need for energy transduction.[274] Also, mutation to the petite state can lead to the loss of flocculence, although certain petite strains retain the capacity to form flocs.[87,115,127,1317] As a pleiotropic effect, flocculence regained may accompany spontaneous reversion.[1318] Apparently, the loss of flocculence due to petite generation is strain variable.

Although functional mitochondria may be required, saturating concentrations of chloramphenicol, a selective inhibitor of mitochondrial protein synthesis, allow nevertheless the development of flocculence.[274] This might simply indicate that the synthesis of mitochondrial proteins is not necessary. The need for cytoplasmic protein synthesis, by contrast, is suggested by the observation that cycloheximide totally blocks flocculence development.[262,274,1304] Once the capacity of a culture to form flocs has been induced, flocculation proceeds undisturbed in the presence of cycloheximide. Gene products synthesized on cytoplasmic ribosomes, therefore, may be required for the induction of competence to form flocs.

That flocculence is a heritable character can be inferred from the fact that flocculent and nonflocculent strains exist. However, the phenotype is rather unstable.[79,109,127,578,1248,1319-1322] A strain established as flocculent in the brewery may, for no apparent reason, lose its flocculent character over the years and become powdery. It may, alternatively, become more and more flocculent, only to lose its flocculence later on. Spontaneous mutation from flocculence to nonflocculence has been quantified.[191] It occurs at the rate of 5.5×10^{-4} to 1.7×10^{-2} per cell division. The backward mutation occurs at a rate lower than 8×10^{-5} per cell division. Besides the instability of the marker, industrial strains are known for their ill-defined ploidy. Moreover, they sporulate poorly even as their spores germinate poorly. All these features of brewery strains tended to postpone the inevitable dissection of flocculence as a genetic system.

Early studies that were purported to unravel the genetic basis of flocculence may now be waived aside as the analyses of chain formation, a totally different phenomenon.[91,578,1247,1323-1327] Nonetheless, they have served to dispel doubts that flocculation is genetically determined. Originally, two dominant genes for flocculence, *FLO1* and *FLO2*, as well as a recessive gene, *flo3*, were identified.[191,1328,1329] Soon after, *FLO4*, another dominant gene, was mapped on chromosome I, about 38 cM from *ade1*.[87,127,1330] Subsequent genetic crosses showed that *FLO1*, *FLO2*, and *FLO4* are allelic markers.[577,1331] As a result, the three dominant loci are now reduced to a single locus designated as *FLO1* (Figure 23, Chapter 3).

The suppression of flocculence seems to have been partly responsible for the initial confusion in the nomenclature. Governed by suppressor genes, the suppression of flocculence, apparently a common phenomenon, may result in the recovery of 3:1 tetrads.[115,1332] One such gene is *fsu1*, which is recessive in certain crosses, but dominant in others. Spontaneous generation of petites can cause also the appearance of unexpected segregation ratios of flocculence genes.[87] As already mentioned, petites may retain the flocculence of their wild-type progenitor, but they tend as a rule to be less flocculent or even nonflocculent altogether.[115,1318,1333-1336] The loss or diminution of flocculence can be due to other mutations, such as that of *ade1*. The flocculence of *Saccharomyces diastaticus*, a brewery contaminant, can be lost as a result of mutation to antibiotic resistance.[1337]

Some success has been achieved from first attempts to identify the gene product of *FLO1*.[1281,1316,1332] As shown by SDS-polyacrylamide gel electrophoretograms, a negligible variance in protein patterns is shown by extracts of flocculent cells and those from nonflocculent mutants. But one polypeptide of about 13 K daltons seems to make the difference. It is not certain at the moment whether the polypeptide is a part of the structures that mediate cell-cell interaction.

Flocculation of brewers' yeast is a good example of a microbial aggregation system that is commonly thought of as nonsexual. Moreover, by most investigators, it is looked upon as a classic example of a homotypic system, but then, nobody has looked really carefully into the possibility that there is some activity, sexual or parasexual, that is associated with the phenomenon. It is possible that a sexual function might have been originally associated, but now that sexual function has been lost through inadvertent selection by brewers over the centuries.[1270] On the other hand, it might be possible, by clever manipulation, to elicit sex out of an apparently nonsexual phenomenon.

VI. OTHER YEAST AGGREGATION SYSTEMS

In the preceding section, flocculation of brewers' yeast was treated in a unified manner as if there is indeed only one phenomenon. The bulk of the literature on the subject seems to treat it in the same manner. In addition to chain formation, which has been excluded by definition, perhaps there are several similar or related phenomena inadvertently lumped together. The foaming of saké and wine yeasts might be altogether different.[1338,1339] Moreover, most flocculation studies have been made with strains of the lager type, but ale strains, it must be recalled, form head instead.[79,1251,1272] Head formation may turn out to be slightly different with respect to details. Apart from the industrial species *Saccharomyces cerevisiae* and *S. uvarum (carlsbergensis),* other species in the genus also exhibit flocculation. One such is *S. diastaticus.*[1337]

Another related phenomenon is the co-flocculation of strain pairs of brewers' yeast.[79,85,1290,1340-1343] By themselves, the strains are nonflocculent or only slightly flocculent. When mixed together, as if in sexual agglutination, cells of strain pairs form flocs and come out of suspension. Two types of strains can be distinguished. The A type is completely nonflocculent on its own. The B type exhibits slight flocculation that is convincing enough, but not compact enough; so the cells remain in suspension. The B type seems more common. The possibility that sexual or parasexual activity occurs between strain pairs during flocculation has not been examined. Although of very minor importance, the co-flocculation of brewers' strains with wild yeasts, and even with bacterial contaminants in the brewery, does take place.[54,158,1258,1344,1345]

The formation of films, rings, and pellicles is commonplace among yeasts of the genera *Candida, Hansenula, Kloekera, Kluyveromyces, Pichia, Saccharomyces,* and *Torulopsis.*[600,1346-1351] In species of *Saccharomyces*, film formation correlates well with the hydrophobicity of the cell surface.[1352,1353] This is specifically true in the case of *S. bayanus*. During the film stage, cells that form films become more hydrophobic than cells in the prefilm stage and cells of filmless strains. Film formation is enhanced when the cells are grown in ethanol, in contrast to glucose, as sole carbon source. When treated with surface-active agents, organic solvents, or NaOH, cells derived from films become less hydrophobic. Fatty acids on the cell surface are readily removed by alkali treatment of whole cells. The recovered amount of surface fatty acids, mainly oleic and palmitoleic acids, is correlated with the degree of hydrophobicity.

Hansenula ciferrii exhibits what appears to be a related phenomemon.[1355-1356] This yeast causes the accumulation of needle-like crystals of tetraacetylphytosphingosine in the culture medium. The cells and their lipid crystal products are so hydrophobic that eventually cells and products form flocs together.

The agglutination of ascospores in all species of *Kluyveromyces* may be related, too, in that ascospores are known to be hydrophobic.[1357] Certain species of other sporogenous genera, such as *Hansenula* and *Schizosaccharomyces*, have also been reported to exhibit ascospore agglutination.[600,601] Ascospores, however, may be presumed to derive in the final analysis from the sexual interaction of haploid cells. However far removed that sexual event might be from sporulation, it suggests the existence of mating types, which could well be expressed prematurely by the meiotic products. Thus, ascospore agglutination is possibly related to sexual agglutination.

Wickerham's short list of yeasts known to be sexually agglutinative included only four species distributed in four genera.[35,62] Aside from *Hansenula wingei*, which merited a separate section by itself earlier, it included *Citeromyces matritensis, Torulopsis globosa*, and *Saccharomyces kluyveri*. A very intense agglutination reaction is observed when cells of strains 9 and 10 of *Citeromyces matritensis* are brought together (Figure 2, Chapter 3). Conjugation follows soon after the agglutination reaction, which is best demonstrated when the strains are pregrown separately as liquid cultures. Cells of *C. matritensis* can be made to mate with those of *Torulopsis globosa*.[63] Indeed, *T. globosa* is now reduced to being merely the imperfect form of the sporogenous *C. matritensis*.[1358]

One of the more interesting yeast aggregation systems is the sexual agglutination in *Saccharomyces kluyveri*.[35] A potentially useful organism for industry, *S. kluyveri* is a vigorous fermenter and a fast grower. When complementary mating types are mixed together, a very strong agglutination reaction occurs, thereby facilitating the recovery of biomass in liquid batch cultures (Figure 3, Chapter 2). According to Wickerham, this organism is found as unisexual diplonts and as unisexual forms believed to be triplonts and tetraplonts.[35,39] These forms are morphologically differentiable from one another and sexually agglutinative. The so-called bisexual forms, however, are not agglutinative. Until recently, *S. kluyveri* has not generated much interest. It is rather intriguing, as already noted earlier, that strain 3 of this species cross-reacts with strain 21 of *Hansenula wingei*.[1111] Also intriguing is that the sex pheromones excreted into the medium by *S. cerevisiae* cross-react with cells of *S. kluyveri*.

The complementary mating types of *S. kluyveri* have been named **α-k** and **a-k**, as in the better known *Saccharomyces* but with the letter **k** added.[1359] Haploid cells designated as mating type **α-k** synthesize **α-k** factor, an excreted pheromone that causes G1 arrest and a morphological change in **a-k** cells. The **α-k** haplonts themselves, as well as the diplonts of the genotype **α-k/a-k**, fail to respond to the pheromone. The **a-k** cells, in turn, excrete **a-k** factor, which brings about a morphological change in **α-k** cells. Interestingly, the **α** factor that is isolated from *S. cerevisiae* is more active in provoking G1 arrest of **a-k** cells of *S. kluyveri* than in provoking the same response from **a** cells of *S. cerevisiae*. With more carefully purified preparations, the comparative cellular responses of the two species are interchanged; now the **a** cells of *S. cerevisiae*, as expected, is more responsive to **α** factor than the **a-k** cells of *S. kluyveri*.

The **α-k** pheromone has been isolated.[1360] It is partially characterized as a small peptide that is very similar to the **α** factor of *S. cerevisiae*. The difference seems to lie in only three amino-acid residues. *S. kluyveri* cells responding to **α** factor from *S. cerevisiae* are thermoresistant as regards sexual agglutination, but sensitive to chemical reduction.[1361] The opposite properties are displayed by the complementary cells that respond to **a** factor. The two species of *Saccharomyces*, despite their pheromonal cross-reactivity, do not seem to mate readily with each other.[1359] Weak sexual reactions between the two species, however, have been reported.[1362]

Sexual agglutination in *Pichia amethionina* is another interesting yeast aggregation system.[1363] The complementary mating types are designated **α** and **a**. Treatment of **α** cells with Zymolyase yields a multivalent agglutinin that brings about the agglutination of **a** cells.[1361] The agglutinin, which is analogous to the 5-agglutinin of *Hansenula wingei*, is a glycoprotein that is resistant to heat but sensitive to chemical reduction. The same properties are exhibited by the intact cells. Treatment of **a** cells with the protease subtilisin yields a small protein that inhibits the agglutination provoked by the isolated agglutinin. Analogous to the 21-factor of *H. wingei*, **a** factor is thermosensitive but resistant to chemical reduction, in contrast to the agglutinin. The isolated factors of *Pichia amethionina* do not cross-react with the 5-agglutinin and the 21-factor of *H. wingei*. Nonetheless, there is a slight cross-reaction between the cells of strain 5 of *H. wingei* and the cells of *P. amethionina*.

Sex-directed flocculation in *Schizosaccharomyces pombe* may be extended to include the other members of the small genus.[483] Reported to be similar is sexual flocculation in *S. japonicus*.[68,1142] The two systems differ only in minor details. Flocculation in *S. japonicus* is observed, for instance, even in a population that is exponentially growing. It is stimulated by yeast extract in the culture medium or by depletion of nitrogen. It is altogether inhibited in total darkness. The chemical nature of the light receptors remains unclear.

A dimorphic fungus with a yeast phase, *Phialophora dermatitidis*, has been reported to exhibit aggregate formation during stationary phase.[248] Conjugating forms are observed after the formation of aggregates consisting of more than 100 thick-walled cells. No aggregates are formed by thin-walled exponentially growing cells.

There are other yeast species known definitely to be sexually agglutinative. Among them are *Hansenula holstii*,[36] *Saccharomyces transvaalensis*,[1364] and *Tremella mesenterica*.[36] Their tendency to form sex-associated agglutinates is only incidentally mentioned in the literature; studies on the agglutination reaction have not been made. In addition to those described or merely mentioned, there are no doubt others. Where sex has been demonstrated, it might be a simple matter of finding the appropriate conditions for identifying a preceding event that may be called agglutination.

Only a few yeast aggregation systems have been well studied. In the complicated series of events that constitute the life cycle of a yeast, cell aggregation may not be the most interesting nor the most conspicuous. Thus, it may remain unstudied and unknown. There is usually some compelling reason, other than its being conspicuous, when it is studied at all. Sexual agglutination in *Hansenula wingei* finds favor because it serves as an uncanny model to the antigen-antibody reaction. Sex-directed flocculation in *Schizosaccharomyces pombe* is a rather convenient marker in the sexual development of a homothallic system. The pheromonal regulation of heterothallic sex is the most attractive feature of sexual agglutination in *Saccharomyces cerevisiae*. In the case of flocculation of brewers' yeast, its prominent role in industrial fermentation serves to make it worthy of study.

It can be safely assumed that cell aggregation is a common phenomenon and that its occurrence among yeasts is no exception. The current list of yeast aggregation systems is not very much longer than a list one could have made a quarter of a century ago. The list will doubtless become much longer, when cell aggregation becomes established as basic structure and function in biology. Where it has been identified in one species, it may just be a matter of extending the identification to other members of the genus. Then, perhaps, a more comprehensive and unifying picture of cell aggregation among microbes will be drawn.

Chapter 6

PUTTING THEM ALL BACK TOGETHER AGAIN

A detailed description of a system is in effect a dissection. That is exactly what was done many times over in the preceding chapters. Initially, we gathered the various systems that make up microbial aggregation, put them all together, and then dissected each of the better studied systems as shown to us predissected in the literature by the original investigators.

Described in detail were a number of fairly well studied bacterial and yeast aggregation systems. The bacterial systems we have covered include mating-aggregate formation in *Escherichia coli*; star formation in rhizobia, pseudomonads, and other bacteria; pellicle formation in fimbriate bacteria; agglutination associated with bacterial transformation; formation of dental plaque and aggregation of oral bacteria; cell aggregation in *Streptococcus faecalis*; cell aggregation in myxobacteria; and other bacterial aggregation systems. The yeast systems include sexual agglutination in *Hansenula wingei*, sex-directed flocculation in *Schizosaccharomyces pombe*, sexual agglutination in *Saccharomyces cerevisiae*, flocculation in the brewery, and other yeast aggregation systems. With regard to these microbial taxa, the monograph has been comprehensive, in that it includes all the well-known systems, be they sexual or nonsexual, so long as they fall within the purview of the subject matter. The other microbial aggregation systems — the cellular slime molds, the filamentous fungi, algae, and protozoa — will be covered in a subsequent volume.

For now, it is time to put the various components all together again. Unlike Humpty-Dumpty after his fall, microbial aggregation after dissection, literal or conceptual, can be put together again, without the need for all the king's men and all the king's horses.

The practical aspects of microbial aggregation we will not delve into again, but what is worth and can bear summarizing is the experimental insight that can be gained from a study of microbial aggregation. Routine experiments call for homogeneous and representative samples of a liquid culture of bacteria or yeast. When such a culture turns clumpy, it becomes unusable because samples from it would most likely be nonrepresentative. When the phenomenon of cell aggregation takes place where it is unwanted, it is often a source of no small irritation. If homogeneous and representative samples of a culture must be obtained at all costs, cultural conditions that favor cell aggregation must be avoided. Otherwise, the whole culture must serve as one sample. On the other hand, when one is looking for a genetic or a developmental system where none has been observed before, the phenomenon of cell aggregation can be experimentally advantageous. A culture may be enriched for suspected sexuality or development by simply favoring aggregation.

Cells multiply by dividing, in a dispersive or nondispersive manner, or in a manner somewhere between the two extremes. In a dispersive manner, the resultant progeny are single cells. In a nondispersive manner, the progeny form a cluster or a multicellular organism. Between the two extremes are small clusters.

Only after a dispersive manner of multiplication can cell aggregation take place. For, as defined in the introductory chapter, the concept of cell aggregation presupposes a state of dispersion. If cells in a culture arrange themselves such as to give an aggregation number much greater than one, then the clusters themselves must come together, or they have to be dispersed into their components prior to what we define as cell aggregation, an addition.

There are many more things that need to be unearthed, not just the details of the mechanisms of the known systems, but also the very existence of the unknown ones. Already the few systems described reveal cell aggregation as a fundamental function and a fundamental structure in the biotic world. Without a study of such functions and structures, biology would be incomplete, like a childless woman.

Cell aggregation is fundamental in the sense that it is common to a variety of microorganisms, prokaryotes and eukaryotes alike. It is found among bacteria, yeasts, filamentous fungi, cellular slime molds, algae, and protozoa. It is also fundamental in another sense: the general mechanism underlying the phenomenon is fairly simple. The event itself comprises the movement of cells toward one another, followed by connections mediated by molecular structures that bear some element of complementarity. The movement, purposive or stochastic, and the connections themselves represent fundamental functions. The resultant structure itself is conceptually nothing more than a group of cells materially connected to one another.

Cell aggregation is fundamental because the coming and the sticking together are a fundamental process of matter, living or nonliving. It is conceptually fundamental, too, for lumping is merely the reverse of splitting; but in a vital sense, it is fundamental, for after intracellular development and interaction with the nonliving component of the environment, what else is there left for a unicellular organism to do but interact with other organisms? It is fundamental because it is a beginning of more complicated events: sex, morphogenesis, development, and survival.

Of course, like any biological function and structure, cell aggregation, simple as it may seem on the surface, is at the same time complex enough. From a structural viewpoint, a cell aggregate represents a higher order of complexity than does a single isolated cell. The multicellular condition is not just a collection of individual cells, but a new synthetic structure. The aggregate is simultaneously more and less than the sum of its components. It has new dimensions and physical properties that do not precisely equal those of cells added together. It is also clear that the vital activities of the individual cells within an aggregate are somewhat different from single isolated cells. There would be functions enhanced, functions inhibited or obliterated, and functions altogether new. There would be functional antagonism and synergism, making extrapolations from the individual to the group difficult and error prone.

That is possibly true for all many-body interactions. Strengths and weaknesses of the individual interacting units may cancel out or reinforce one another. In union may be strength, but in the same union, a weakness may be amplified rather than canceled. The amplification may cause a particular weakness or strength to exceed the threshold for the precipitation of a morphogenic catastrophe.[1365]

The close-up details of the interactions in each of the systems are varied and much more complex than the generalizations made at a distance. So are the events related to cell aggregation, the events that come before and after and during its unfolding, the genetic bases, the molecular structures, the signals, and the regulation. These things are complex. It will take some time to simplify them; thus, the continuity of research along these lines is assured. The subject matter is just beginning to thicken.

Although, conceptually, cell aggregation is the reverse of dispersion and cell division taken together, one is not mechanistically the reverse of the other in the manner of a film segment played backwards. Granted, in most cases, the coming together is stochastic and so is the dispersive process, but the molecular connections of aggregated cells are quite different from the molecular connections of progeny and progenitor before dispersion. Moreover, it is most unlikely that progeny and progenitor will be found adjacent to each other in an aggregate, in contrast to a clone-cluster.

Although cell fusion, if it occurs at all, after cell aggregation is presumed to be enzymic in the same manner that scission after cell division is enzymic, the details are not symmetrical and the processes are not reverses of each other. Moreover, at the time of aggregation, the cells are morphologically and physiologically different from the cells right after cell division. As Heraclitus once put it: ''All things flow, nothing abides. Into the same river one cannot step twice.'' Consequently, in most systems, an aggregate of a given species can be visually distinguished from a cluster-clone, not just by the morphology of the group as a whole, but also by the morphology of the individual cells.

Of course, many microbial aggregation systems are considered reversible. A good example is sex-directed flocculation of fission yeast. The reverse of aggregation is none other than deaggregation, not scission and dispersion. Even then, the temporal mirror-image is cracked. An aggregate reconstituted from deaggregated cells is not the same as the original aggregate.

Perhaps, it is not surprising that cell aggregation is not the reverse of cell dispersion and scission. Perhaps, rarely ever are multi-step, multi-body processes inside or between cells precisely reversible in a mechanistic way. Protein synthesis and proteolysis are reverse processes only conceptually. So are permeation and excretion, so are birth and death.

Because cell aggregation, by definition, presumes a dispersive state prior to the phenomenon, it must be in itself a developmental event, an inducible segment in the life cycle of an organism. Whether the trigger is internal or external, chemical or physical, the result is a gathering that leads to intimate cell-cell interactions. In many examples, the cell-cell interactions usher in further developmental events, whether these be copulation, conjugation, cell fusion, sporulation, fruiting-body formation, or morphogenesis. The immediately scorable event, however, is simply a group of cells that have become materially connected to one another.

It is obvious that in most situations, the dispersion of cells is the way for the species to colonize the face of the earth. Once in a while, however, comes a time when the cells ought to band together to survive, to perform a particular function not available to single isolated cells, to form a multicellular unit. This capacity to band together when that time comes remains in the cell's repertoire for survival. No doubt it is somehow encoded in the genome.

What cell aggregation accomplishes that obviously a single cell cannot accomplish is the realization of its sexuality. Morphogenesis, differentiation, survival, and development, all these can and do proceed after sex. These processes can also occur without sexual interaction. For purposes of sex, it is more effective for any form of life, including human, to be initially single, unbound, and dispersed. That is why sex cells, even among higher forms, are generated in a dispersive manner. Sex cells, however, retain or acquire the competence to come together in a fashion that is rather specific and selective.

Sexual aggregation must be heterotypic to be of any advantage, for sex ought to bring about genic recombination that ultimately leads to a gain of functions. Indeed, by operational definition of many systems, a sexual interaction at the cellular level must be fruitful — fruitful to be scored as such. Otherwise, a "blush unseen", a "sweetness on the desert air" it remains.

Moreover, aggregated cells, even in a homotypic system, would be influenced by their position within the aggregate. The peripheral cells, for instance, must be presumed to exist in an immediate milieu different from that of cells in the core. This positional effect could easily be envisaged to lead to differentiation.

The multicellularity of an aggregate and the arrangement of the cells constitute by themselves a morphogenic process. Further morphogenesis can be due to differentiation and development. In turn, development itself may be triggered by morphogenesis and differentiation. Thus, morphogenesis, differentiation, and development — overlapping, malleable concepts that can be applied to concomitant or associated events — are expected consequences of even the more primitive forms of cell aggregation. In whatever terms, the functions and structures bring to the species a life cycle that is richer and larger.

Survival of the species, of course, can be the ultimate reward of sex. Or it may simply be due to development without prior sexual interaction. Too, it is plausible that the chances for survival of the individual cells are enhanced simply by their being together, that is, even in the absence of sex or morphogenesis or development. There has been no outright test of the survival value of cell aggregation per se, except possibly that of pellicle formation in fimbriate bacteria. Perhaps, only because enhanced survival of cells through aggregation appears to be glaringly obvious. The counsel of Ben Franklin seems apropos even to cells: "We must all hang together, else we shall all hang separately." A ball of cells could survive adverse conditions, like flash elevated temperatures and desiccation, better than a collection of isolated cells. Indeed, enhanced survival is assumed where sex and morphogenic development in a species have not been observed, where otherwise microbial aggregation is prosaic and uneventful.

If cell aggregation enhances the survivability of the species, then there is reason enough for its being conserved to, perhaps, evolve in time. If it does not enhance nor detract, then it may just be allowed to persist, but if it contributes in one way or another to the perishability of the individual and of the species, then it may be on its way out. Its fairly wide distribution among extant microbial taxa suggests all three possibilities.

There are not many examples of microbial aggregation leading to the formation of a multicellular organism. Among bacteria and yeasts, only the fruiting body of myxobacteria stands out. Indeed, the evolutionary origins of multicellular organisms cannot be assigned to cell aggregation, but rather to a nondispersive mode of cell division. In other words, whereas cell aggregation does lead to the formation of a multicellular unit, this unit usually falls short of integrating its components into a functional whole.

The multicellular condition can be achieved in two different ways (Figure 1). The first, more direct and apparently more primitive, is through cell division without the subsequent complete separation of the progeny. It requires no other function. The immediate result is a cluster of cells. This way may be viewed as the passive path to multicellularity.

The other path, more active and a little more complicated, calls for other functions. Cell division must be followed by scission, the complete material separation of the progeny. The separation is presumed to be an enzymic process, but the dispersion can be assumed to be largely mechanical. Sometime afterwards, the separated cells must come together and stick together. The coming together and the sticking together constitute what has been defined as cell aggregation.

Variations and combinations of the two ways do occur, but the two ways, in essence, cover all biological systems. The more direct way is used far more often by most living forms to attain multicellularity. Neither, however, is sufficient guarantee toward the formation of a multicellular organism. The juxtaposition of cells is just a beginning. A chain of streptococci is not one organism, but several. On the other hand, a slug of cellular slime mold may be treated quite properly as a single organism. For the development of such an organism, the integration and coordination of functions of the many cells into a cohesive unit must be achieved.

What the second path lacks in simplicity and efficiency, it makes up in better opportunities for pattern formation and for cellular recombination. A randomly dispersed group of cells made to come together in a random way has more opportunities to arrange themselves in a greater variety of ways than a cell cluster arisen as a consequence of cell division without opportunity for intraclonal rearrangement. Thus, a given cell can be adjacent to, and interact more directly with, any cell in an aggregate, not just its sib or its immediate progeny or progenitor. Moreover, interactions need not be limited to members of a clone, thus making room for better mixing of genotypes.

Note again that the dispersive path is for the species to follow the Biblical exhortation to go forth and multiply. If dispersive multiplication were the only path, there would be no multicellular organisms, save those arisen from cell aggregation. It is clear that nondispersive multiplication is the more effective way toward the generation of a multicellular organism. Higher forms of life seem to have retained both pathways.

Mechanistically, both pathways are found in most higher organisms. A tree, for example, grows from a seed, the cells dividing without dispersion of progeny, which eventually make up the roots, the trunk, the branches, and the leaves. However, the seed from which the tree arose derives from two cells that have arisen separately from progenitors that divided and dispersed. Somehow the separated cells eventually came together, stuck together, and fused to form the seed. During germination and subsequent growth, the seed becomes a

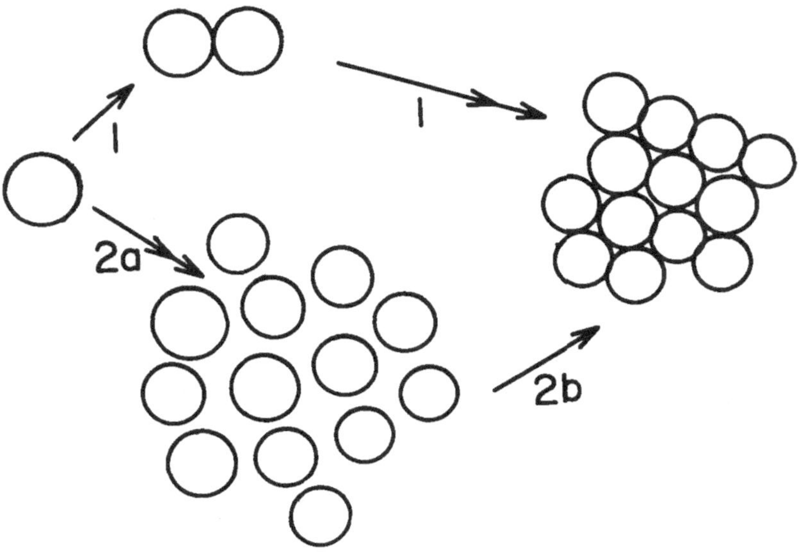

FIGURE 1. The paths to multicellularity. (1) Cell division without separation. (2a) Cell division followed by scission (complete material separation). (2b) Cell aggregation.

seedling and then a tree by the direct route to multicellularity. Only the cells differentiated and fated to become the sex cells and then the seed go the second, more energy-consuming, route. The same paths are traveled in the generation and regeneration of human life.

The first path is the way of somatic cells of higher organisms, the way of all flesh, so to speak. The formation of gametes follows the second, which forms the basis of sex at the cellular level. Of course, certain somatic cells, such as those found in the blood stream, are generated in a dispersive manner. Of course, there can be sexual interactions even among cells in a clone. Sex in a clone, however, is less likely to generate new forms. The second path is more than likely to have given rise in the course of evolution to the present diversity of life.

Recall that our definition of cell aggregation excludes the mathematically minimal union of two cells, but only as an expedient; that is, the exclusion is there to make the subject matter manageable. Mechanistically, the gathering and the sticking together of cells below the cutoff aggregation number of ten are similar to the processes that fall conveniently within the confines of our definition. A fruitful sexual encounter requires only a pair of cells, although, in some instances, *ménages à trois* (or of a higher number) are not too uncommon. Nevertheless, many conjugation systems, including fertilization among metazoa, usually involve interactions of much more than just a pair of cells. The fertilization of the human ovum, for example, is preceded by initial events that involve a myriad of sperm, normally only one of which succeeds.

The roots of sociobiology are found in the interaction of cells. In turn, cellular interactions, within cells and among cells, derive from chemical and physical reactions, molecular and atomic. Atoms, in turn, are the interactions of elementary particles. Thus, all phenomena, living and nonliving, are reducible to interactions.

All interactions are surface phenomena in that they must involve surfaces. To find the site of interaction is a matter of locating where the surfaces of contact are. The surfaces are the bounds of the interacting entities. They are defined materially and conceptually. On the matter of cell-cell interactions, the site of interaction is the cell surface. In most microbial systems, the cell surface means the cell wall plus any appendages attached to it. The wall sets the bounds of the cell. Thus, a cell-cell interaction is a mural event. Cell aggregation is principally an interaction of cell walls.

Nevertheless, cell aggregation, as we have already seen, is more than a mural phenomenon. It includes events other than wall-wall interactions. Because of its species specificity, not only does it arise from deep in the heart of a cell, the genome, it also extends to touch in a labyrinthine way the genomes of other cells. By origin and by extension, it becomes genetic expression and interaction. By virtue of its being a collection of cell-cell interactions, cell aggregation is a social phenomenon.

A cell aggregate is a community of cells, a society in a microcosm. It may not be coordinated and integrated into a unified and functional whole, but the interactions of component cells allow it to be treated, at least conceptually, as a community. It is not a new organism. Rather, it is an alternative for unicellular organisms, a halfway house toward the truly multicellular condition. It is a suprastructure, beyond unicellularity, but usually short of and not quite a multicellular organism. It is simply a consortium of interacting, intercommunicating cells hanging together. It may be viewed as a league of cells, like a league of nations, each cell in many ways independently sufficient and sufficiently independent, but united by a common affinity for one another toward a common end, be it survival or a patently social function, such as sex or intercellular development.

Locked in the genome of a solitary cell and awaiting expression is the will to socialize, to form a social structure with other cells of its own kind or even of another kind. Dispersed and isolated from its progenitor and sib and progeny, an individual cell moves inexorably in time toward events other than those occurring within itself. It reaches out to consort with other cells to form an agglutinate, a floc, a clump, an aggregate.

The actual journey of the cells toward one another may be endogenously orchestrated, like chickens coming home to roost, salmon going up the river. Or the cells may be brought together unceremoniously by stochastic forces, or even by the investigator's hand. Regardless, cell-cell interactions and connections are forged, a beginning. Thence, perhaps, there may occur cell fusion, sex, differentiation, morphogenesis, development. From there, a fruiting body, transformation, recombination, a zygote, meiosis, sporulation, germination, a cycle completed, a regeneration, rarely ever a truly multicellular organism, but doubtless a community. As well, microbial aggregation may be nothing more than a temporal refuge from forces that subvert life, the mere survival of individuals and of the species, a tribal ritual of returning, an awaiting. Perhaps, even an end. Whatever the fate of the members or of the group, process becomes performance, in concert, live.

Chapter 7

REFERENCES

1. **Lipton, J.,** *An Exaltation of Larks or, the Venereal Game,* Grossman Publishers, New York, 1968.
2. **Doherty, J.,** Animal group terminology, in *Information Please Almanac 1981,* Information Please Publishing, New York, 1980, 663.
3. **Curtis, A. S. G.,** *The Cell Surface: its Molecular Role in Morphogenesis,* Academic Press, London, 1967.
4. **Lilien, J. E.,** Toward a molecular explanation for specific cell adhesion, in *Current Topics in Developmental Biology,* Moscona, A. A. and Monroy, A., Eds., Academic Press, New York, 1969, 169.
5. **Silvestri, L. G., Ed.,** *Cell Interaction,* North-Holland, Amsterdam, 1972.
6. **Edwards, J.,** Intercellular adhesion, in *New Techniques in Biophysics and Cell Biology,* Vol. 1, Pain, R. H. and Smith, B. J., Eds., John Wiley & Sons, London, 1973, chap. 1.
7. **Moscona, A. A., Ed.,** *The Cell Surface in Development,* John Wiley & Sons, New York, 1974.
8. **Hoffman, H.,** Morphogenesis of bacterial aggregations, *Annu. Rev. Microbiol.,* 18, 111, 1964.
9. **Morris, E. O.,** Aggregation of unicells: yeasts, in *The Fungi: an Advanced Treatise,* Vol. 2, Ainsworth, G. C. and Sussman, A. S., Academic Press, New York, 1966, chap. 2.
10. **Harris, R. H. and Mitchell, R.,** The role of polymers in microbial aggregation, *Annu. Rev. Microbiol.,* 27, 27, 1973.
11. **Reissig, J. L.,** Decoding of regulatory signals at the microbial surface, *Curr. Top. Microbiol. Immunol.,* 67, 43, 1974.
12. **Kalckar, H. M.,** Galactose metabolism and cell "sociology", *Science,* 150, 305, 1965.
13. **Atkinson, B. and Daoud, I. S.,** Microbial flocs and flocculation in fermentation process engineering, *Adv. Biochem. Eng.,* 4, 41, 1976.
14. **Ottow, J. C. G.,** Ecology, physiology, and genetics of fimbriae and pili. *Annu. Rev. Microbiol.,* 29, 79, 1975.
15. **Smith, H.,** Microbial surfaces in relation to pathogenicity, *Bacteriol. Rev.,* 41, 475, 1977.
16. **Crandall, M.,** Mating-type interactions in micro-organisms, in *Receptors and Recognition,* Ser. A, Vol. 3, Cuatrecasas, P. and Greaves, M. F., Eds., Chapman and Hall, London, 1977, chap. 2.
17. **Reissig, J. L., Ed.,** *Microbial Interactions,* Chapman and Hall, London, 1977.
18. **Burnett, J. H.,** *Fundamentals of Mycology,* 1st ed., Edward Arnold, London, 1968.
19. **Burnett, J. H.,** *Fundamentals of Mycology,* 2nd ed., Edward Arnold, London, 1976.
20. **Hawker, L. E.,** *The Physiology of Reproduction in Fungi,* Cambridge University Press, Cambridge, 1959.
21. **Garrett, S. D.,** Inoculum potential, in *Plant Pathology: an Advanced Treatise,* Vol. 3, Horsfall, J. G. and Dimond, A. E., Eds., Academic Press, New York, 1960, chap. 2.
22. **Butler, G. M.,** Vegetative structures, in *The Fungi: an Advanced Treatise,* Vol. 2, Ainsworth, G. C. and Sussman, A. S., Eds., Academic Press, New York, 1966, chap. 3.
23. **Willetts, H. J.,** The morphogenesis and possible evolutionary origins of fungal sclerotia, *Biol. Rev. Cambridge Philos. Soc.,* 47, 515, 1972.
24. **Chet, I. and Henis, Y.,** Sclerotial morphogenesis in fungi, *Annu. Rev. Phytopathol.,* 13, 169, 1975.
25. **Carlile, M. J. and Gooday, G. W.,** Cell fusion in myxomycetes and fungi, in *Membrane Fusion,* Poste, G. and Nicolson, G. L., Eds., North-Holland, Amsterdam, 1978, chap. 5.
26. **Bonner, J. T.,** *The Cellular Slime Molds,* 1st ed., Princeton University Press, Princeton, N. J., 1959.
27. **Calleja, G. B. and Johnson, B. F.,** *The Effects of Nutritional Factors on Microbial Aggregation,* NRCC Contribution No. 16394, Canada Institute for Scientific and Technical Information (CISTI), Ottawa, 1979.
28. **Achtman, M.,** Mating aggregates in *Escherichia coli* conjugation, *J. Bacteriol.,* 123, 505, 1975.
29. **Pakula, R., Ray, P., and Spencer, L. R.,** Some characteristics of streptococci competent for uptake of deoxyribonucleic acid, *Can. J. Microbiol.,* 16, 345, 1970.
30. **Tomasz, A. and Zanati, E.,** Appearance of a protein "agglutinin" on the sphaeroplast membrane of pneumococci during induction of competence, *J. Bacteriol.,* 105, 1213, 1971.
31. **Erickson, R. J. and Copeland, J. C.,** Structure and replication of chromosomes in competent cells of *Bacillus subtilis, J. Bacteriol.,* 109, 1075, 1972.
32. **Miyake, A.,** Induction of conjugation by cell-free fluid in the ciliate *Blepharisma, Proc. Jpn. Acad.,* 44, 837, 1968.
33. **Miyake, A. and Nobili, R.,** Mating reaction and its daily rhythm in *Euplotes crassus, J. Protozool.,* 21, 584, 1974.
34. **Pringle, J. R. and Mor, J.-R.,** Methods for monitoring the growth of yeast cultures and for dealing with the clumping problem, in *Methods in Cell Biology,* Vol. 11, Prescott, D. M., Ed., Academic Press, New York, 1975, chap. 7.
35. **Wickerham, L. J.,** Sexual agglutination of heterothallic yeasts in diverse taxonomic areas, *Science,* 128, 1504, 1958.

36. **Wickerham, L. J.**, Yeast taxonomy in relation to ecology, genetics, and phylogeny, *Antonie van Leeuwenhoek J. Microbiol. Serol.*, 35 (Suppl. Yeast Symp. 1969), 31, 1969.
37. **Gibbons, R. J. and van Houte, J.**, Bacterial adherence in oral microbial ecology, *Annu. Rev. Microbiol.*, 29, 19, 1975.
38. **Wickerham, L. J. and Burton, K. A.**, Heterothallism in *Saccharomyces rouxii, J. Bacteriol.*, 80, 492, 1960.
39. **Wickerham, L. J. and Dworschack, R. G.**, Extracellular invertase production by sexually agglutinative mating types of *Saccharomyces kluyveri, Science*, 131, 985, 1960.
40. **Buller, A. H. R.**, *Researches on Fungi*, Vol. 4, Longmans Green, London, 1931.
41. **Buller, A. H. R.**, *Researches on Fungi*, Vol. 5, Longmans Green, London, 1934.
42. **Brefeld, O.**, *Dictyostelium mucoroides*. Ein neuer Organismus aus der Myxomyceten, *Abh. Senckenb. Naturforsch. Ges.*, 7, 85, 1869.
43. **van Tieghem, P.**, Sur quelques Myxomycètes à plasmode agrégé, *Bull. Soc. Bot. Fr.*, 27, 317, 1880.
44. **Thaxter, R.**, On the Myxobacteriaceae, a new order of Schizomycetes, *Bot. Gaz.*, 17, 389, 1892.
45. **Berthold, G.**, Die geschlechtliche Fortpflanzung der eigentlichen Phaeosporeen, *Mitt. Zool. Stat. Neapel*, 2, 401, 1881.
46. **Pasteur, L.**, *Etudes sur la Bière*, Gauthier Villars, Paris, 1876, p. 196.
47. **Seyffert, H.**, Einiges über Reinzucht-Hefen und ihre Ernährung, *Z. Gesamte Brauwes.*, 19, 318, 1896.
48. **Kusserow, R.**, *Brennerei Ztg.*, 14, 318, 1897; cited by Jansen, H. E., Flocculation of yeasts, in *The Chemistry and Biology of Yeasts*, Cook, A. H., Ed., Academic Press, New York, 1958, chap. 13.
49. **Lange, H.**, Ueber den Einfluss verschiedenartiger Stickstoffernährung auf die Hefe, *Wochenschr. Brau.*, 16, 49, 1899.
50. **Lange, H.**, Einfluss mechanischer Beimengungen auf die grobflockige Abscheidung der Hefe. (Bruchbildung.), *Jahrb. Vers. Lehranst. Brau. Berlin*, 4, 19, 1901.
51. **Delbrück, M.**, Die Zurückführung des physiologischen Zustandes der Hefe auf die Veränderungen ihres Enzymgehaltes während der Gährung und Lagerung; ihr Einfluss auf Vergähurngsgrad und Bruch, *Jahrb. Vers. Lehranst. Brau. Berlin*, 4, 295, 1901.
52. **Holderer, H.**, Discussion, *Jahrb. Vers. Lehranst. Brau. Berlin*, 4, 306, 1901.
53. **Lindner, P.**, Discussion, *Jahrb. Vers. Lehranst. Brau. Berlin*, 4, 309, 1901.
54. **Barendrecht, H. P.**, Die Agglutination von Hefe, *Centralbl. Bakteriol. Parasitenk. Infektionskr. Abt. 2*, 7, 623, 1901.
55. **Brefeld, O.**, *Botanische Untersuchungen über Schimmelpilze*, Vol. 3, Felix, Leipzig, 1877.
56. **de Bary, A.**, *Comparative Morphology and Biology of the Fungi, Mycetozoa and Bacteria*, Oxford Clarendon Press, London, 1887.
57. **Reinhardt, M. O.**, Das Wachsthum der Pilzhyphen. Ein Beitrag zur Kenntniss des Flächenwachsthums vegetabilischer Zellmembranen, *Jahrb. Wiss. Bot.*, 23, 479, 1892.
58. **Sonneborn, T. M.**, Sex, sex inheritance and sex determination in *Paramecium aurelia, Proc. Natl. Acad. Sci. U.S.A.*, 23, 378, 1937.
59. **Bütschli, O.**, Studien über die erster Enwicklungsvorgange des Einzelle, die Zellteilung und der Konjugation der Infusorien, *Abh. Senckenb. Naturforsch. Ges.*, 10, 1, 1876.
60. **Maupas, E.**, Le rajeunissement karyogamique chez les ciliés, *Arch. Zool. Exp. Gen. 2e Ser.*, 7, 149, 1889.
61. **Hertwig, R.**, Über die Konjugation der Infusorien, *Abh. Bayer. Akad. Wiss.*, 17, 150, 1889.
62. **Wickerham, L. J.**, Influence of agglutination on zygote formation in *Hansenula wingei*, a new species of yeast, *C. R. Trav. Lab. Carlsberg Ser. Physiol.*, 26, 423, 1956.
63. **Wickerham, L. J.**, Genus 1. *Citeromyces* Santa Maria, in *The Yeasts: a Taxonomic Study*, Lodder, J., Ed., North-Holland, Amsterdam, 1970, chap. 4.
64. **Leupold, U.**, Die Vererbung von Homothallie und Heterothallie bei *Schizosaccharomyces pombe, C. R. Trav. Lab. Carlsberg Ser. Physiol.*, 24, 381, 1950.
65. **Calleja, G. B.**, Flocculation in *Schizosaccharomyces pombe, J. Gen. Microbiol.*, 64, 247, 1970.
66. **Calleja, G. B. and Johnson, B. F.**, Flocculation in a fission yeast: an initial step in the conjugation process, *Can. J. Microbiol.*, 17, 1175, 1971.
67. **Egel, R.**, Physiological aspects of conjugation in fission yeast, *Planta*, 98, 89, 1971.
68. **Itoh, S., Takahashi, S., Tsuboi, M., Shimoda, C., and Hayashibe, M.**, Effect of light on sexual flocculation in *Schizosaccharomyces japonicus, Plant Cell Physiol.*, 17, 1355, 1976.
69. **Sakai, K. and Yanagishima, N.**, Mating reaction in *Saccharomyces cerevisiae*. I. Cell agglutination related to mating, *Arch. Mikrobiol.*, 75, 260, 1971.
70. **Lindegren, C. C. and Lindegren, G.**, A new method for hybridizing yeast, *Proc. Natl. Acad. Sci. U.S.A.*, 29, 306, 1943.
71. **Levi, J. D.**, Mating reaction in yeast, *Nature (London)*, 177, 753, 1956.
72. **Duntze, W., MacKay, V., and Manney, T. R.**, *Saccharomyces cerevisiae*: a diffusible sex factor, *Science*, 168, 1472, 1970.

73. **Lederberg, J. and Tatum, E. L.,** Gene recombination in *Escherichia coli, Nature (London),* 158, 558, 1946.
74. **Wickerham, L. J. and Burton, K. A.,** A simple technique for obtaining mating types in heterothallic diploid yeasts, with special reference to their uses in the genus *Hansenula, J. Bacteriol.,* 67, 303, 1954.
75. **Leupold, U.,** Methodisches zur Genetik von *Schizosaccharomyces pombe, Schweiz. Z. Allgem. Pathol. Bakteriol.,* 20, 535, 1957.
76. **Calleja, G. B. and Johnson, B. F.,** A comparison of quantitative methods for measuring yeast flocculation, *Can. J. Microbiol.,* 23, 68, 1977.
77. **Eddy, A. A.,** Flocculation characteristics of yeasts. I. Comparative survey of various strains of *Saccharomyces cerevisiae, J. Inst. Brew. London,* 61, 307, 1955.
78. **Townsend, B. B.,** Nutritional factors influencing the production of sclerotia by certain fungi, *Ann. Bot. (N.S.),* 21, 153, 1957.
79. **Hough, J. S.,** Characterizing the principal components of pitching yeasts, *J. Inst. Brew. London,* 63, 483, 1957.
80. **Hunt, D. E. and Carpenter, P. L.,** Sterols as components in the mating reaction of *Hansenula wingei, J. Bacteriol.,* 86, 845, 1967.
81. **Mill, P. J.,** The nature of the interactions between flocculent cells in the flocculation of *Saccharomyces cerevisiae, J. Gen. Microbiol.,* 35, 61, 1964.
82. **Gibbons, R. J. and Fitzgerald, R. J.,** Dextran-induced agglutination of *Streptococcus mutans,* and its potential role in the formation of microbial dental plaques, *J. Bacteriol.,* 98, 341, 1969.
83. **Takahashi, M. and Hiwatashi, K.,** Disappearance of mating reactivity in *Paramecium caudatum* upon repeated washing, *J. Protozool.,* 17, 667, 1970.
84. **Calleja, G. B.,** Role of mitochondria in the sex-directed flocculation of a fission yeast, *Arch. Biochem. Biophys.,* 154, 382, 1973.
85. **Stewart, G. G., Russell, I., and Garrison, I. F.,** Further studies on flocculation and co-flocculation in brewer's yeast strains, *Am. Soc. Brew. Chem. Proc.,* p. 100, 1973.
86. **Wiese, L.,** Nature of sex specific glycoprotein agglutinins in *Chlamydomonas, Ann. N.Y. Acad. Sci.,* 234, 383, 1974.
87. **Stewart, G. G., Garrison, I. F., Goring, T. E., Meleg, M., Pipasts, P., and Russell, I.,** Biochemical and genetic studies on yeast flocculation, *Kemia-Kemi,* 3, 465, 1976.
88. **Bonner, J. T.,** Development in the cellular slime molds: the role of cell division, cell size, and cell number, in *Developing Cell Systems and Their Controls,* Rudnick, D., Ed., Ronald Press, New York, 1960, 3.
89. **Parker, D. S., Kaufman, W. J., and Jenkins, D.,** Physical conditioning of activated sludge floc, *J. Water Pollut. Control Fed.,* 43, 1817, 1971.
90. **Manney, T. R. and Meade J. H.,** Cell-cell interactions during mating in *Saccharomyces cerevisiae,* in *Microbial Interactions,* Riessig, J. L., Ed., Chapman and Hall, London, 1977, chap. 7.
91. **Gilliland, R. B.,** The flocculation characteristics of brewing yeasts during fermentation, Eur. Brew. Conv. Proc. Congr., Brighton, 1951, 35.
92. **Hartong, B. D.,** Flocculation of yeast in the brewery, Eur. Brew. Conv. Proc. Congr., Brighton, 1951, 110.
93. **de Clerk, J. and van Roey, G.,** Courbe de fermentation et floculation de la levure, Eur. Brew. Conv. Proc. Congr., Brighton, 1951, 120.
94. **Brock, T. D.,** Mating reaction in the yeast *Hansenula wingei.* Preliminary observations and quantitation, *J. Bacteriol.,* 75, 697, 1958.
95. **Woof, J. B.,** An approach to yeast sedimentation characteristics, *J. Inst. Brew. London,* 68, 315, 1962.
96. **Ginterová, A. and Janotková, O.,** A simple method for characterizing yeast sedimentation, *J. Inst. Brew. London,* 71, 417, 1965.
97. **Stanley, S. O. and Rose, A. H.,** On the clumping of *Corynebacterium xerosis* as affected by temperature, *J. Gen. Microbiol.,* 48, 9, 1967.
98. **Crandall, M. A. and Brock, T. D.,** Molecular basis of mating in yeast *Hansenula wingei, Bacteriol. Rev.,* 32, 139, 1968.
99. **Calleja, G. B.,** On the nature of the forces involved in the sex-directed flocculation of a fission yeast, *Can. J. Microbiol.,* 20, 797, 1974.
100. **Shimoda, C. and Yanagishima, N.,** Mating reaction in *Saccharomyces cerevisiae.* VI. Effect of 2-deoxyglucose on conjugation, *Plant Cell Physiol.,* 15, 767, 1974.
101. **Osumi, M., Shimoda, C., and Yanagishima, N.,** Mating reaction in *Saccharomyces cerevisiae.* V. Changes in the fine structure during the mating reaction, *Arch. Microbiol.,* 97, 27, 1974.
102. **Taylor, N. W. and Orton, W. L.,** Calcium in flocculence of *Saccharomyces cerevisiae, J. Inst. Brew. London,* 81, 53, 1975.
103. **Burns, J. A.,** Yeast flocculation, *J. Inst. Brew. London,* 43, 31, 1937.
104. **Neilsen, N.,** Untersuchungen über die Senkungsgeschwindigkeit der Hefe, *C. R. Trav. Lab. Carlsberg Ser. Physiol.,* 22, 61, 1937.

105. **Lund, A.,** Flocculation as a racial character in bottom fermentation yeast, Eur. Brew. Conv. Proc. Congr., Brighton, 1951, 10.
106. **Helm, E., Nøhr, B., and Thorne, R. S. W.,** The measurement of yeast flocculence and its significance in brewing, *Wallerstein Lab. Commun.*, 16, 315, 1953.
107. **Greenshields, R. N., Yates, J., Sharp. F., and Davies, T. M. C.,** Studies in tower continuous fermentation. I. Methods for the characterization of flocculent yeasts, *J. Inst. Brew. London*, 78, 236, 1972.
108. **Jansen, H. E. and Mendlik, F.,** A study on yeast flocculation, Eur. Brew. Conv. Proc. Congr., Brighton, 1951, 59.
109. **Chester, V. E.,** A study of changes in flocculence in a single cell culture of a strain of *Saccharomyces cerevisiae, Proc. R. Soc. London Ser. B.*, 157, 223, 1963.
110. **Mill, P. J.,** The effect of nitrogenous substances on the time of flocculation of *Saccharomyces cerevisiae, J. Gen. Microbiol.*, 35, 53, 1964.
111. **Porter, A. M. and Macaulay, R. J.,** Studies on flocculation. I. A relationship between the pH and calcium content of the growth medium, *J. Inst. Brew. London*, 71, 175, 1965.
112. **Ault, R. G., Hampton, A. N., Newton, R., and Roberts, R. H.,** Biological and biochemical aspects of tower fermentation, *J. Inst. Brew. London*, 75, 261, 1969.
113. **Greenshields, R. N., Coote, S. D. J., Yates, J., and Smith, E. L.,** The characteristics and kinetics of flocculent yeasts in continuous tower fermentation systems, in *Proc. 4th Int. Symp. on Yeasts*, Part I, Klaushofer, H. and Sleytr, U. B., Eds., Hochschule f. Bodenkultur, Wien, 1974, 99.
114. **Patel, G. B. and Ingledew, W. M.,** A flocculation test for moderately flocculent brewing yeast, *J. Inst. Brew. London*, 81, 123, 1975.
115. **Holmberg, S. and Keilland-Brandt, M. C.,** A mutant of *Saccharomyces cerevisiae* temperature sensitive for flocculation. Influence of oxygen and respiratory deficiency on flocculence, *Carlsberg Res. Commun.*, 43, 37, 1978.
116. **Greenshields, R. N. and Smith, E. L.,** Tower-fermentation systems and their applications, *Chem. Eng. (London)*, 249, 182, 1971.
117. **Newell, P. C.,** How cells communicate: the system used by slime moulds, *Endeavour (N.S.)*, 1, 63, 1977.
118. **Sussman, M.,** A developmental analysis of cellular slime mold aggregation, in *Chemical Basis of Development*, McElroy, W. D. and Glass, B., Eds., Johns Hopkins Press, Baltimore, 1958, 264.
119. **Miyake, A.,** Cell interaction in conjugation of ciliates, *Curr. Top. Microbiol. Immunol.*, 64, 49, 1974.
120. **Johnson, B. F., Yoo, B. Y., and Calleja, G. B.,** Cell division in yeasts: movement of organelles associated with cell plate growth of *Schizosaccharomyces pombe, J. Bacteriol.*, 115, 358, 1973.
121. **Calleja, G. B., Yoo, B. Y., and Johnson, B. F.,** Fusion and erosion of cell walls during conjugation in the fission yeast *(Schizosaccharomyces pombe), J. Cell Sci.*, 25, 139, 1977.
122. **Hicks, J. B. and Herskowitz, I.,** Interconversion of yeast mating types. I. Direct observation of the action of the homothallism (*HO*) gene, *Genetics*, 83, 245, 1976.
123. **Egel, R.,** "Flip-flop" control and transposition of mating-type genes in fission yeast, in *DNA: Insertion Elements, Plasmids, and Episomes*, Bukhari, A. I., Shapiro, J. A., and Adhya, S. L., Eds., Cold Spring Harbor Laboratory, Cold Spring Harbor, N. Y., 1977, 447.
124. **Calleja, G. B., Johnson, B. F., Zuker, M., and James, A. P.,** The mating system of a homothallic fission yeast, *Mol. Gen. Genet.*, 172, 1, 1979.
125. **Crandall, M. and Caulton, J. H.,** Induction of haploid glycoprotein mating factors in diploid yeasts, in *Methods in Cell Biology*, Vol. 12, Prescott, D. M., Ed., Academic Press, New York, 1975, 185.
126. **Sena, E. P., Radin, D. N., Welch, J., and Fogel, S.,** Synchronous mating in yeasts, in *Methods in Cell Biology*, Vol. 11, Prescott, D. M., Ed., Academic Press, New York, 1975, chap. 4.
127. **Stewart, G. G. and Russell, I.,** The identification, characterization, and mapping of a gene for flocculation in *Saccharomyces* sp., *Can. J. Microbiol.*, 23, 441, 1977.
128. **Hartwell, L. H.,** Synchronization of haploid yeast cell cycles, a prelude to conjugation, *Exp. Cell Res.*, 76, 111, 1973.
129. **Bücking-Throm, E., Duntze, W., Hartwell, L. H., and Manney, T. R.,** Reversible arrest of haploid yeast cells at the initiation of DNA synthesis by a diffusible sex factor, *Exp. Cell Res.*, 76, 99, 1973.
130. **Egel, R. and Egel-Mitani, M.,** Premeiotic DNA synthesis in fission yeast, *Exp. Cell Res.*, 88, 127, 1974.
131. **Streiblová, E. and Wolf, A.,** Role of cell wall topography in conjugation of *Schizosaccharomyces pombe, Can. J. Microbiol.*, 21, 1399, 1975.
132. **Streiblová, E.,** Control of conjugation during the cell cycle of *Schizosaccharomyces pombe, Folia Microbiol. (Prague)*, 21, 194, 1976.
133. **Schmeisser, E. T., Baumgartel, D. M., and Howell, S. H.,** Gametic differentiation in *Chlamydomonas reinhardii*. Cell dependence and rates of attainment of mating competency, *Dev. Biol.*, 31, 31, 1973.
134. **Calleja, G. B.,** Glucose effect and the effect of cyclic AMP on flocculation and sporulation of *Schizosaccharomyces pombe*, in *Proceedings of the Fourth International Symposium on Yeasts*, Part I, Klaushofer, H. and Sleytr, U. B., Eds., Hochschule f. Bodenkultur, Wien, 1974, 73.

135. **Montenecourt, B. S., Kuo, S.-C., and Lampen, J. O.,** *Saccharomyces* mutants with invertase formation resistant to repression by hexoses, *J. Bacteriol.*, 114, 233, 1973.
136. **Schamhart, D. H. J., ten Berge, A. M. A., and van de Poll, K. W.,** Isolation of a catabolite repression mutant of yeast as a revertant of a strain that is maltose negative in the respiratory-deficient state, *J. Bacteriol.*, 121, 747, 1975.
137. **Kates, J. R. and Jones, R. F.,** The control of gametic differentiation in liquid cultures of *Chlamydomonas*, *J. Cell. Comp. Physiol.*, 63, 157, 1964.
138. **Cohen, L. W.,** The basis for the circadian rhythm of mating in *Paramecium bursaria*, *Exp. Cell. Res.*, 37, 360, 1965.
139. **Imafuku, M.,** Adaptation of the circadian rhythm of mating reactivity to abnormal light dark cycles in *Paramecium bursaria*, *J. Interdiscip. Cycle Res.*, 6, 141, 1975.
140. **Shaffer, B. M.,** Integration in aggregating cellular slime moulds, *Q. J. Microsc. Sci.*, 99, 103, 1958.
141. **Doi, S. and Yoshimura, M.,** Temperature-sensitive loss of sexual agglutinability in *Saccharomyces cerevisiae*, *Arch. Microbiol.*, 114, 287, 1977.
142. **Doi, S. and Yoshimura, M.,** Temperature-dependent conversion of sexual agglutinability in *Saccharomycs cerevisiae*, *Mol. Gen. Genet.*, 162, 251, 1978.
143. **Calleja, G. B. and Johnson, B. F.,** Temperature sensitivity of flocculation induction, conjugation and sporulation in fission yeast, *Antonie van Leeuwenhoek J. Microbiol.*, 45, 391, 1979.
144. **Hirsch, H. M.,** Environmental factors influencing the differentiation of protoperithecia and their relation to tyrosinase and melanin formation in *Neurospora crassa*, *Physiol. Plant.*, 7, 72, 1954.
145. **Bowman, D. H.,** Sporidial fusion in *Ustilago maydis*, *J. Agric. Res.*, 72, 233, 1946.
146. **Hawker, L. E.,** Environmental influences on reproduction, in *The Fungi: an Advanced Treatise*, Vol. 2, Ainsworth, G. C. and Sussman, A. S., Eds., Academic Press, New York, 1966, 435.
147. **Fowell, R. R.,** Sporulation and hybridization of yeasts, in *The Yeasts*, Vol. 1, Rose, A. H. and Harrison, J. S., Eds., Academic Press, London, 1969, 303.
148. **Lu, B. C.,** Dark dependence of meiosis at elevated temperatures in the basidiomycete *Coprinus lagopus*, *J. Bacteriol.*, 111, 833, 1972.
149. **Lüers, H. and Heuss, R.,** Zur Kenntnis der Bruchbildung der Hefen, *Z. Gesamte Brauwes.*, 44, 18, 1922.
150. **Geys, K.,** Über die Bruchbildung der Hefe und ihre Beeinflussung durch die Reinzucht, *Z. Gesamte Brauwes.*, 45, 51, 1922.
151. **Winslow, C. E. A. and Fleeson, E. H.,** The influence of electrolytes upon the electrophoretic migration of bacteria and yeast cells, *J. Gen. Physiol.*, 8, 195, 1926.
152. **Stockhausen, F.,** Elektrische Ladung der Hefe und Wasserstoffionenkonzentration, Einfluss auf Gärfähigkeit und Flockenbildungsvermögen, *Wochenschr. Brau.*, 44, 121, 1927.
153. **Koch, R.,** Beiträge zur Kenntnis der Bruchbildung der Brauereihefen, *Wochenschr. Brau.*, 45, 175, 1928.
154. **McCandlish, D. and Hagues, G.,** Some practical advantages to be derived from the neutralisation of brewing liquor, *J. Inst. Brew. London*, 35, 61, 1929.
155. **de Clerk, J.,** Le problème de l'atténuation, *Bull. Assoc. Anc. Eleves Ec. Super. Brass. Univ. Louvain*, 30, 1, 1930.
156. **Moldawskaya, E. A.,** Über Besonderheiten der elektrischen Ladung der Hefezellen, *Biochem. Z.*, 257, 480, 1933.
157. **Malkow, A., Petina, A., and Zwetkowa, N.,** Studien über Flockenbildung (Agglutination) der Hefe, *Zentralbl. Bakteriol. Parasitenk. Infektionskr. Abt. 2*, 88, 193, 1933.
158. **Malkow, A.,** Studien über Flockenbildung (Agglutination) der Hefe. II. Mitteilung, *Zentralbl. Bakteriol. Parasitenk. Infektionskr. Abt. 2*, 90, 212, 1934.
159. **Hennig, K. and Ay, H.,** Die kataphoretische Untersuchung der Wein- und Brauereihefen. Über die kataphoretischen Eigenschaften der Hefe, *Biochem. Z.*, 299, 123, 1938.
160. **Silbereisen, K.,** Die elektrische Ladung der Hefe und ihre Bedeutung für die Agglutination (Bruchbildung), *Wochenschr. Brau.*, 55, 153, 1938.
161. **Metz, C. B.,** Mating substances and the physiology of fertilization in ciliates, in *Sex in Microorganisms*, Wenrich, D. H., Ed., American Association for the Advancement of Science, Washington, D.C., 1954, 284.
162. **Yanagishima, N., Shimoda, C., and Kitano, S.,** Sex-specific factors responsible for sexual cell agglutination in *Saccharomyces cerevisiae*, in *Proc. 1st Int. Congr. of IAMS*, Vol. 1, Hasegawa, T., Ed., Science Council of Japan, Tokyo, 1975, 265.
163. **Calleja, G. B.,** The effects of cyclic AMP and actinomycin D on an inducible system in a fission yeast, *Kalikasan Philipp. J. Biol.*, 1, 63, 1972.
164. **Lodish, H. F., Margolskee, J. P., and Blumberg, D. D.,** Transcriptional and translational control of protein synthesis during differentiation of *Dictyostelium discoideum*, in *Differentiation and Development*, Miami Winter Symposia, Vol. 15, Ahmad, F., Russell, T. R., Schultz, J., and Werner, R., Eds., Academic Press, New York, 1978, 169.

165. **Miyake, A. and Honda, H.,** Cell union and protein synthesis in conjugation of *Blepharisma, Exp. Cell Res.*, 100, 31, 1976.
166. **Taylor, N. W. and Orton, W. L.,** Association constant of the sex-specific agglutinin in the yeast, *Hansenula wingei, Biochemistry*, 9, 2931, 1970.
167. **Taylor, N. W. and Orton, W. L.,** Cooperation among the active binding sites in the sex-specific agglutinin from the yeast, *Hansenula wingei, Biochemistry*, 10, 2043, 1971.
168. **Yen, P. H. and Ballou, C. E.,** Composition of a specific intercellular agglutination factor, *J. Biol. Chem.*, 248, 8316, 1973.
169. **Yen, P. H. and Ballou, C. E.,** Partial characterization of the sexual agglutination factor from *Hansenula wingei* Y-2340 type 5 cells, *Biochemistry*, 13, 2428, 1974.
170. **Crandall, M., Lawrence, L. M., and Saunders, R. M.,** Molecular complementarity of yeast glycoprotein mating factors, *Proc. Natl. Acad. Sci. U.S.A.*, 71, 26, 1974.
171. **Shimoda, C., Kitano, S., and Yanagishima, N.,** Mating reaction in *Saccharomyces cerevisiae*. VII. Effect of proteolytic enzymes on sexual agglutinability and isolation of crude sex-specific substances responsible for sexual agglutination, *Antonie van Leeuwenhoek J. Microbiol. Serol.*, 41, 513, 1975.
172. **Shimoda, C. and Yanagishima, N.,** Mating reaction in *Saccharomyces cerevisiae*. VIII. Mating-type-specific substances responsible for sexual cell agglutination, *Antonie van Leeuwenhoek J. Microbiol. Serol.*, 41, 521, 1975.
173. **Sakurai, A., Tamura, S., Yanagishima, N., and Shimoda, C.,** Isolation of a peptidyl factor controlling sexual agglutination in *Saccharomyces cerevisiae, Proc. Jpn. Acad.*, 51, 291, 1975.
174. **Sakurai, A., Tamura, S., Yanagishima, N., and Shimoda, C.,** Isolation of a peptidyl factor, α substance-I_A, inducing sexual agglutinability in *Saccharomyces cerevisiae, Agric. Biol. Chem.*, 41, 389, 1977.
175. **Wiese, L. and Hayward, P. C.,** On sexual agglutination and mating type substances in isogamous dioecious chlamydomonads. III. The sensitivity of sex cell contact to various enzymes, *Am. J. Bot.*, 59, 530, 1972.
176. **Mclean, R. J., Laurendi, C. J., and Brown, R. M., Jr.,** The relationship of gamone to the mating reaction in *Chlamydomonas moewusii, Proc. Natl. Acad. Sci. U.S.A.*, 71, 2610, 1974.
177. **Bergman, K., Goodenough, R. U., Goodenough, D. A., Jawitz, J., and Martin, H.,** Gametic differentiation in *Chlamydomonas reinhardtii*. II. Flagellar membranes and the agglutination reaction, *J. Cell Biol.*, 67, 606, 1975.
178. **Snell, W. J.,** Mating in *Chlamydomonas*: a system for the study of specific cell adhesion. I. Ultrastructural and electrophoretic analyses of flagellar surface components involved in adhesion, *J. Cell Biol.*, 68, 48, 1976.
179. **Clark, M. A.,** Control of mating type expression in *Paramecium multimicronucleatum*, syngen 2, *J. Cell Physiol.*, 79, 1, 1972.
180. **Kitamura, K. and Hiwatashi, K.,** Mating-reactive membrane vesicles from cilia of *Paramecium caudatum, J. Cell Biol.*, 69, 736, 1976.
181. **Sherman, F. and Lawrence, C. W.,** *Saccharomyces*, in *Handbook of Genetics*, Vol. 1, King, R. C., Ed., Plenum Press, New York, 1974, chap. 24.
182. **Mortimer, R. K.,** Genetic map of *Saccharomyces cerevisiae*, in Abstr. and Program 9th Int. Conf. on Yeast Genetics and Molecular Biology, University of Rochester, Rochester, N.Y., 1978, back cover and inside back.
183. **Gutz, H., Heslot, H., Leupold, U., and Loprieno, N.,** *Schizosaccharomyces pombe*, in *Handbook of Genetics*, Vol. 1, King, R. C., Ed., Plenum Press, New York, 1974, chap. 25.
184. **Kohli, J., Hottinger, H., Munz, P., Strauss, A., and Thuriaux, P.,** Genetic mapping in *Schizosaccharomyces pombe* by mitotic and meiotic analysis and induced haploidization, *Genetics*, 87, 471, 1977.
185. **Levine, R. P.,** *Chlamydomonas reinhardi*, in *Handbook of Genetics*, Vol. 2, King, R. C., Ed., Plenum Press, New York, 1974, chap. 17.
186. **Preer, J. R., Jr.,** Genetics of the protozoa, in *Research in Protozoology*, Vol. 3, Chen, T. T., Ed., Pergamon Press, Oxford, 1969, 130.
187. **Sonneborn, T. M.,** *Paramecium aurelia*, in *Handbook of Genetics*, Vol. 2, King, R. C., Ed., Plenum Press, New York, 1974, chap. 20.
188. **Low, K. B.,** *Escherichia coli* K12 F-prime factors, old and new, *Bacteriol. Rev.*, 36, 587, 1972.
189. **Achtman, M.,** Genetics of the F sex factor in Enterobacteriaceae, *Curr. Top. Microbiol. Immunol.*, 60, 79, 1973.
190. **Low, K. B.,** F-prime factors of *Escherichia coli*, in *Handbook of Genetics*, Vol. 1, King, R. C., Ed., Plenum Press, New York, 1974, chap. 7.
191. **Lewis, C. W., Johnston, J. R., and Martin, P. A.,** The genetics of yeast flocculation, *J. Inst. Brew. London*, 82, 158, 1976.
192. **Lancaster, J. H., Goldschmidt, E. P., and Wyss, O.,** Characterization of conjugation factors in *Escherichia coli* cell walls. I. Inhibition of recombination by cell walls and cell extracts, *J. Bacteriol.*, 89, 1478, 1965.

193. **Ou, J. T. and Anderson, T. F.,** Effect of Zn^{2+} on bacterial conjugation: inhibition of mating pair formation, *J. Bacteriol.*, 111, 177, 1972.
194. **Curtiss, R., III, Caro, L. G., Allison, D. P., and Stallions, D. R.,** Early stages of conjugation in *Escherichia coli, J. Bacteriol.*, 100, 1091, 1969.
195. **Curtiss, R., III, Charamella, L. J., Stallions, D. R., and Mays, J. A.,** Parental functions during conjugation in *Escherichia coli* K-12, *Bacteriol. Rev.*, 32, 320, 1968.
196. **Ou, J. T.,** Effect of Zn^{2+} on bacterial conjugation: increases in ability of F^- cells to form mating pairs, *J. Bacteriol.*, 115, 648, 1973.
197. **Heumann, W.,** Conjugation in starforming *Rhizobium lupini, Mol. Gen. Genet.*, 102, 132, 1968.
198. **Mayer, F. and Schmitt, R.,** Electronenmikroskopische, diffraktometrische und disc-elektrophoretische Untersuchungen an Fimbrien des sternbildenden Bodenbakteriums *Pseudomonas echinoides* und einer nicht-sternbildenden Mutante, *Arch. Mikrobiol.*, 79, 311, 1971.
199. **Mayer, F., Kall, S., and Schmitt, R.,** Untersuchung der morphologischen Grundlagen für einen möglichen Übertragungsweg der DNA bei der Konjugation sternbildender Bodenbakterien, *Z. Allgem. Mikrobiol.*, 14, 221, 1974.
200. **Moll, G. and Ahrens, R.,** Ein neuer Fimbrientyp, *Arch. Mikrobiol.*, 70, 361, 1970.
201. **Duguid, J. P. and Gillies, R. R.,** Fimbriae and adhesive properties in dysentery bacilli, *J. Pathol. Bacteriol.*, 74, 397, 1957.
202. **Duguid, J. P.,** Fimbriae and adhesive properties in *Klebsiella* strains, *J. Gen. Microbiol.*, 21, 271, 1959.
203. **Old, D. C., Corneil, I., Gibson, L. F., Thomson, A. D., and Duguid, J. P.,** Fimbriation, pellicle formation and the amount of growth of salmonellas in broth, *J. Gen. Microbiol.*, 51, 1, 1968.
204. **Duguid, J. P.,** The function of bacterial fimbriae, *Arch. Immunol. Ther. Exp.*, 16, 173, 1968.
205. **Old, D. C. and Duguid, J. P.,** Selective outgrowth of fimbriate bacteria in static liquid medium, *J. Bacteriol.*, 103, 447, 1970.
206. **Old, D. C. and Duguid, J. P.,** Selection of fimbriate transductants of *Salmonella typhimurium* dependent on motility, *J. Bacteriol.*, 107, 655, 1971.
207. **Frøholm, L. O. and Bøvre, K.,** The fimbriated state and competence in genetic transformation of three *Moraxella* species, in *Bacterial Transformation*, Archer, L. J., Ed., Academic Press, London, 1973, 125.
208. **Tweedy, J. M., Park, R. W. A., and Hodgkiss, W.,** Evidence for the presence of fimbriae (pili) on *Vibrio* species, *J. Gen. Microbiol.*, 51, 235, 1968.
209. **Pakula, R. and Ray, P.,** Requirements for transformation of the Challis streptococcus, *Can. J. Microbiol.*, 16, 499, 1970.
210. **Tomasz, A. and Mosser, J. L.,** On the nature of the pneumococcal activator substance, *Proc. Natl. Acad. Sci. U.S.A.*, 55, 58, 1966.
211. **Tomasz, A.,** Cellular metabolism in genetic transformation of pneumococci: requirement for protein synthesis during induction of competence, *J. Bacteriol.*, 101, 860, 1970.
212. **Kohoutová, M.,** The mechanism of binding of the competence substance to the surface of non-competent cells in pneumococcus, in *Bacterial Transformation*, Archer, L. J., Ed., Academic Press, London, 1973, 89.
213. **Tomasz, A. and Hotchkiss, R. D.,** Regulations of the transformability of pneumococcal cultures by macromolecular cell products, *Proc. Natl. Acad. Sci. U.S.A.*, 51, 480, 1964.
214. **Young, F. E. and Spizizen, J.,** Incorporation of deoxyribonucleic acid in the *Bacillus subtilis* transformation system, *J. Bacteriol.*, 86, 392, 1963.
215. **Horvath, S.,** Competence in *Bacillus subtilis* transformation system, *J. Gen. Microbiol.*, 51, 85, 1968.
216. **Spizizen, J.,** Genetic activity of deoxyribonucleic acid in the reconstitution of biosynthetic pathways, *Fed. Proc. Fed. Am. Soc. Exp. Biol.*, 18, 957, 1959.
217. **Young, F. E. and Spizizen, J.,** Physiological and genetic factors affecting transformation of *Bacillus subtilis, J. Bacteriol.*, 81, 823, 1961.
218. **Anagnostopoulos, C. and Spizizen, J.,** Requirements for transformation in *Bacillus subtilis, J. Bacteriol.*, 81, 741, 1961.
219. **Wilson, G. A. and Bott, K. F.,** Nutritional factors influencing the development of competence in the *Bacillus subtilis* transformation system, *J. Bacteriol.*, 95, 1439, 1968.
220. **Bott, K. F. and Wilson, G. A.,** Metabolic and nutritional factors influencing the development of competence for transfection of *Bacillus subtilis, Bacteriol. Rev.*, 32, 370, 1968.
221. **Spizizen, J. and Anagnostopoulos, C.,** Transformation of indole and sucrose characters in *Bacillus subtilis*, strain 168, *Bacteriol. Proc.*, p. 185, 1960.
222. **Bott, K. F. and Wilson, G. A.,** Development of competence in the *Bacillus subtilis* transformation system, *J. Bacteriol.*, 94, 562, 1967.
223. **Portolés, A., Lopez, R., and Tapia, A.,** An approach to the study of competence regulation in *B. subtilis* growing in a chemostat, in *Bacterial Transformation*, Archer, L. J., Ed., Academic Press, London, 1973, 65.

224. **Stewart, C. R.,** Enhancement and inhibition of transformation in *Bacillus subtilis, J. Bacteriol.*, 95, 2428, 1968.
225. **Kelstrup, J. and Funder-Nielsen, T. D.,** Adhesion of dextran to *Streptococcus mutans, J. Gen. Microbiol.*, 81, 485, 1974.
226. **Germaine, G. R. and Schachtele, C. F.,** *Streptococcus mutans* dextransucrase: mode of interaction with high-molecular-weight dextran and role in cellular aggregation, *Infect. Immun.*, 13, 365, 1976.
227. **Gibbons, R. J. and van Houte, J.,** Selective bacterial adherence to oral epithelial surfaces and its role as an ecological determinant, *Infect. Immun.*, 3, 567, 1971.
228. **Freedman, M. L. and Tanzer, J. M.,** Dissociation of plaque formation from glucan-induced agglutination in mutants of *Streptococcus mutans, Infect. Immun.*, 10, 189, 1974.
229. **Ellwood, D. C.,** Chemostat studies of oral bacteria, in *Microbial Aspects of Dental Caries*, Stiles, H. M., Loesche, W. J., and O'Brien, T. C., Eds., Information Retrieval, Washington, D.C., 1976, 785.
230. **Schachtele, C. F., Harlander, S. K., Fuller, D. W., Zollinger, P. K., and Leung, W.-L. S.,** Bacterial interference with sucrose-dependent adhesion of oral streptococci, in *Microbial Aspects of Dental Caries*, Stiles, H. M., Loesche, W. J., and O'Brien, T. C., Eds., Information Retrieval, Washington, D. C., 1976, 401.
231. **McCabe, M. M. and Smith, E. E.,** Relationship between cell-bound dextransucrase and the agglutination of *Streptococcus mutans, Infect. Immun.*, 12, 512, 1975.
232. **Clark, J. B.,** Slime as a possible factor in cell clumping in *Nocardia corallina, J. Bacteriol.*, 75, 400, 1958.
233. **Dworkin, M.,** Nutritional regulation of morphogenesis in *Myxococcus xanthus, J. Bacteriol.*, 86, 67, 1963.
234. **Quehl, A.,** Untersuchungen über die Myxobakterien, *Centralbl. Bakteriol. Parasitenk, Infektionskr. Abt. 2*, 16, 9, 1906.
235. **Vahle, C.,** Vergleichende Untersuchungen über die Myxobakteriazeen und Bakteriazeen, sowie die Rhodobakteriazeen und Spirillazeen, *Centralbl. Bakteriol. Parasitenk. Infektionskr. Abt. 2*, 25, 178, 1910.
236. **Oetker, H.,** Untersuchungen über die Ernährung einiger Myxobakterien, *Arch. Mikrobiol.*, 19, 206, 1953.
237. **Kühlwein, H.,** Weitere Untersuchungen an Myxobakterien, *Arch. Mikrobiol.*, 19, 365, 1953.
238. **McVittie, A. and Zahler, S. A.,** Chemotaxis in *Myxococcus, Nature (London)*, 194, 1299, 1962.
239. **Leadbetter, E. R.,** Control of growth and morphogenesis in some *Myxococcus* species, *Nature (London)*, 200, 1127, 1963.
240. **McCurdy, H. D., Jr.,** Growth and fruiting body formation of *Chondromyces crocatus* in pure culture, *Can. J. Microbiol.*, 10, 935, 1964.
241. **Hemphill, H. E. and Zahler, S. A.,** Nutritional induction and suppression of fruiting in *Myxococcus xanthus* FBa, *J. Bacteriol.*, 95, 1018, 1968.
242. **Dworkin, M.,** Nutritional regulation of fruiting body formation in *Myxococcus xanthus, Bacteriol. Proc.*, p. 116, 1962.
243. **Jennings, J.,** Association of a steroid and a pigment with a diffusible fruiting factor in *Myxococcus virescens, Nature (London)*, 190, 190, 1961.
244. **Fluegel, W.,** Induced fruiting in myxobacteria, *Proc. Minn. Acad. Sci.*, 31, 114, 1964.
245. **Fluegel, W.,** Fruiting chemotaxis in *Myxococcus fulvus* (myxobacteria), *Proc. Minn. Acad. Sci.*, 30, 120, 1963.
246. **Brock, T. D.,** Biochemical and cellular changes occurring during conjugation in *Hansenula wingei, J. Bacteriol.*, 90, 1019, 1965.
247. **Crandall, M. and Caulton, J. H.,** Induction of glycoprotein mating factors in diploid yeast of *Hansenula wingei* by vanadium salts or chelating agents, *Exp. Cell Res.*, 82, 159, 1973.
248. **Oujezdsky, K. B. and Szaniszlo, P. J.,** Conjugation in the dimorphic chromomycosis fungus *Phialophora dermatitidis, J. Bacteriol.*, 114, 1356, 1973.
249. **Jakob, H.,** Technique de synchronisation de la formation des zygotes chez la levure *Saccharomyces cerevisiae, C.R. Acad. Sci.*, 254, 3909, 1962.
250. **Sena, E. P., Radin, D. N., and Fogel, S.,** Synchronous mating in yeast, *Proc. Natl. Acad. Sci. U.S.A.*, 70, 1373, 1973.
251. **Lee, E.-H., Lusena, C. V., and Johnson, B. F.,** A new method of obtaining zygotes in *Saccharomyces cerevisiae, Can. J. Microbiol.*, 21, 802, 1975.
252. **Sakurai, A., Tamura, S., Yanagishima, N., and Shimoda, C.,** Structure of the peptidyl factor inducing sexual agglutination in *Saccharomyces cerevisiae, Agric. Biol. Chem.*, 40, 1057, 1976.
253. **Shimoda, C., Yanagishima, N., Sakurai, A., and Tamura, S.,** Mating reaction in *Saccharomyces cerevisiae*. IX. Regulation of sexual cell agglutinability of *a* type cells by a sex factor produced by α type cells, *Arch. Microbiol.*, 108, 27, 1976.
254. **Yanagishima, N., Yoshida, K., Hamada, K., Hagiya, M., Kawanabe, Y., Sakurai, A., and Tamura, S.,** Regulation of sexual agglutinability in *Saccharomyces cerevisiae* of *a* and α types by sex-specific factors produced by their respective opposite mating types, *Plant Cell Physiol.*, 17, 439, 1976.

255. **Sakurai, A., Tamura, S., Yanagishima, N., and Shimoda, C.,** Structure of a peptidyl factor, α substance-I_A, inducing sexual agglutinability in *Saccharomyces cerevisiae, Agric. Biol. Chem.*, 41, 395, 1977.
256. **Biliński, T., Litwińska, J., Żuk, J., and Gajewski, W.,** Synchronization of zygote production in *Saccharomyces cerevisiae, J. Gen. Microbiol.*, 79, 285, 1973.
257. **Windisch, S.,** Flocculation of brewing yeasts, *Brew. Dig.*, 43, 62, 1968.
258. **Stewart, G. G.,** Yeast flocculation: practical implications and experimental findings, *Brew. Dig.*, 50, 42, 1975.
259. **Piendl, A.,** Brauereitechnologische Faktoren und Bruchbildung der Hefen, *Eur. Brew. Conv. Proc. Congr.*, 12, 381, 1969.
260. **Morimoto, K., Shimazu, T., Fujii, T., and Horie, Y.,** Some substances in malt inducing earley flocculation of yeast. I. Preliminary investigation on high molecular weight substances in malt and wort, *Rep. Res. Lab. Kirin Brew. Co.*, 18, 63, 1975.
261. **Fujii, T. and Horie, Y.,** Some substances in malt inducing earley flocculation of yeast. II. Further investigation on a substance isolated from wort inducing early flocculation, *Rep. Res. Lab. Kirin Brew. Co.*, 18, 75, 1975.
262. **Baker, D. A. and Kirsop, B. H.,** Flocculation in *Saccharomyces cerevisiae* as influenced by wort composition and by actidione, *J. Inst. Brew. London*, 78, 454, 1972.
263. **Kijima, M.,** Effect of wort protein on yeast flocculation, *J. Inst. Brew. London*, 60, 223, 1954.
264. **Kudo, S.,** On a polysaccharide as a new factor stimulating the flocculation of yeast in the primary fermentation of beer, *Proc. Jpn. Acad.*, 28, 93, 1952.
265. **Kudo, S.,** Flockung der Brauerei-Hefe: Über den Mechanismus der Hefeflockung dürch Würze-Gummi, der bei der Hauptgärung die Bruchbildung fördert, *Rep. Kirin Brew. Co.*, 3, 25, 1960.
266. **St. Johnston, J. H.,** The termination of primary fermentation. I. Experimental data obtained from flocculation and fermentation tests, *Wallerstein Lab. Commun.*, 16, 39, 1953.
267. **Lindquist, W.,** On the mechanism of yeast flocculation, *J. Inst. Brew. London*, 59, 59, 1953.
268. **Kudo, S. and Kigima, M.,** Über die Bruchbildung der Brauerei Hefe. IV. Mitteil. Über die Identifizierung des Bruch verursachenden Stoff, "Bärmigen", und das Hefebruch-Phänomen durch verwandte Verbindungen, *J. Agric. Chem. Soc. Jpn.*, 27, 809, 1952.
269. **Kudo, S. and Kijima, M.,** Studies of yeast flocculation: identification of a yeast-flocculating agent "Bärmigen" and yeast flocculation caused by the related substances, *Rep. Kirin Brew. Co.*, 3, 33, 1960.
270. **Eddy, A. A.,** Role of humic acid in the flocculation of yeast, *J. Inst. Brew. London*, 62, 320, 1956.
271. **Stewart, G. G., Russell, I., and Garrison, I. F.,** Some considerations of the flocculation characteristics of ale and lager yeast strains, *J. Inst. Brew. London*, 81, 248, 1975.
272. **Eddy, A. A.,** Flocculation characteristics of yeasts. II. Sugars as dispersing agents, *J. Inst. Brew. London*, 61, 313, 1955.
273. **Patel, G. B. and Ingledew, W. M.,** Glycogen — a physiological determinant of yeast flocculation?, *Can. J. Microbiol.*, 21, 1614, 1975.
274. **Nishihara, H., Toraya, T., and Fukui, S.,** Induction of floc-forming ability in brewer's yeast, *J. Ferment. Technol.*, 54, 356, 1976.
275. **Eddy, A. A.,** Flocculation characteristics of yeasts. III. General role of flocculating agents and special characteristics of a yeast flocculated by alcohol, *J. Inst. Brew. London*, 61, 318, 1955.
276. **Lyons, T. P. and Hough, J. S.,** The role of yeast cell walls in brewing, *Brew. Dig.*, 45, 52, 1970.
277. **Jayatissa, P. M. and Rose, A. H.,** Role of wall phosphomannan in flocculation of *Saccharomyces cerevisiae, J. Gen. Microbiol.*, 96, 165, 1976.
278. **Nishihara, H., Toraya, T., and Fukui, S.,** Factors affecting flocculation of brewer's yeast, *J. Ferment. Technol.*, 54, 351, 1976.
279. **Taylor, N. W. and Orton, W. L.,** Effect of alkaline-earth metal salts on flocculence in *Saccharomyces cerevisiae, J. Inst. Brew. London*, 79, 294, 1973.
280. **Masschelein, C. and Devreux, A.,** Floculence et structure de la paroi cellulaire de la levure, Eur. Brew. Conv. Proc. Congr., Copenhagen, 1957, 194.
281. **Harris, J. O.,** Possible mechanisms of yeast flocculation, *J. Inst. Brew. London*, 65, 5, 1959.
282. **Stewart, G. G. and Goring, T. E.,** Effect of some monovalent and divalent metal ions on the flocculation of brewers yeast strains, *J. Inst. Brew. London*, 82, 341, 1976.
283. **Katô, S. and Nishikawa, N.,** Studies on flocculation of brewers' yeast. I. A new method for determination of flocculating power and a comparative survey of various strains, *Bull. Brew. Sci.*, 3, 39, 1957.
284. **Eddy, A. A.,** Comparison of the respective electrophoretic and flocculation characteristics of different strains of *Saccharomyces, J. Inst. Brew. London*, 64, 139, 1958.
285. **Chester, V. E.,** Aspects of yeast flocculation. Some effects of toxic reagents on a flocculent brewing yeast, in *Surface Activity and the Microbial Cell*, S.C.I. Monograph 19, Society of Chemical Industry, London, 1965, 59.
286. **Fisher, D. J.,** Flocculation — some observations on the surface charges of yeast cells, *J. Inst. Brew. London*, 81, 107, 1975.

287. **Bonner, J. T.,** Epigenetic development in the cellular slime moulds, *Symp. Soc. Exp. Biol.*, 17, 341, 1963.
288. **Wright, B. E.,** Biochemistry of Acrasiales, in *Biochemistry and Physiology of Protozoa*, Vol. 3, Hunter, S. H., Ed., Academic Press, New York, 1964, 341.
289. **Gregg, J. H.,** Organization and synthesis in the cellular slime molds, in *The Fungi: an Advanced Treatise*, Vol. 2, Ainsworth, G. C. and Sussman, A. S., Eds., Academic Press, New York, 1966, 235.
290. **Bonner, J. T.,** *The Cellular Slime Molds*, 2nd ed., Princeton University Press, Princeton, N.J., 1967.
291. **Gerisch, G.,** Cell aggregation and differentiation in *Dictyostelium*, *Curr. Top. Dev. Biol.*, 3, 157, 1968.
292. **Bonner, J. T.,** Aggregation and differentiation in the cellular slime molds, *Annu. Rev. Microbiol.*, 25, 75, 1971.
293. **Newell, P. C.,** The development of the cellular slime mould *Dictyostelium discoideum*: a model system for the study of cellular differentiation, *Essays Biochem.*, 7, 87, 1971.
294. **Konijn, T. M.,** Cyclic AMP as a first messenger, *Adv. Cyclic Nucleotide Res.*, 1, 17, 1972.
295. **Gustafson, G. L. and Wright, B. E.,** Analysis of approaches used in studying differentiation of the cellular slime mold, *Crit. Rev. Microbiol.*, 1, 453, 1972.
296. **Garrod, D. and Ashworth, J. M.,** Development of the cellular slime mould *Dictyostelium discoideum*, in *Microbial Differentiation*, Symp. Soc. Gen. Microbiol., 23, Ashworth, J. M. and Smith, J. E., Eds., Cambridge University Press, Cambridge, U.K., 1973, 407.
297. **Killick, K. A. and Wright, B. E.,** Regulation of enzyme activity during differentiation in *Dictyostelium discoideum*, *Annu. Rev. Microbiol.*, 28, 139, 1974.
298. **Loomis, W. F.,** *Dictyostelium discoideum: a Developmental System*, Academic Press, New York, 1974.
299. **Jacobson, A. and Lodish, H. F.,** Genetic control of development of the cellular slime mold, *Dictyostelium discoideum*, *Annu. Rev. Genet.*, 9, 145, 1975.
300. **Newell, P. C.,** Cellular communication during aggregation of the slime mold *Dictyostelium*, in *Microbiology — 1975*, Schlessinger, D., Ed., American Society for Microbiology, Washington, D.C., 1975, 426.
301. **Gerisch, G. and Malchow, D.,** Cyclic AMP receptors and the control of cell aggregation in *Dictyostelium*, *Adv. Cyclic Nucleotide Res.*, 7, 49, 1976.
302. **Sussman, M.,** The genesis of multicellular organization and the control of gene expression in *Dictyostelium discoideum*, *Prog. Mol. Subcell. Biol.*, 4, 103, 1976.
303. **Gregg, J. H.,** Oxygen utilization in relation to growth and morphogenesis of the slime mold *Dictyostelium discoideum*, *J. Exp. Zool.*, 114, 173, 1950.
304. **Takeuchi, I. and Tazawa, M.,** Studies on the morphogenesis of the slime mould, *Dictyostelium discoideum*, *Cytologia*, 20, 157, 1955.
305. **Bradley, S. G., Sussman, M., and Ennis, H. L.,** Environmental factors affecting the aggregation of the cellular slime mold, *Dictyostelium discoideum*, *J. Protozool.*, 3, 33, 1956.
306. **Bonner, J. T., Kelso, A. P., and Gillmor, R. G.,** A new approach to the problem of aggregation in the cellular slime molds, *Biol. Bull. (Woods Hole, Mass.)*, 130, 28, 1966.
307. **Konijn, T. M.,** Effect of bacteria on chemotaxis in the cellular slime molds, *J. Bacteriol.*, 99, 503, 1969.
308. **Sussman, M., Alexander, S., Boschwitz, C., Brackenbury, R., Cohen, A., and Schindler, J.,** Morphogenetic feedback and gene expression in *Dictyostelium discoideum*, in *Developmental Biology*, McMahon, D. and Fox, C. F., Eds., W. A. Benjamin, Menlo Park, Calif., 1975, 89.
309. **Krichevsky, M. I. and Wright, B. E.,** Environmental control of the course of development in *Dictyostelium discoideum*, *J. Gen. Microbiol.*, 32, 195, 1963.
310. **Krichevsky, M. I. and Love, L. L.,** Adenine inhibition of the rate of sorocarp formation in *Dictyostelium discoideum*, *J. Gen. Microbiol.*, 37, 293, 1964.
311. **Yanagisawa, K. O., Tanaka, Y., and Yanagisawa, K.,** Effect of glucose and adenosine 3′,5′-cyclic monophosphate on the early development of *Dictyostelium discoideum*, *J. Biochem. (Tokyo)*, 75, 1321, 1974.
312. **Hirschberg, E. and Merson, G.,** Effect of test compounds on the aggregation and culmination of the slime mold *Dictyostelium discoideum*, *Cancer Res.*, 3 (Suppl.), 76, 1955.
313. **Pan, P., Hall, E. M., and Bonner, J. T.,** Determination of the active portion of the folic acid molecule in cellular slime mold chemotaxis, *J. Bacteriol.*, 122, 185, 1975.
314. **Krichevsky, M. I. and Love, L. L.,** The uptake and utilization of histidine by washed amoebae in the course of development in *Dictyostelium discoideum*, *J. Gen. Microbiol*,. 34, 483, 1964.
315. **Hirschberg, E. and Rusch, H. P.,** Effects of compounds of varied biochemical action on the aggregation of a slime mold, *Dictyostelium discoidem*, *J. Cell. Comp. Physiol.*, 36, 105, 1950.
316. **Konijn, T. M., van de Meene, J. G. C., Bonner, J. T., and Barkeley, D. S.,** The acrasin activity of adenosine-3′,5′-cyclic phosphate, *Proc. Natl. Acad. Sci. U.S.A.*, 58, 1152, 1967.
317. **Bonner, J. T., Barkley, D. S., Hall, E. M., Konijn, T. M., Mason, J. W., O'Keefe, G., III, and Wolfe, P. B.,** Acrasin, acrasinase, and the sensitivity to acrasin in *Dictyostelium discoideum*, *Dev. Biol.*, 20, 72, 1969.

318. **Konijn, T. M., van de Meene, J. G. C., Chang, Y. Y., Barkley, D. S., and Bonner, J. T.,** Identification of adenosine-3′,5′-monophosphate as the bacterial attractant for myxamoebae of *Dictyostelium discoideum, J. Bacteriol.*, 99, 510, 1969.
319. **Chi, Y.-Y. and Francis, D.,** Cyclic AMP and calcium exchange in a cellular slime mold, *J. Cell. Physiol.*, 77, 169, 1971.
320. **Mason, J. W., Rasmussen, H., and Dibella, F.,** 3′5′ AMP and Ca^{2+} in slime mold aggregation, *Exp. Cell Res.*, 67, 156, 1971.
321. **Robertson, A., Drage, D. J., and Cohen, M. H.,** Control of aggregation in *Dictyostelium discoideum* by an external periodic pulse of cyclic adenosine monophosphate, *Science*, 175, 333, 1972.
322. **McMahon, D.,** A cell-contact model for cellular position determination in development, *Proc. Natl. Acad. Sci. U.S.A.*, 70, 2396, 1973.
323. **Gerisch, G., Malchow, D., and Hess, B.,** Cell communication and cyclic-AMP regulation during aggregation of the slime mold, *Dictyostelium discoideum*, in *Biochemistry of Sensory Functions*, Jaenicke, L., Ed., Springer-Verlag, Berlin, 1974, 279.
324. **Darmon, M., Brachet, P., and Pereira da Silva, L. H.,** Chemotactic signals induce cell differentiation in *Dictyostelium discoideum, Proc. Natl. Acad. Sci. U.S.A.*, 72, 3163, 1975.
325. **Gerisch, G., Fromm, H., Huesgen, A., and Wick, U.,** Control of cell-contact sites by cyclic AMP pulses in differentiating *Dictyostelium* cells, *Nature (London)*, 255, 547, 1975.
326. **Gerisch, G., Malchow, D., Huesgen, A., Nanjundiah, V., Roos, W., Wick, U., and Hülser, D.,** Cyclic-AMP reception and cell recognition in *Dictyostelium discoideum*, in *Developmental Biology*, McMahon, D. and Fox, C. F., Eds., W. A. Benjamin, Menlo Park, Calif., 1975, 76.
327. **Mato, J. M. and Konijn, T. M.,** Chemotaxis and binding of cyclic AMP in cellular slime molds, *Biochim. Biophys. Acta*, 385, 173, 1975.
328. **Mato, J. M., Losada, A., Nanjundiah, V., and Konijn, T. M.,** Signal input for a chemotactic response in the cellular slime mold *Dictyostelium discoideum, Proc. Natl. Acad. Sci. U.S.A.*, 72, 4991, 1975.
329. **Shaffer, B. M.,** Secretion of cyclic AMP induced by cyclic AMP in the cellular slime mould *Dictyostelium discoideum, Nature (London)*, 255, 549, 1975.
330. **George, R. P.,** Disruption of multicellular organization in the cellular slime molds by cyclic AMP, *Cell Differ.* 5, 293, 1977.
331. **Mato, J. M. and Konijn, T. M.,** Enhanced cell aggregation in *Dictyostelium discoideum* by ATP activation of cyclic AMP receptors, *Dev. Biol.*, 47, 233, 1975.
332. **Mato, J. M. and Konijn, T. M.,** The activation of cell aggregation by phosphorylation in *Dictyostelium discoideum, Exp. Cell Res.*, 99, 328, 1976.
333. **Pan, P., Hall, E. M., and Bonner, J. T.,** Folic acid as second chemotatic substance in the cellular slime moulds, *Nature New Biol.*, 237, 181, 1972.
334. **Wright, B. E.,** Effect of steroids on aggregation in the slime mold *Dictyostelium discoideum, Bacteriol. Proc.*, p. 115, 1958.
335. **Heftmann, E., Wright, B. E., and Liddel, G. U.,** Identification of a sterol with acrasin activity in a slime mold, *J. Am. Chem. Soc.*, 81, 6525, 1959.
336. **Brachet, P. and Klein, C.,** Inhibition of growth and cellular aggregation of *Dictyostelium discoideum* by steroid compounds, *Exp. Cell Res.*, 93, 159, 1975.
337. **Klein, C. and Brachet, P.,** Effects of progesterone and EDTA on cyclic AMP and phosphodiesterase in *Dictyostelium discoideum, Nature (London)*, 254, 432, 1975.
338. **Brachet, P. and Klein, C.,** Cell responsiveness to cAMP during the aggregation phase of *Dictyostelium discoideum, Differentiation*, 8, 1, 1977.
339. **Brachet, P.,** Effet d'un ionophore sur l'aggrégation de *Dictyostelium discoideum, C.R. Acad. Sci. Ser. D*, 282, 377, 1976.
340. **Gerisch, G.,** Zellkontaktbildung vegatitiver und aggregationsreifer Zellen von *Dictyostelium discoideum, Naturwissenschaften*, 48, 436, 1961.
341. **Gingell, D. and Garrod, D. R.,** Effect of EDTA on electrophoretic mobility of slime mould cells and its relationship to current theories of cell adhesion, *Nature (London)*, 221, 192, 1969.
342. **Born, G. V. R. and Garrod, D.,** Photometric demonstration of aggregation of slime mould cells showing effects of temperature and ionic strength, *Nature (London)*, 220, 616, 1968.
343. **Gingell, D., Garrod, D. R., and Palmer, J. F.,** Divalent cations and cell adhesion, in *Symp. Calcium and Cell Function*, Cuthbert, A., Ed., MacMillan, New York, 1969, 59.
344. **Hohl, H.-R. and Raper, K. B.,** Nutrition of cellular slime molds. II. Growth of *Polysphondylium pallidum* in axenic culture, *J. Bacteriol.*, 85, 199, 1963.
345. **Wurster, B., Pan, P., Tyan, G.-G., and Bonner, J. T.,** Preliminary characterization of the acrasin of the cellular slime mold *Polysphondylium violaceum*. *Proc. Natl. Acad. Sci. U.S.A.*, 73, 795, 1976.
346. **Pan, P.,** Chemical messengers in the development of cellular slime molds, in *Developmental Biology*, McMahon, D. and Fox, C. F., Eds., W. A. Benjamin, Menlo Park, Calif., 1975, 421.

347. **Clark, M. A.,** Possible role for acetylcholine in aggregation centre spacing in *Polysphondylium violaceum, Nature (London),* 266, 170, 1977.
348. **Hostak, M. B. and Raper, K. B.,** The induction of cell aggregation in *Acytostelium* by alkaloids, *Bacteriol. Proc.,* p. 58, 1960.
349. **Butler, G. M.,** The development and behaviour of mycelial strands in *Merulius lacrymans* (Wulf.) Fr. I. Strand development during growth from a food-base through a non-nutrient medium, *Ann. Bot. (N.S.),* 21, 523, 1957.
350. **Butler, G. M.,** The development and behaviour of mycelial strands in *Merulius lacrymans* (Wulf.) Fr. II. Hyphal behaviour during strand formation, *Ann. Bot. (N.S.),* 22, 219, 1958.
351. **Garrett, S. D.,** A study of violet root rot: factors affecting production and growth of mycelial strands in *Helicobasidium purpureum* Pat., *Trans. Br. Mycol. Soc.,* 29, 114, 1946.
352. **Garrett, S. D.,** Function of the mycelial strands in substrate colonization by the cultivated mushroom, *Psalliota hortensis, Trans. Br. Mycol. Soc.,* 37, 51, 1954.
353. **Mathew, K. T.,** Morphogenesis of mycelial strands in the cultivated mushroom, *Agaricus bisporus, Trans. Br. Mycol. Soc.,* 44, 285, 1961.
354. **Snider, P. J.,** Stages of development in rhizomorphic thalli of *Armillaria mellea, Mycologia,* 51, 693, 1959.
355. **Garrett, S. D.,** Rhizomorph behaviour in *Armillaria mellea* (Vahl) Quél. I. Factors controlling rhizomorph initiation by *A. mellea* in pure culture, *Ann. Bot. (N.S.),* 17, 63, 1953.
356. **Townsend, B. B.,** Morphology and development of fungal rhizomorphs, *Trans. Br. Mycol. Soc.,* 37, 222, 1954.
357. **Macdonald, J. A. and Cartter, M. A.,** The rhizomorphs of *Marasmius androsaceus* Fries, *Trans. Br. Mycol. Soc.,* 44, 72, 1961.
358. **Chet, I.,** Sclerotial morphogenesis in *Sclerotium rolfsii,* in *Abstr. 2nd Int. Mycol. Congr.,* Bigelow, H. E. and Simmons, E. G., Eds., University of South Florida, Tampa, 1977, 93.
359. **Wheeler, B. E. J. and Waller, J. M.,** The production of sclerotia by *Sclerotium rolfsii.* II. The relationship between mycelial growth and initiation of sclerotia, *Trans. Br. Mycol. Soc.,* 48, 303, 1965.
360. **Henis, Y., Chet, I., and Avizohar-Hershenzon, Z.,** Nutritional and mechanical factors involved in mycelial growth and production of sclerotia by *Sclerotium rolfsii* in artifical medium and amended soil, *Phytopathology,* 55, 87, 1965.
361. **Boyle, L. W.,** The ecology of *Sclerotium rolfsii* with emphasis on the role of saprophytic media, *Phytopathology,* 51, 117, 1961.
362. **Wheeler, B. E. J. and Sharan, N.,** The production of sclerotia by *Sclerotium rolfsii.* I. Effects of varying the supply of nutrients in an agar medium, *Trans. Br. Mycol. Soc.,* 48, 291, 1965.
363. **Tucker, C. M.,** Investigations with *Sclerotium delphinii, Mo. Agric. Exp. Stn. Bull.,* 340, 29, 1934.
364. **Misra, A. P. and Haque, S. Q.,** Factors affecting the growth and sclerotial production in *Sclerotium rolfsii* Sacc., causing storage rot of potato, *Proc. Natl. Acad. Sci. India Sect. B,* 56, 157, 1962.
365. **Okon, Y., Chet, I., and Henis, Y.,** Lactose-induced synchronous sclerotium formation in *Sclerotium rolfsii* and its inhibition by ethanol, *J. Gen. Microbiol.,* 71, 465, 1972.
366. **Henis, Y., Okon, Y., and Chet, I.,** The relationship between early hyphal branching and formation of sclerotia in *Sclerotium rolfsii, J. Gen. Microbiol.,* 79, 147, 1973.
367. **Liu, T. M. E. and Wu, L. C.,** The effect of amino acids on the growth and morphogenesis of *Sclerotium rolfsii* Sacc., *Plant Prot. Soc. Bull. (Taipei),* 13, 87, 1971.
368. **Chet, I., Henis, Y., and Mitchell, R.,** The morphogenetic effect of sulfur-containing amino acids, glutathione and iodoacetic acid on *Sclerotium rolfsii* Sacc., *J. Gen. Microbiol.,* 45, 541, 1966.
369. **Chet, I. and Henis, Y.,** The control mechanism of sclerotial formation in *Sclerotium rolfsii* Sacc., *J. Gen. Microbiol.,* 54, 231, 1968.
370. **Henis, Y. and Chet, I.,** Developmental biology of sclerotia of *Sclerotium rolfsii, Can. J. Bot.,* 46, 947, 1968.
371. **Trevethick, J. and Cooke, R. C.,** Effects of some metabolic inhibitors and sulfur-containing amino acids on sclerotium formation in *Sclerotium rolfsii, S. delphinii* and *Sclerotinia sclerotiorum, Trans. Br. Mycol. Soc.,* 57, 340, 1971.
372. **Chet, I. and Henis, Y.,** The response of two types of *Sclerotium rolfsii* to factors affecting sclerotium formation, *J. Gen. Microbiol.,* 73, 483, 1972.
373. **Higgins, B. B.,** Physiology and parasitism of *Sclerotium rolfsii* Sacc., *Phytopathology,* 17, 417, 1927.
374. **Wheeler, B. E. J.,** Effect of ethanol on production of sclerotia by *Sclerotium rolfsii, Trans. Br. Mycol. Soc.,* 59, 453, 1972.
375. **Robbins, W. J. and Kavanagh, F.,** Vitamin B_1 or its intermediates and growth of certain fungi, *Am. J. Bot.,* 25, 229, 1938.
376. **Melhuish, J. H., Jr. and Bean, G. A.,** Effect of dimethyl sulfoxide on the sclerotia of *Sclerotium rolfsii, Can. J. Microbiol.,* 17, 429, 1971.

377. **Lentz, P. L. and McKay, H. H.,** Sclerotial formation in *Corticium olivascens* and other hymenomycetes, *Mycopathol. Mycol. Appl.*, 40, 1, 1970.

378. **Goujon, M.,** Mise en evidence, dans le mycélium du *Corticium rolfsii* (Sacc.) Curzi, d'un facteur morphogénétique responsable de l'apparition des sclérotes, *C. R. Acad. Sci. Ser. D*, 267, 409, 1968.

379. **Goujon, M.,** Nature et synthèse du facteur morphogène responsable de l'apparition des sclérotes chez le *Corticium rolfsii* (Sacc.) Curzi, *C. R. Acad. Sci. Ser. D*, 269, 2195, 1969.

380. **Goujon, M.,** Mécanismes physiologiques de la formation des sclérotes chez le *Corticium rolfsii* (Sacc.) Curzi, *Physiol. Veg.*, 8, 349, 1970.

381. **Geiger, J.-P. and Goujon, M.,** Mise en évidence, dans les extraits de thalle, d'un facteur morphogène responsable de l'apparition des sclérotes du *Corticium rolfsii* (Sacc.) Curzi, *C. R. Acad. Sci. Ser. D*, 271, 41, 1970.

382. **Tanrikut, S. and Vaughan, E. K.,** Studies on the physiology of *Sclerotinia sclerotiorum*, *Phytopathology*, 41, 1099, 1951.

383. **Bedi, K. S.,** Studies on *Sclerotinia sclerotiorum* (Lib.) de Bary. I. Some chemical factors affecting the formation of sclerotia, *Proc. Natl. Acad. Sci. India Sect. B*, 26, 112, 1956.

384. **Wang, S.-Y. C. and Le Tourneau, D.,** Carbon sources, growth, sclerotium formation and carbohydrate composition of *Sclerotinia sclerotiorum*, *Arch. Mikrobiol.*, 80, 219, 1971.

385. **Démétriadès, S. D.,** Etudes sur la biologie du *Sclerotinia sclerotiorum* (Lib.) Massee. II. L'utilisation de diverses sources de carbone, *Ann. Inst. Phytopathol. Benaki*, 6, 51, 1952.

386. **Wang, S.-Y. C. and Le Tourneau, D.,** Amino acids as nitrogen sources for growth and sclerotium formation in *Sclerotinia sclerotiorum*, *Trans. Br. Mycol. Soc.*, 59, 509, 1972.

387. **Vega, R. R. and Le Tourneau, D.,** The effect of zinc on the formation of sclerotia by *Sclerotinia sclerotiorum*, *Phytopathology*, 60, 1537, 1970.

388. **Lilly, V. G. and Barnett, H. L.,** Growth rates, vitamin deficiencies, and sclerotia formation by some Sclerotiniaceae, *Proc. W. Va. Acad. Sci.*, 20, 69, 1949.

389. **Rai, J. N., Tewari, J. P., and Sinha, A. K.,** Effect of environmental conditions on sclerotia and cleistothecia production in *Aspergillus*, *Mycopathol. Mycol. Appl.*, 31, 209, 1967.

390. **Heath, L. A. F. and Eggins, H. O. W.,** Effects of light, temperature and nutrients on the production of conidia and sclerotia by forms of *Aspergillus japonicus*, *Experientia*, 21, 385, 1965.

391. **Rudolph, E. D.,** The effect of some physiological and environmental factors on sclerotial aspergilli, *Am. J. Bot.*, 49, 71, 1962.

392. **Moore, E. J.,** The ontogeny of the sclerotia of *Pyronema domesticum*, *Mycologia*, 54, 312, 1962.

393. **Brinkerhoff, L. A.,** The influence of temperature, aeration, and soil microflora on microsclerotial development of *Verticillium albo-atrum* in abscised cotton leaves, *Phytopathology*, 59, 805, 1969.

394. **Heale, J. B. and Isaac, I.,** Environmental factors in the production of dark resting structures in *Verticillium alboatrum*, *V. dahliae* and *V. tricorpus*, *Trans. Br. Mycol. Soc.*, 48, 39, 1965.

395. **Brandt, W. H.,** Manganese stimulation of melanin synthesis and microsclerotia development in *Verticillium*, *Plant Physiol.*, 37 (Suppl.), xxx, 1962.

396. **Brandt, W. H. and Reese, J. E.,** Morphogenesis in *Verticillium:* a self-produced, diffusible morphogenetic factor, *Am. J. Bot.*, 51, 922, 1964.

397. **Wyllie, T. D. and DeVay, J. E.,** Growth characteristics of several isolates of *Verticillium albo-atrum* and *Verticillium nigrescens* from cotton, *Phytopathology*, 60, 907, 1970.

398. **Kaiser, W. J.,** Biotin requirement of the *Verticillium* wilt fungus, *Phytopathology*, 54, 481, 1964.

399. **Loughheed, T. C.,** The effect of nutrition on synnemata formation in *Hirsutella gigantea* Petch, *Can. J. Bot.*, 39, 865, 1961.

400. **Taber, W. A.,** Nutritional factors affecting the morphogenesis of the synnema, in *Recent Advances in Botany*, Vol. 4, International Botanical Congress (1959), Montreal, 1961, 289.

401. **Taber, W. A.,** Studies on *Isaria cretacea:* morphogenesis of the synnema and endogenous nutrition, *Can. J. Microbiol.*, 6, 53, 1960.

402. **Taber, W. A. and Vining, L. C.,** Studies on *Isaria cretacea:* nutritional and morphological characteristics of two strains and morphogenesis of the synnema, *Can. J. Microbiol.*, 5, 513, 1959.

403. **Carlile, M. J., Lewis, B. G., Mordue, E. M., and Northover, J.,** The development of coremia. I. *Penicillium claviforme*, *Trans. Br. Mycol. Soc.*, 44, 129, 1961.

404. **Carlile, M. J., Dickens, J. S. W., Mordue, E. M., and Schipper, M. A. A.,** The development of coremia. II. *Penicillium isariiforme*, *Trans. Br. Mycol. Soc.*, 45, 457, 1962.

405. **Carlile, M. J., Dickens, J. S. W., and Schipper, M. A. A.,** The development of coremia. III. *Penicillium clavigerum,* with some observations on *P. expansum* and *P. italicum*, *Trans. Br. Mycol. Soc.*, 45, 462, 1962.

406. **Tinnell, W. H., Jefferson, B. L., and Benoit, R. E.,** The organic nitrogen exigency and the effects of manganese on coremia production in *Penicillium clavigerum* and *Penicillium claviforme*, *Can. J. Microbiol.*, 20, 91, 1974.

407. **Tinnell, W. H., Jefferson, B. L., and Benoit, R. E.,** Manganese-mediated morphogenesis in *Penicillium claviforme* and *Penicillium clavigerum, Can. J. Microbiol.*, 23, 209, 1977.
408. **Magrou, J., Marneffe, H., and Mariat, F.,** Action morphogène de facteurs vitaminique sur certains champignons, *Ann. Inst. Pasteur Paris*, 80, 443, 1951.
409. **Lewin, R. A.,** Control of sexual activity in *Chlamydomonas* by light, *J. Gen. Microbiol.*, 15, 170, 1956.
410. **Stifter, I.,** Untersuchungen über einige Zusammenhänge zwischen Stoffwechsel- und Sexualphysiologie an dem Flagellaten *Chlamydomonas eugametos, Arch. Protistenk.*, 104, 364, 1959.
411. **Wiese, L.,** Algae, in *Fertilization: Comparative Morphology, Biochemistry, and Immunology*, Vol. 2, Metz, C. B. and Monroy, A., Eds., Academic Press, New York, 1969, 135.
412. **Förster, H. and Wiese, L.,** Gamonwirkungen bei *Chlamydomonas eugametos, Z. Naturforsch.*, 96, 548, 1954.
413. **Wiese, L.,** On sexual agglutination and mating-type substances (gamones) in isogamous heterothallic chlamydomonads. I. Evidence of the identity of the gamones with the surface components responsible for sexual flagellar contact, *J. Phycol.*, 1, 46, 1965.
414. **Coleman, A. W.,** Sexuality, in *Physiology and Biochemistry of Algae*, Lewin, R. A., Ed., Academic Press, New York, 1962, 711.
415. **Sager, R. and Granick, S.,** Nutritional studies with *Chlamydomonas reinhardi, Ann. N. Y. Acad. Sci.*, 56, 831, 1953.
416. **Sager, R. and Granick, S.,** Nutritional control of sexuality in *Chlamydomonas reinhardi, J. Gen. Physiol.*, 37, 729, 1954.
417. **Bernstein, E. and Jahn, T. L.,** Certain aspects of the sexuality of two species of *Chlamydomonas, J. Protozool.*, 2, 81, 1955.
418. **Tsubo, Y.,** Observations on sexual reproduction in a *Chlamydomonas, Bot. Mag.*, 69, 1, 1956.
419. **Tsubo, Y.,** On the mating reaction of a *Chlamydomonas*, with special references to clumping and chemotaxis, *Bot. Mag.*, 70, 327, 1957.
420. **Trainor, F. R.,** Control of sexuality in *Chlamydomonas chlamydogama, Am. J. Bot.*, 45, 621, 1958.
421. **Trainor, F. R.,** A comparative study of sexual reproduction in four species of *Chlamydomonas, Am. J. Bot.*, 46, 65, 1959.
422. **Kates, J. R. and Jones, R. F.,** Pattern of CO_2 fixation during vegetative development and gametic differentiation in *Chlamydomonas reinhardtii, J. Cell. Physiol.*, 67, 101, 1966.
423. **Kates, J. R., Chiang, K. S., and Jones, R. F.,** Studies on DNA replication during synchronized vegetative growth and gametic differentiation in *Chlamydomonas reinhardtii, Exp. Cell Res.*, 49, 121, 1968.
424. **Chiang, K.-S., Kates, J. R., Jones, R. F., and Sueoka, N.,** On the formation of a homogeneous zygotic population in *Chlamydomonas reinhardtii, Dev. Biol.*, 22, 655, 1970.
425. **Jones, R. F.,** Physiological and biochemical aspects of growth and gametogenesis in *Chlamydomonas reinhardtii, Ann. N. Y. Acad. Sci.*, 175, 648, 1970.
426. **Honeycutt, R. C. and Margulies, M. M.,** Control of ribosome content, gamete formation and amino acid uptake in wildtype and arg-1 *Chlamydomonas reinhardtii, Biochim. Biophys. Acta*, 281, 399, 1972.
427. **Martin, N. C. and Goodenough, U. W.,** Gametic differentiation in *Chlamydomonas reinhardtii.* I. Production of gametes and their fine structure, *J. Cell Biol.*, 67, 587, 1975.
428. **Greenwood, H. L.,** The effects of actinomycin D, gibberellin A7, and cyclic AMP on mating of *Chlamydomonas reinhardtii, J. Protozool.*, 20, 305, 1973.
429. **Lorch, S. K. and Karlander, E. P.,** Gametogenesis in *Chlamydomonas eugametos.* I. Light requirements, *Plant Physiol.*, 51, 1046, 1973.
430. **Lewin, R. A.,** The genetics of *Chlamydomonas moewusii* Gerloff, *J. Genet.*, 51, 543, 1953.
431. **Lewin, R. A.,** Sex in unicellular algae, in *Sex in Microorganisms*, Wenrich, D. H., Ed., American Association for the Advancement of Science, Washington, D.C., 1954, 100.
432. **Wiese, L. and Jones, R. F.,** Studies on gamete copulation in heterothallic chlamydomonads, *J. Cell. Comp. Physiol.*, 61, 265, 1963.
433. **Tsubo, Y.,** Sexual reproduction of *Chlamydomonas* as affected by ionic balance in the medium, *Bot. Mag.*, 74, 442, 1961.
434. **Jones, R. F. and Wiese, L.,** Studies on the mating reaction in *Chlamydomonas, J. Gen. Physiol.*, 46, 362A, 1962.
435. **Lerche, W.,** Untersuchungen über Entwicklung und Fortpflanzung in der Gattung *Dunaliella, Arch. Protistenk.*, 88, 236, 1937.
436. **Grell, K. G.,** Sexual reproduction in protozoa, in *Research in Protozoology*, Vol. 2, Chen, T.-T., Ed., Pergamon Press, Oxford, 1967, 147.
437. **Rayburn, W. R.,** Sexual reproduction in *Pandorina unicocca, J. Phycol.*, 10, 258, 1974.
438. **Coleman, A. W.,** Sexual isolation in *Pandorina morum, J. Protozool.*, 6, 249, 1959.
439. **Thiadens, A. J. H. and Zeuthen, E.,** Meiosis and sporulation induced in sporophytes of *Ulva mutabilis* ("slender") with synchronous mitosis, *Planta*, 72, 60, 1967.

440. **Bråten, T.,** The ultrastructure of fertilization and zygote formation in the green alga *Ulva mutabilis* Føyn, *J. Cell Sci.*, 9, 621, 1971.

441. **Nordby, Ø. and Hoxmark, R. C.,** Changes in cellular parameters during synchronous meiosis in *Ulva mutabilis* Føyn, *Exp. Cell Res.*, 75, 321, 1972.

442. **Weisz, P. B.,** Multiconjugation in *Blepharisma*, *Biol. Bull. (Woods Hole, Mass.)*, 98, 242, 1950.

443. **Miyake, A. and Beyer, J.,** Cell interaction by means of soluble factors (gamones) in conjugation of *Blepharisma intermedium*, *Exp. Cell Res.*, 76, 15, 1973.

444. **Miyake, A. and Beyer, J.,** Blepharmone: a conjugation-inducing glycoprotein in the ciliate *Blepharisma*, *Science*, 185, 621, 1974.

445. **Miyake, A. and Bleyman, L. K.,** Gamones and mating types in the genus *Blepharisma* and their possible taxonomic application, *Genet. Res. Camb.*, 27, 267, 1976.

446. **Kubota, T., Tokoroyama, T., Tsukuda, Y., Koyama, H., and Miyake, A.,** Isolation and structure determination of blepharismin, a conjugation initiating gamone in the ciliate *Blepharisma*, *Science*, 179, 400, 1973.

447. **Miyake, A.,** Conjugation of the ciliate *Blepharisma:* a possible model system for biochemistry of sensory mechanisms, in *Biochemistry of Sensory Functions*, Jaenicke, L., Ed., Springer-Verlag, Berlin, 1974, 299.

448. **Braun, V. and Miyake, A.,** Composition of blepharmone, a conjugation-inducing glycoprotein of the ciliate *Blepharisma*, *FEBS Lett.*, 53, 131, 1975.

449. **Honda, H. and Miyake, A.,** Taxis to a conjugation-inducing substance in the ciliate *Blepharisma*, *Nature (London)*, 257, 678, 1975.

450. **Katashima, R.,** Mating types in *Euplotes eurystomus*, *J. Protozool.*, 6, 75, 1959.

451. **Sonneborn, T. M.,** Methods in the general biology and genetics of *Paramecium aurelia*, *J. Exp. Zool.*, 113, 87, 1950.

452. **Wenrich, D. H.,** Sex in protozoa: a comparative review, in *Sex in Microorganisms*, Wenrich, D. H., Ed., American Association for the Advancement of Science, Washington, D. C., 1954, 134.

453. **Sonneborn, T. M.,** Breeding systems, reproductive methods, and species problems in protozoa, in *The Species Problem*, Publ. 50, Mayr, E., Ed., American Association for the Advancement of Science, Washington, D.C., 1957, 155.

454. **Nanney, D. L.,** Ciliate genetics: patterns and programs of gene action, *Annu. Rev. Genet.*, 2, 121, 1968.

455. **Hiwatashi, K.,** *Paramecium*, in *Fertilization: Comparative Morphology, Biochemistry, and Immunology*, Vol. 2, Metz, C. B. and Monroy, A., Eds., Academic Press, New York, 1969, 255.

456. **Gibson, I.,** Interacting genetic systems in *Paramecium*, *Adv. Morphog.*, 8, 159, 1970.

457. **van Wagtendonk, W. J. and Soldo, A. T.,** Methods used in the axenic cultivation of *Paramecium aurelia*, in *Methods in Cell Physiology*, Vol. 4, Prescott, D. M., Ed., Academic Press, New York, 1970, 117.

458. **Sonneborn, T. M.,** Methods in *Paramecium* research, in *Methods in Cell Physiology*, Vol. 4, Prescott, D. M., Ed., Academic Press, New York, 1970, 241.

459. **Hanson, E. D.,** Methods in the cellular and molecular biology of *Paramecium*, in *Methods in Cell Biology*, Vol. 8, Prescott, D. M., Ed., Academic Press, New York, 1974, 319.

460. **Karakashian, M. W.,** The rhythm of mating in *Paramecium aurelia* syngen 3, *J. Cell. Physiol.*, 71, 197, 1969.

461. **Barnett, A.,** A circadian rhythm of mating type reversals in *Paramecium multimicronucleatum*, syngen 2, and its genetic control, *J. Cell. Physiol.*, 67, 239, 1966.

462. **Hiwatashi, K.,** Studies on the conjugation of *Paramecium caudatum*. III. Some properties of the mating type substances, *Sci. Rep. Tôhoku Univ. 4th Ser. Biol.*, 18, 270, 1950.

463. **Hiwatashi, K.,** Studies on the conjugation of *Paramecium caudatum*. VI. On the nature of the union of conjugation, *Sci. Rep. Tôhoku Univ. 4th Ser. Biol.*, 21, 207, 1955.

464. **Hiwatashi, K.,** Induction of conjugation by ethylenediamine tetraacetic acid (EDTA) in *Paramecium caudatum*, *Sci. Rep. Tôhoku Univ. 4th Ser. Biol.*, 25, 81, 1959.

465. **Hallet, M.M.,** Effet du chlorure de calcium sur le déterminisme génétique du type sexuel chez *Paramecium aurelia* (souche 60, syng. 1), *Protistologica*, 8, 387, 1972.

466. **Miyake, A.,** Physiological analysis of the life cycle of the protozoa. III. Artificial induction of selfing conjugation by chemical agents in *Paramecium caudatum*, *Physiol. Ecol.*, 7, 14, 1956.

467. **Miyake, A.,** Induction of conjugation by chemical agents in *Paramecium caudatum*, *J. Inst. Polytechnics Osaka City Univ. Ser. D*, 9, 251, 1958.

468. **Miyake, A.,** Induction of conjugation by chemical agents in *Paramecium*, *J. Exp. Zool.*, 167, 359, 1968.

469. **Miyake, A.,** Chemical induction of conjugation. I. General account, in *Methods in Cell Physiology*, Vol. 4, Prescott, D. M., Ed., Academic Press, New York, 1970, 289.

470. **Hiwatashi, K.,** Induction of conjugation in *P. caudatum* by EDTA, $MnCl_2$, and $CoCl_2$, in *Methods in Cell Physiology*, Vol. 4, Prescott, D. M., Ed., Academic Press, New York, 1970, 291.

471. **Hiwatashi, K. and Kasuga, T.,** Artificial induction of conjugation by manganese ion in *Paramecium caudatum*, *J. Protozool.*, 7 (Suppl.), 20, 1960.

472. **Cronkite, D. L.,** A role of calcium ions in chemical induction of mating in *Paramecium tetraurelia, J. Protozool.*, 23, 431, 1976.
473. **Miyake, A. and Okamoto, C.,** Induction of conjugation by Rb and Cs in *Paramecium caudatum, Physiol. Ecol.*, 8, 9, 1958.
474. **Zweibaum, J.,** La conjugaison et la différenciation sexuelle chez les infusoires (Enriques et Zweibaum). V. Les conditions nécessaires et suffisantes pour la conjugaison du *Paramaecium caudatum, Arch. Protistenk.*, 26, 275, 1912.
475. **Cronkite, D. L.,** Genetics of chemical induction of conjugation in *Paramecium aurelia, Genetics*, 76, 703, 1974.
476. **Band, R. N. and Mohrlok, S. H.,** An analysis of clumping in the soil amoeba *Mayorella palestinensis, J. Protozool.*, 16, 35, 1969.
477. **Warren, A. J., Warren, W. D., and Cox, E. C.,** Genetic complexity of aggregation in the cellular slime mold, *Proc. Natl. Acad. Sci. U.S.A.*, 72, 1041, 1975.
478. **Williams, K. L. and Newell, P. C.,** A genetic study of aggregation in the cellular slime mould *Dictyostelium discoideum* using complementation analysis, *Genetics*, 82, 287, 1976.
479. **Coukell, M. B.,** Evidence against mutational "hot spots" at aggregation loci in *Dictyostelium discoideum, Mol. Gen. Genet.*, 151, 269, 1977.
480. **Racker, E.,** *Mechanisms in Bioenergetics*, Academic Press, New York, 1965, 145.
481. **Bucher, T. and Sies, H., Eds.,** *Inhibitors: Tools in Cell Research*, Springer-Verlag, Heidelberg, 1970.
482. **Kröning, A. and Egel, R.,** Autolytic activities associated with conjugation and sporulation in fission yeast, *Arch. Microbiol.*, 99, 241, 1974.
483. **Calleja, G. B., Johnson, B. F., and Yoo, B. Y.,** The cell wall as sex organelle in fission yeast, in *Sexual Interactions in Eucaryotic Microbes*, O'Day, D. H. and Horgen, P. A., Eds., Academic Press, New York, 1981, 225.
484. **Miyake, A. and Honda, H.,** Cell union and protein synthesis in conjugation of *Blepharisma, Exp. Cell Res.*, 100, 31, 1976.
485. **Brock, T. D.,** Physiology of the conjugation process in the yeast *Hansenula wingei, J. Gen. Microbiol.*, 26, 487, 1961.
486. **Wiese, L. and Metz, C. B.,** On the trypsin sensitivity of gamete contact at fertilization as studied with living gametes in *Chlamydomonas, Biol. Bull. (Woods Hole, Mass.)*, 136, 483, 1969.
487. **Wiese, L. and Hayward, P. C.,** On sexual agglutination and mating type substances in isogamous dioecious chlamydomonads. III. The sensitivity of sex cell contact of various enzymes, *Am. J. Bot.*, 59, 530, 1972.
488. **Eddy, A. A.,** Part of the yeast surface apparently involved in flocculation, *J. Inst. Brew. London*, 64, 19, 1958.
489. **Takeuchi, I., Okamoto, K., Tasaka, M., and Takemoto, S.,** Regulation of cell-differentiation in slime-mold development, *Bot. Mag. Spec. Issue*, 1, 47, 1978.
490. **Wiese, L. and Shoemaker, D. W.,** On sexual agglutination and mating-type substances (gamones) in isogamous heterothallic chlamydomonads. II. The effect of concanavalin A upon the mating-type reaction, *Biol. Bull. (Woods Hole, Mass.)*, 138, 88, 1970.
491. **Gerisch, G.,** Membrane sites implicated in cell adhesion: their developmental control in *Dictyostelium discoideum*, in *International Cell Biology 1976—1977*, Brinkley, B. R. and Porter, K. R., Eds., Rockefeller University Press, New York, 1977, 36.
492. **Karlson, P. and Lüscher, M.,** 'Pheromones': a new term for a class of biologically active substances, *Nature (London)*, 183, 55, 1959.
493. **van den Ende, H.,** *Sexual Interactions in Plants: the Role of Specific Substances in Sexual Reproduction*, Academic Press, London, 1976.
494. **Gooday, G. W.,** Chemotaxis and chemotropism in fungi and algae, in *Primitive Sensory and Communication Systems*, Carlile M. J., Ed., Academic Press, London, 1975, 155.
495. **Gooday, G. W.,** Fungal sex hormones, *Annu. Rev. Biochem.*, 43, 35, 1974.
496. **Duntze, W., Stötzler, D., Bücking-Throm, E., and Kalbitzer, S.,** Purification and partial characterization of α-factor, a mating-type specific inhibitor of cell reproduction from *Saccharomyces cerevisiae, Eur. J. Biochem.*, 35, 357, 1973.
497. **Stötzler, D. and Duntze, W.,** Isolation and characterization of four related peptides exhibiting α-factor activity from *Saccharomyces cerevisiae, Eur. J. Biochem.*, 65, 257, 1976.
498. **Stötzler, D., Kiltz, H.-H., and Duntze, W.,** Primary structure of α-factor peptides from *Sacharomyces cerevisiae, Eur. J. Biochem.*, 69, 397, 1976.
499. **MacKay, V. and Manney, T. R.,** Mutations affecting sexual conjugation and related processes in *Saccharomyces cerevisiae*. I. Isolation and phenotypic characterization of non-mating mutants, *Genetics*, 76, 255, 1974.
500. **Miyake, A.,** Cell communication, cell union, and initiation of meiosis in ciliate conjugation, in *Current Topics in Developmental Biology*, Vol. 12, Moscona, A. A. and Monroy, A., Eds., Academic Press, New York, 1978, chap. 3.

501. **Tokoroyama, T., Hori, S., and Kubota, T.,** Synthesis of blepharismone, a conjugation-inducing gamone in ciliate *Blepharisma, Proc. Jpn. Acad.*, 49, 461, 1973.
502. **Kimball, R. F.,** Mating types in *Euplotes, Am. Nat.*, 73, 57, 1939.
503. **Kimball, R. F.,** The nature and inheritance of mating types in *Euplotes patella, Genetics*, 27, 269, 1942.
504. **Katashima, R.,** Breeding system of *Euplotes patella* in Japan, *Jpn. J. Zool.*, 13, 39, 1961.
505. **Bonner, J. T.,** Evidence for the formation of cell aggregates by chemotaxis in the development of the slime mold *Dictyostelium discoideum, J. Exp. Zool.*, 106, 1, 1947.
506. **Konijn, T. M.,** Chemotaxis in the cellular slime molds. I. The effect of temperature, *Dev. Biol.*, 12, 487, 1965.
507. **Pauling, L.,** Molecular basis of biological specificity, *Nature (London)*, 248, 769, 1974.
508. **Greaves, M. F.,** *Cellular Recognition*, Chapman and Hall, London, 1975.
509. **Greaves, M. F.,** Cell surface receptors: a biological perspective, in *Receptors and Recognition*, Ser. A, Vol. 1, Cuatrecasas, P. and Greaves, M. F., Eds., Chapman and Hall, London, 1976, chap. 1.
510. **Monod, J., Changeux, J.-P., and Jacob, F.,** Allosteric proteins and cellular control systems, *J. Mol. Biol.*, 6, 306, 1963.
511. **Monod, J., Wyman, J., and Changeux, J.-P.,** On the nature of allosteric transitions: a plausible model, *J. Mol. Biol.*, 12, 88, 1965.
512. **Koshland, D. E., Jr. and Neet, K. E.,** The catalytic and regulatory properties of enzymes, *Annu. Rev. Biochem.*, 37, 359, 1968.
513. **Taylor, N. W.,** Specific, soluble factor involved in sexual agglutination of the yeast *Hansenula wingei, J. Bacteriol.*, 87, 863, 1964.
514. **Brock, T. D.,** The purification and characterization of an intracellular sex-specific mannan protein from yeast, *Proc. Natl. Acad. Sci. U.S.A.*, 54, 1104, 1965.
515. **Crandall, M., Egel, R., and MacKay, V. L.,** Physiology of mating in three yeasts, *Adv. Microb. Physiol.*, 15, 307, 1977.
516. **Crandall, M.,** Mating-type interactions in yeasts, in *Cell-Cell Recognition*, Curtis, A. S. G., Ed., Cambridge University Press, Cambridge, U.K., 1978, 105.
517. **Taylor, N. W. and Orton, W. L.,** Sexual agglutination in yeast. VII. Significnce of the 1.7S component from reduced 5-agglutinin, *Arch. Biochem. Biophys.*, 126, 912, 1968.
518. **Sing, V., Yeh, Y.-F., and Ballou, C. E.,** Isolation of a *Hansenula wingei* mutant with an altered sexual agglutinin, in *Surface Membrane Receptors*, Bradshaw, R. A., Frasier, W. A., Merrell, R. C., Gottlieb, D. I., and Hogue-Angeletti, R. A., Eds., Plenum Press, New York, 1976, 87.
519. **Calleja, G. B., Johnson, B. F., and Yoo, B. Y.** Wall-bound protein mediates cell-cell interaction in fission yeast during mating and differentiation, *Abstr. 11th Int. Congr. Biochem., Toronto*, 1979, 511.
520. **Yanagishima, N.,** Sexual cell agglutination in *Saccharomyces cerevisiae:* sexual cell recognition and its regulation, *Bot. Mag. Spec. Issue*, 1, 61, 1978.
521. **Yoshida, K., Hagiya, M., and Yanagishima, N.,** Isolation and purification of the sexual agglutination substance of mating type *a* cells in *Saccharomyces cerevisiae, Biochem. Biophys. Res. Commun.*, 71, 1085, 1976.
522. **Hagiya, M., Yoshida, K., and Yanagishima, N.,** The release of sex-specific substances responsible for sexual agglutination from haploid cells of *Saccharomyces cerevisiae, Exp. Cell Res.*, 104, 263, 1977.
523. **Frazier, W. A.,** The role of cell surface components in the morphogenesis of the cellular slime molds, *Trends Biochem. Sci.*, 1, 130, 1976.
524. **Rosen, S. D., Kafka, J. A., Simpson, D. L., and Barondes, S. H.,** Developmentally regulated, carbohydrate-binding protein in *Dictyostelium discoideum, Proc. Natl. Acad. Sci. U.S.A.*, 70, 2554, 1973.
525. **Simpson, D. L., Rosen, S. D., and Barondes, S. H.,** Discoidin, a developmentally regulated carbohydrate-binding protein from *Dictyostelium discoideum.* Purification and characterization, *Biochemistry*, 13, 3487, 1974.
526. **Frazier, W. A., Rosen, S. D., Reitherman, R. W., and Barondes, S. H.,** Purification and comparison of two developmentally regulated lectins from *Dictyostelium discoideum:* discoidin I and II, *J. Biol. Chem.*, 250, 7714, 1975.
527. **Siu, C.-H., Lerner, R. A., Ma, G., Firtel, R. A., and Loomis, W. F.,** Developmentally regulated proteins of the plasma membrane of *Dictyostelium discoideum:* the carbohydrate-binding protein, *J. Mol. Biol.*, 100, 157, 1976.
528. **Reitherman, R. W., Rosen, S. D., Frazier, W. A., and Barondes, S. H.,** Cell surface species-specific high affinity receptors for discoidin: developmental regulation in *Dictyostelium discoideum, Proc. Natl. Acad. Sci. U.S.A.*, 72, 3541, 1975.
529. **Beug, H., Katz, F. E., Stein, A., and Gerisch, G.,** Quantitation of membrane sites in aggregating *Dictyostelium* cells by use of tritiated univalent antibody, *Proc. Natl. Acad. Sci. U.S.A.*, 70, 3150, 1973.
530. **Beug, H., Katz, F. E., and Gerisch, G.,** Dynamics of antigenic membrane sites relating to cell aggregation in *Dictyostelium discoideum, J. Cell Biol.*, 56, 647, 1973.

531. **Gerisch, G., Beug, H., Malchow, D., Schwarz, H., and Stein, A.,** Receptors for intercellular signals in aggregating cells of the slime mold, *Dictyostelium discoideum,* in *Biology and Chemistry of Eucaryotic Cell Surfaces,* Vol. 7, Miami Winter Symposia, Lee, E. Y. C. and Smith, E. E., Eds., Academic Press, New York, 1974, 49.
532. **Heusgen, A. and Gerisch, G.,** Solubilized contact site A from cell membranes of *Dictyostelium discoideum, FEBS Lett.,* 56, 46, 1975.
533. **Geltosky, J. E., Siu, C.-H., and Lerner, R. A.,** Glycoproteins of the plasma membrane of *Dictyostelium discoideum* during development, *Cell,* 8, 391, 1976.
534. **Simpson, D. L., Rosen, S. D., and Barondes, S. H.,** Pallidin. Purification and characterization of a carbohydrate-binding protein from *Polysphondylium pallidum* implicated in intercellular adhesion, *Biochim. Biophys. Acta,* 412, 109, 1975.
535. **Rosen, S. D., Haywood, P. L. and Barondes, S. H.,** Inhibition of intercellular adhesion in a cellular slime mould by univalent antibody against a cell-surface lectin, *Nature (London),* 263, 425, 1976.
536. **Cohen, L. W. and Siegel, R. W.,** The mating-type substances of *Paramecium bursaria, Genet. Res. Camb.,* 4, 143, 1963.
537. **Miyake, A.,** Induction of conjugation by cell-free preparations in *Paramecium multimicronucleatum, Science,* 146, 1583, 1964.
538. **Curtiss, R., III,** Bacterial conjugation, *Annu. Rev. Microbiol.,* 23, 69, 1969.
539. **Brinton, C. C., Jr.,** The properties of sex pili, the viral nature of 'conjugal' genetic transfer systems, and some possible approaches to the control of bacterial drug resistance, *Crit. Rev. Microbiol.,* 1, 105, 1971.
540. **Tomoeda, M., Inuzuka, M., and Date, T.,** Bacterial sex pili, *Prog. Biophys. Mol. Biol.,* 30, 23, 1975.
541. **Yoo, B. Y., Calleja, G. B., and Johnson, B. F.,** Electron microscopy of pre-conjugal flocculating fission yeast cells, *Proc. Can. Fed. Biol. Soc.,* 14, 62, 1971.
542. **Calleja, G. B., Johnson, B. F., and Yoo, B. Y.,** Wall-wall interaction during conjugation in fission yeast, *J. Cell Biol.,* 70, 132a, 1976.
543. **Poon, N. H. and Day, A. W.,** Fungal fimbriae. I. Structure, origin, and synthesis, *Can. J. Microbiol.,* 21, 537, 1975.
544. **Day, A. W., Poon, N. H., and Stewart, G. G.,** Fungal fimbriae. III. The effect on flocculation in *Saccharomyces, Can. J. Microbiol.,* 21, 558, 1975.
545. **Goodenough, U. W.,** Mating interactions in *Chlamydomonas,* in *Microbial Interactions,* Reissig, J. L., Ed., Chapman and Hall, London, 1977, chap. 8.
546. **Takahashi, M., Takeuchi, N., and Hiwatashi, K.,** Mating agglutination of cilia detached from complementary mating types of *Paramecium, Exp. Cell Res.,* 87, 415, 1974.
547. **Brown, R. M., Johnson, C., and Bold, M. C.,** Electron and phase-contrast microscopy of sexual reproduction in *Chlamydomonas moewusii, J. Phycol.,* 4, 100, 1968.
548. **Ringo, D. L.,** Flagellar motion and fine structure of the flagellar apparatus in *Chlamydomonas, J. Cell Biol.,* 33, 543, 1967.
549. **Jennings, H. S.,** Genetics of *Paramecium bursaria.* I. Mating types and groups, their interrelations and distribution: mating behavior and self-sterility, *Genetics,* 24, 202, 1939.
550. **Hiwatashi, K.,** Locality of mating reactivity on the surface of *Paramecium caudatum, Sci. Rep. Tôhoku Univ. 4th Ser.,* 27, 93, 1961.
551. **Brinton, C. C., Jr.,** The structure, function, synthesis and genetic control of bacterial pili and a molecular model for DNA and RNA transport in Gram negative bacteria, *Trans. N.Y. Acad. Sci.,* 277, 1003, 1960.
552. **Sussman, M. and Rossomando, E. F.,** Cellular slime molds, in *Handbook of Genetics,* Vol. 2, King, R. C., Ed., Plenum Press, New York, chap. 18.
553. **Newell, P. C.,** Genetics of the cellular slime molds, *Annu. Rev. Genet.,* 12, 69, 1978.
554. **Coukell, M. B.,** Parasexual genetic analysis of aggregation deficient mutants of *Dictyostelium discoideum, Mol. Gen. Genet.,* 142, 119, 1975.
555. **Warren, A. J., Warren, W. D., and Cox, E. C.,** Genetic and morphological study of aggregation in cellular slime mold *Polysphondylium violaceum, Genetics,* 83, 25, 1976.
556. **Erdos, G. W., Raper, K. B., and Vogen, L. K.,** Mating types and macrocyst formation in *Dictyostelium discoideum, Proc. Natl. Acad. Sci. U.S.A.,* 70, 1828, 1973.
557. **Clark, M. A., Francis, D., and Eisenberg, R.,** Mating types in the cellular slime molds, *Biochem. Biophys. Res. Commun.,* 52, 672, 1973.
558. **Clark, M. A.,** Syngenic division of cellular slime mold *Polysphondylium violaceum, J. Protozool.,* 21, 755, 1974.
559. **Erdos, G. W., Raper, K. B., and Vogen, L. K.,** Sexuality in the cellular slime mold *Dictyostelium giganteum, Proc. Natl. Acad. Sci. U.S.A.,* 72, 970, 1975.
560. **Wiese, L. and Wiese, W.,** Sex-cell contact in *Chlamydomonas,* a model for cell-recognition, in *Cell-Cell Recognition,* Curtis, A. S. G., Ed., Cambridge University Press, Cambridge, U.K., 1978, 83.
561. **Goodenough, U. W., Hwang, C., and Martin, H.,** Isolation and genetic analysis of mutant strains of *Chlamydomonas reinhardi* defective in gametic differentiation, *Genetics,* 82, 169, 1976.

562. **Lindegren, C. C.,** *The Yeast Cell: its Genetics and Cytology*, Educational Publishers, St. Louis, Mo., 1949.
563. **Hawthorne, D. C. and Mortimer, R. K.,** Chromosome mapping in *Saccharomyces:* centromere linked genes, *Genetics*, 45, 1085, 1960.
564. **Fogel, S. and Mortimer, R. K.,** Recombination in yeast, *Annu. Rev. Genet.*, 5, 219, 1971.
565. **Winge, Ø. and Roberts, C.,** A gene for diploidization in yeast, *C. R. Trav. Lab. Carlsberg Ser. Physiol.*, 24, 341, 1949.
566. **Takahashi, T.,** Complementary genes controlling homothallism in *Saccharomyces*, *Genetics*, 43, 705, 1958.
567. **Takano, I. and Oshima, Y.,** Mutational nature of an allele-specific conversion of the mating type by the homothallic gene *HOα* in *Saccharomyces*, *Genetics*, 65, 421, 1970.
568. **Oshima, Y. and Takano, I.,** Mating types in *Saccharomyces:* their convertibility and homothallism, *Genetics*, 67, 327, 1971.
569. **Harashima, S., Nogi, Y., and Oshima, Y.,** The genetic system controlling homothallism in *Saccharomyces* yeasts, *Genetics*, 77, 639, 1974.
570. **Hicks, J. B., Strathern, J. N., and Herskowitz, I.,** The cassette model of mating-type interconversion, in *DNA: Insertion Elements, Plasmids, and Episomes*, Bukhari, A. I., Shapiro, J. A., and Adhya, S. L., Eds., Cold Spring Harbor Laboratory, Cold Spring Harbor, N. Y., 1977, 457.
571. **Leupold, U.,** Studies on recombination in *Schizosaccharomyces pombe*, *Cold Spring Harbor Symp. Quant. Biol.*, 23, 161, 1958.
572. **Meade, J. H. and Gutz, H.,** Mating-type mutations in *Schizosaccharomyces pombe:* isolation of mutants and analysis of strains with an h^- or h^+ phenotype: *Genetics*, 83, 259, 1976.
573. **Sermonti, G.,** Bacteria, in *Fertilization: Comparative Morphology, Biochemistry, and Immunology*, Vol. 2, Metz, C. B. and Monroy, A., Eds., Academic Press, New York, 1969, chap. 2.
574. **Finnegan, D. J.,** *Bacterial Conjugation*, Meadowfield Press, Durham, England, 1976.
575. **Achtman, M. and Skurray, R.,** A redefinition of the mating phenomenon in bacteria, in *Microbial Interactions*, Reissig, J. L., Ed., Chapman and Hall, London, 1977, chap. 6.
576. **Novick, R. P., Clowes, R. C., Cohen, S. N., Curtiss, R., III, Datta, N., and Falkow, S.,** Uniform nomenclature for bacterial plasmids: a proposal, *Bacteriol. Rev.*, 40, 168, 1976.
577. **Russell, I., Stewart, G. G., Martin, P. A., Johnston, J. R., and Reader, H. P.,** Genetic control of yeast flocculation, *Yeast Newsl.*, 27, 95, 1978.
578. **Thorne, R. S. W.,** Some aspects of yeast flocculence, Eur. Brew. Conv. Proc. Congr., Brighton, 1951, 21.
579. **Anderson, E. and Martin, P. A.,** The sporulation and mating of brewing yeasts, *J. Inst. Brew. London*, 81, 242, 1975.
580. **O'Melia, C. R.,** Coagulation and flocculation, in *Physicomechanical Processes for Water Quality Control*, Weber, W. J., Jr., Ed., Interscience, New York, 1972, 61.
581. **La Mer, V. K. and Healy, T. W.,** Adsorption-flocculation reactions of macromolecules at the solid-liquid interface, *Rev. Pure Appl. Chem.*, 13, 112, 1963.
582. **Yusa, M.,** Mechanisms of pellet flocculation, *Int. J. Miner. Process.*, 4, 293, 1977.
583. **Stevenson, D. G.,** Flocculation and floc behaviour, *J. Inst. Water Eng. London*, 26, 155, 1972.
584. **Overbeek, J. T. G.,** Kinetics of flocculation, in *Colloid Science*, Vol. 1, Kruyt, H. R., Ed., Elsevier, Amsterdam, 1952, 291.
585. **Vold, M. J.,** Computer simulation of floc formation in a colloidal suspension, *J. Colloid Sci.*, 18, 684, 1963.
586. **Goel, N. S., Campbell, R. D., Gordon, R., Rosen, R., and Ycas, M.,** Self-sorting of isotropic cells, *J. Theor. Biol.*, 28, 423, 1970.
587. **Leith, A. G. and Goel, N. S.,** Self-sorting of anisotropic cells, *J. Theor. Biol.*, 28, 469, 1971.
588. **Antonelli, P. L., Rogers, T. D., and Willard, M. A.,** Geometry and the exchange principle in cell aggregation kinetics, *J. Theor. Biol.*, 41, 1, 1973.
589. **Mostow, G., Ed.,** *Mathematical Models of Cell Rearrangement*, Yale University Press, New Haven, 1975.
590. **Rogers, T. D. and Sampson, J. R.,** A random walk model of cellular kinetics, *Int. J. Bio-Med. Comput.*, 8, 45, 1977.
591. **Cohen, M. H. and Robertson, A.,** Wave propagation in the early stages of aggregation of cellular slime molds, *J. Theor. Biol.* 31, 101, 1971.
592. **Cohen, M. H. and Robertson, A.,** Chemotaxis and the early stages of aggregation of cellular slime molds, *J. Theor. Biol.*, 31, 119, 1971.
593. **Robertson, A. and Cohen, M. H.,** Control of developing fields, *Annu. Rev. Biophys. Bioeng.*, 1, 409, 1972.
594. **Nanjundiah, V.,** Chemotaxis, signal relaying and aggregation morphology, *J. Theor. Biol.*, 42, 63, 1973.
595. **Robertson, A. and Cohen, M. H.,** Quantitative analysis of the development of the cellular slime molds. II, *Lect. Math. Life Sci.*, 6, 44, 1974.

596. **Durston, A. J.,** Pacemaker activity during aggregation in *Dictyostelium discoideum, Dev. Biol.,* 37, 225, 1974.
597. **Novak, B. and Seeling, F. F.,** Phase-shift model for the aggregation of amoebae: a computer study, *J. Theor. Biol.,* 56, 301, 1976.
598. **Parnas, H. and Segel, L. A.** Computer evidence concerning the chemotactic signal in *Dictyostelium discoideum, J. Cell Sci.,* 25, 191, 1977.
599. **MacKay, S. A.,** Computer simulation of aggregation in *Dictyostelium discoideum, J. Cell Sci.,* 33, 1, 1978.
600. **Lodder, J., Ed.,** *The Yeasts: a Taxonomic Study,* 2nd ed., North-Holland, Amsterdam, 1970.
601. **Wickerham, L. J.,** Genus 7. *Hansenula* H. et P. Sydow, in *The Yeasts: a Taxonomic Study,* 2nd ed., Lodder, J., Ed., North-Holland, Amsterdam, 1970, 226.
602. **Barbour, I. G.,** *Issues in Science and Religion,* SCM Press, London, 1966.
603. **Ayala, F. J.,** Biology as an autonomous science, *Am. Sci.,* 56, 207, 1968.
604. **Popper, K. R.,** Scientific reduction and the essential incompleteness of all science, in *Studies in the Philosophy of Biology,* Ayala, F. J. and Dobzhansky, T., Ed., University of California Press, Berkeley, 1974, 259.
605. **Smith, J. M.,** What use is sex?, *J. Theor. Biol.,* 30, 319, 1971.
606. **Williams, G. C.,** *Sex and Evolution,* Princeton University Press, Princeton, N.J., 1975.
607. **Murray, R. G. E.,** Microbial structure as an aid to microbial classification and taxonomy, *Spisy Prirodoved. Fak. Univ. J. E. Purkyne Brne,* 43, 249, 1968.
608. **Lederberg, J. and Tatum, E. L,** Novel genotypes in mixed cultures of biochemical mutants of bacteria, *Cold Spring Harbor Symp. Quant. Biol.,* 21, 150, 1946.
609. **Jacob, F. and Wollman, E. L.,** Les épisomes, éléments génétiques ajouté, *C. R. Acad. Sci.,* 247, 154, 1958.
610. **Lederberg, J.,** Conjugal pairing in *Escherichia coli, J. Bacteriol.,* 71, 497, 1956.
611. **Anderson, T. F. and Maze, R.,** Analyse de la descendance de zygotes formé par conjugaison chez *Escherichia coli* K12, *Ann. Inst. Pasteur Paris,* 93, 194, 1957.
612. **Anderson, T. F.,** Recombination and segregation in *Escherichia coli, Cold Spring Harbor Symp. Quant. Biol.,* 23, 47, 1958.
613. **Valentine, R. C., Silverman, P. M., Ippen, K. A., and Mobach, H.,** The F-pilus of *Escherichia coli, Adv. Microb. Physiol.,* 3, 1, 1969.
614. **Crawford, E. M. and Gesteland, R. F.,** The adsorption of bacteriophage R-17, *Virology,* 22, 165, 1964.
615. **Caro, L. G. and Schnös, M.,** The attachment of the male-specific bacteriophage f1 to sensitive strains of *Escherichia coli, Proc. Natl. Acad. Sci. U.S.A.,* 56, 126, 1966.
616. **Beard, J. P., Howe, T. G. B., and Richmond, M. H.,** Purification of sex pili from *Escherichia coli* carrying a derepressed F-like R factor, *J. Bacteriol.,* 111, 814, 1972.
617. **Achtman, M., Willetts, N., and Clark, A. J.,** Beginning a genetic analysis of conjugational transfer determined by the F factor in *Escherichia coli* by isolation and characterization of transfer-deficient mutants, *J. Bacteriol.,* 106, 529, 1971.
618. **Novotny, C. and Fives-Taylor, P.,** Retraction of F pili, *J. Bacteriol.,* 117, 1306, 1974.
619. **Henning, U., Hoehn, B., and Sonntag, I.,** Cell envelope and shape of *Escherichia coli* K12. The ghost membrane, *Eur. J. Biochem.,* 39, 27, 1973.
620. **Chai, T. and Foulds, J.,** Demonstration of a missing outer membrane protein in *tolG* mutants of *Escherichia coli, J. Mol. Biol.,* 85, 465, 1974.
621. **Lugtenberg, B., Meijers, J., Peters, R., van den Hoek, P., and van Alphen, L.,** Electrophoretic resolution of the "major outer membrane protein" of *Escherichia coli* K-12 into four bands, *FEBS Lett.,* 58, 254, 1975.
622. **Schnaitman, C. A.,** Outer membrane proteins of *Escherichia coli.* III. Evidence that the major protein of *Escherichia coli* 0111 outer membrane consists of four distinct polypeptide species, *J. Bacteriol.,* 118, 442, 1974.
623. **Reithmeier, R. A. F. and Bragg, P. D.,** Purification and characterization of a heat-modifiable protein from the outer membrane of *Escherichia coli, FEBS Lett.,* 41, 195, 1974.
624. **Hindennach, I. and Henning, U.,** The major proteins of the *Escherichia coli* outer cell envelope membrane, *Eur. J. Biochem.,* 59, 207, 1975.
625. **Garten, W., Hindennach, I., and Henning, U.,** The major proteins of the *Escherichia coli* outer cell envelope membrane. Characterization of proteins II* and III, comparison of all proteins, *Eur. J. Biochem.,* 59, 215, 1975.
626. **Foulds, J.,** Chromosomal location of the *tolG* locus for tolerance to bacteriocin JF246 in *Escherichia coli* K-12, *J. Bacteriol.,* 117, 1354, 1974.
627. **Henning, U., Hindennach, I., and Haller, I.,** The major proteins of the *Esherichia coli* outer cell envelope membrane: evidence for the structural gene of protein II*, *FEBS Lett.,* 61, 46, 1976.

628. **Manning, P. A., Puspurs, A., and Reeves, P.,** Outer membrane of *Escherichia coli* K-12: isolation of mutants with altered protein 3A by using host-range mutants of bacteriophage K3, *J. Bacteriol.*, 127, 1080, 1976.
629. **Manning, P. A. and Reeves, P.,** Outer membrane of *Escherichia coli:* demonstration of the temperature sensitivity of a mutant in one of the major outer membrane proteins, *Biochem. Biophys. Res. Commun.*, 72, 694, 1976.
630. **Havekes, L. M., Lugtenberg, B. J. J., and Hoekstra, W. P. M.,** Conjugation deficient *E. coli* K12 F^- mutants with heptose-less lipopolysaccharide, *Mol. Gen. Genet.*, 146, 43, 1976.
631. **Havekes, L. M. and Hoekstra, W. P. M.,** Characterization of an *Escherichia coli* K-12 F^- con^- mutant, *J. Bacteriol.*, 126, 593, 1976.
632. **Havekes, L., Tommassen, J., Hoekstra, W., and Lugtenberg, B.,** Isolation and characterization of *Escherichia coli* K-12 F^- mutants defective in conjugation with an I-type donor, *J. Bacteriol.*, 129, 1, 1977.
633. **Sharp, P. A., Hsu, M., Ohtsubo, E., and Davidson, N.,** Electron microscope heteroduplex studies of sequence relations among plasmids of *Escherichia coli.* I. Structure of F prime factors, *J. Mol. Biol.*, 71, 471, 1972.
634. **Kline, B. C. and Helinski, D. R.,** F_1 sex factor of *Escherichia coli.* Size and purification in the form of a strand-specific relaxation complex of supercoiled deoxyribonucleic acid and protein, *Biochemistry,* 10, 4975, 1971.
635. **Achtman, M., Willetts, N., and Clark, A. J.,** A conjugational complementation analysis of transfer-deficient mutants of F*lac* in *Escherichia coli, J. Bacteriol.*, 110, 831, 1972.
636. **Minkley, E. G., Jr., Polen, S., Brinton, C. C., Jr., and Ippen-Ihler, K.,** Identification of the structural gene for F-pilin, *J. Mol. Biol.*, 108, 111, 1976.
637. **Willets, N. S. and Achtman, M.,** A genetic analysis of transfer by the *Escherichia coli* sex factor F, using P1 transductional complementation, *J. Bacteriol.*, 110, 843, 1972.
638. **Willetts, N.,** Mapping loci for surface exclusion and incompatibility on the F factor of *Escherichia coli* K-12, *J. Bacteriol.*, 118, 778, 1974.
639. **Achtman, M. and Helmuth, R.,** The F factor carries an operon of more than 15 × 10^6 daltons coding for deoxyribonucleic acid transfer and surface exclusion, in *Microbiology — 1974,* Schlessinger, D., Ed., American Society for Microbiology, Washington, D. C., 1974, 95.
640. **Minkley, E. G., Jr. and Ippen-Ihler, K.,** Identification of a membrane protein associated with expression of the surface exclusion region of the F transfer operon, *J. Bacteriol.*, 129, 1613, 1977.
641. **Helmuth, R. and Achtman, M.,** Operon structure of DNA transfer cistrons on the F sex factor, *Nature (London),* 257, 652, 1975.
642. **Achtman, M.,** Transfer-positive J-independent revertants of the F factor in *Escherichia coli* K12, *Genet. Res. Camb.*, 21, 67, 1973.
643. **Willetts, N.,** The transcriptional control of fertility in F-like plasmids, *J. Mol. Biol.*, 112, 141, 1977.
644. **Willetts, N.,** The genetics of transmissible plasmids, *Annu. Rev. Genet.*, 6, 257, 1972.
645. **Gasson, M. J. and Willetts, N. S.,** Five control systems preventing transfer of *Escherichia coli* K-12 sex factor F, *J. Bacteriol.*, 122, 518, 1975.
646. **Meynell, E., Meynell, G. G., and Datta, N.,** Phylogenetic relationships of drug-resistance factors and other transmissible bacterial plasmids, *Bacteriol. Rev.*, 32, 55, 1968.
647. **Finnegan, D. J. and Willetts, N. S.,** Two classes of F*lac* mutants insensitive to transfer inhibition by an F-like R factor, *Mol. Gen. Genet.*, 111, 256, 1971.
648. **Finnegan, D. J. and Willetts, N. S.,** The site of action of the F transfer inhibitor, *Mol. Gen. Genet.*, 127, 307, 1973.
649. **Jazwinski, S. M., Marco, R., and Kornberg, A.,** A coat protein of the bacteriophage M13 virion participates in membrane-oriented synthesis of DNA, *Proc. Natl. Acad. Sci. U.S.A.*, 70, 205, 1973.
650. **Singer, S. J.,** The molecular organization of membranes, *Annu. Rev. Biochem.*, 43, 805, 1974.
651. **Lederberg, J., Cavalli, L. L., and Lederberg, E. M.,** Sex compatability in *Escherichia coli, Genetics,* 37, 720, 1952.
652. **Alfoldi, L., Jacob, F., and Wollman, E. L.,** Zygose létale dans les croisements entre souches colicinogènes et non colicinogènes d' *Escherichia coli, C. R. Acad. Sci.*, 244, 2974, 1957.
653. **Nagel de Zwaig, R., Anton, D. N., and Puig, J.,** The genetic control of colicinogenic factors E_2, I and V, *J. Gen. Microbiol.*, 29, 473, 1962.
654. **Clowes, R. C.,** Colicin factors and episomes, *Genet. Res. Camb.*, 4, 162, 1963.
655. **Gross, J. D.,** Cellular damage associated with multiple mating in *Escherichia coli, Genet. Res. Camb.*, 4, 463, 1963.
656. **Skurray, R. A. and Reeves, P.,** Characterization of lethal zygosis associated with conjugation in *Escherichia coli* K-12, *J. Bacteriol.*, 113, 58, 1973.
657. **Skurray, R. A. and Reeves, P.,** Physiology of *Escherichia coli* K-12 during conjugation: altered recipient functions associated with lethal zygosis, *J. Bacteriol.*, 114, 11, 1973.

658. **Skurray, R. A. and Reeves, P.**, F factor-mediated immunity to lethal zygosis in *Escherichia coli* K-12, *J. Bacteriol.*, 117, 100, 1974.
659. **Skurray, R. A., Willetts, N., and Reeves, P.**, Effect of *tra* mutation on F fac‑ ‑pecified immunity to lethal zygosis, *Mol. Gen. Genet.*, 146, 161, 1976.
660. **Calleja, G. B., Yoo, B. Y., and Johnson, B. F.**, Conjugation-induced lysis ‑ ‑zosaccharomyces pombe*, *J. Bacteriol.*, 130, 512, 1977.
661. **Johnson, B. F., Calleja, G. B., and Yoo, B. Y.**, A model for controlled autolyis during differential morphogenesis of fission yeast, in *Eucaryotic Microbes as Model Developmental Systems*, O'Day, D. H. and Horgen, P. A., Eds., Marcel Dekker, New York, 1977, 212.
662. **Schweizer, M. and Henning, U.**, Action of a major outer cell envelope membrane protein in conjugation of *Escherichia coli* K-12, *J. Bacteriol.*, 129, 1651, 1977.
663. **van Alphen, L., Havekes, L., and Lugtenberg, B.**, Major outer membrane protein *d* of *Escherichia coli* K12. Purification and in vitro activity on bacteriophage K3 and F-pilus mediated conjugation, *FEBS Lett.*, 75, 285, 1977.
664. **Davidson, N., Deonier, R. C., Hu, S., and Ohtsubo, E.**, Electron microscope duplex studies of sequence relations among plasmids of *Escherichia coli*. X. Deoxyribonucleic acid sequence organization of F and F-primes, and the sequences involved in Hfr formation, in *Microbiology — 1974*, Schlessinger, D., Ed., American Society for Microbiology, Washington, D.C., 1975, 56.
665. **Helmuth, R. and Achtman M.**, Operon structure of DNA transfer cistrons on the F sex factor, *Nature (London)*, 257, 652, 1975.
666. **Skurray, R. A., Nagaishi, H., and Clark, A. J.**, Molecular cloning of DNA from F sex factor of *Escherichia coli* K-12, *Proc. Natl. Acad. Sci. U.S.A.*, 73, 64, 1976.
667. **Achtman, M.**, A physical analysis of mating in *Escherichia coli*, in *Plasmids: Medical and Theoretical Aspects*, Mitsuhashi, S., Rosival, L., and Krčméry, V., Eds., Avicenum, Czechoslovak Medical Press, Prague, 1977, 117.
668. **Ou, J. T. and Anderson, T. F.**, Role of pili in bacterial conjugation, *J. Bacteriol.*, 102, 648, 1970.
669. **Folkhard, W., Leonard, K. R., Malsey, S., Marvin, D. A., Dubochet, J., Engel, A., Achtman, M., and Helmuth, R.**, X-ray diffraction and electron microscope studies on the structure of bacterial F pili, *J. Mol. Biol.*, 130, 145, 1979.
670. **Date, T., Inuzuka, M., and Tomoeda, M.**, Purification and characterization of F pili from *Escherichia coli*, *Biochemistry*, 16, 5579, 1977.
671. **Helmuth, R. and Achtman, M.**, Cell-cell interactions in conjugating *Escherichia coli:* purification of F pili with biological activity, *Proc. Natl. Acad. Sci. U.S.A.*, 75, 1237, 1978.
672. **Achtman, M., Morelli, G., and Schwuchow, S.**, Cell-cell interactions in conjugating *Escherichia coli:* role of F pili and fate of mating aggregates, *J. Bacteriol.*, 13, 1053, 1978.
673. **Heumann, W.**, Versuche zur Rekobination sternbildender Bakterien, *Naturwissenschaften*, 14, 330, 1960.
674. **Knösel, D.**, Prüfung von Bakterien auf Fähigkeit zur Sternbildung, *Zentralbl. Bakteriol. Parasitenk. Infektionskr. Hyg. Abt. 2*, 116, 70, 1962.
675. **Heumann, W.**, Der Sexualcyclus sternbildender Bakterien, *Arch. Mikrobiol.*, 24, 362, 1956.
676. **Heumann, W.**, Kreuzungsversuche mit sternbildenden Bakterien, *Arch. Mikrobiol.*, 36, 244, 1960.
677. **Heumann, W.**, Die Methodik der Kreuzung sternbildender Bakterien, *Biol. Zentralbl.*, 81, 341, 1962.
678. **Heumann, W.**, Das Genetische Verhalten sternbildender Bakterien, *Naturwissenschaften*, 18, 430, 1962.
679. **Heumann, W.**, Genetische Untersuchung sternbildender Bakterien, *Z. Verebungsl.*, 93, 441, 1962.
680. **Heumann, W., Pühler, A., and Wagner, E.**, The two transfer regions of the *Rhizobium lupini* conjugation. I. Fertility factor elimination and one way transfer, *Mol. Gen. Genet.*, 113, 308, 1971.
681. **Heumann, W., Pühler, A., and Wagner, E.**, The two transfer regions of the *Rhizobium lupini* conjugation. II. Genetic characterization of the transferred chromosomal segments, *Mol. Gen. Genet.*, 126, 267, 1973.
682. **Heumann, W. and Springer, R.**, Formation of merodiploid clones by conjugation in *Rhizobium lupini*, *Mol. Gen. Genet.*, 150, 73, 1977.
683. **Moll, G., Ahrens, R., and Rheinheimer, G.**, Electronenoptische Untersuchungen über sternbildende Bakterien aus der Ostsee, *Kiel. Meeresforsch.*, 23, 137, 1967.
684. **Marx, R. and Heumann, W.**, Über Geisselfeinstrukturen und Fimbrien bei zwei *Pseudomonas*-Stämmen, *Arch. Mikrobiol.*, 43, 245, 1962.
685. **Heumann, W. and Marx, R.**, Feinstruktur und Funktion der Fimbrien bei dem sternbildenden Bakterium *Pseudomonas echinoides*, *Arch. Mikrobiol.*, 47, 325, 1964.
686. **Ahrens, R., Moll, G., and Rheinheimer, G.**, Die Rolle der Fimbrien bei der eigenartigen Sternbildung von *Agrobacterium luteum*, *Arch. Mikrobiol.*, 63, 321, 1968.
687. **Mayer, F.**, Elektronenmikroskopische Untersuchung der Fimbrienkontraktion bei dem sternbildenden Bodenbakterium *Pseudomonas echinoides*, *Arch. Mikrobiol.*, 76, 166, 1971.
688. **Pühler, A. and Burkardt, H.-J.**, Fertility inhibition in *Rhizobium lupini* by the resistance plasmid RP4, *Mol. Gen. Genet.*, 162, 163, 1978.

689. **Spriger, R. and Heumann, W.**, Genetical analysis of merodiploidy in *Rhizobium lupini, Mol. Gen. Genet.*, 165, 57, 1978.
690. **Duguid, J. P., Smith, I. W., Dempster, G., and Edmunds, P. N.**, Non-flagellar filamentous appendages ("fimbriae") and haemagglutinating activity in *Bacterium coli, J. Pathol. Bacteriol.*, 70, 335, 1955.
691. **Duguid, J. P. and Wilkinson, J. F.**, Environmentally induced changes in bacterial morphology, in *Microbial Reaction to Environment,* 11th Symp. Soc. Gen. Microbiol., Meynell, G. G. and Gooder, H., Eds., Cambridge University Press, Cambridge, U.K., 1961, 69.
692. **Old, D. C. and Payne, S. B.**, Antigens of the type-2 fimbriae of salmonellae: "cross-reacting material" (CRM) of type-1 fimbriae, *J. Med. Microbiol.*, 4, 215, 1971.
693. **Duguid, J. P. and Campbell, I.**, Antigens of the type-1 fimbriae of salmonellae and other enterobacteria, *J. Med. Microbiol.*, 2, 535, 1969.
694. **Gillies, R.R. and Duguid, J. P.**, The fimbrial antigens of *Shigella flexneri, J. Hyg. Camb.*, 56, 303, 1958.
695. **Anderson, T. F.**, On the mechanism of adsorption of bacteriophages on host cells, in *The Nature of the Bacterial Surface*, Miles, A. A. and Pirie, N. W., Eds., Blackwell, Oxford, U.K., 1949, 76.
696. **Houwink, A. L.**, Discussion, in *The Nature of the Bacterial Surface*, Miles, A. A. and Pirie, N. W., Eds., Blackwell, Oxford, U.K., 1949, 92.
697. **Houwink, A. L. and van Iterson, W.**, Electron microscopical observations on bacterial cytology. II. A study on flagellation, *Biochim. Biophys. Acta*, 5, 10, 1950.
698. **Brinton, C. C., Jr.**, Non-flagellar appendages of bacteria, *Nature (London)*, 183, 782, 1959.
699. **Duguid, J. P.**, Note on the terms "fimbriae" and "pili", *J. Pathol. Bacteriol.*, 92, 137, 1966.
700. **Duguid, J. P. and Anderson, E. S.**, Terminology of bacterial fimbriae, or pili, and their types, *Nature (London)*, 215, 89, 1967.
701. **Sokatch, J. R.**, Roles of appendages and surface layers in adaptation of bacteria to their environment, in *The Bacteria: a Treatise on Structure and Function,* Vol. 7, Sokatch, J. R. and Ornston, L. N., Eds., Academic Press, New York, 1979, chap. 5.
702. **Dettori, R., Maccacaro, G. A., and Piccinin, G. L.**, Sex-specific bacteriophages of *Escherichia coli* K-12, *G. Microbiol.*, 9, 141, 1961.
703. **Duguid, J. P. and Gillies, R. R.**, Fimbriae and haemagglutinating activity in *Salmonella, Klebsiella, Proteus* and *Chromobacterium, J. Pathol. Bacteriol.*, 75, 519, 1958.
704. **Duguid, J. P., Anderson, E. S., and Campbell, I.**, Fimbriae and adhesive properties in salmonellae, *J. Pathol. Bacteriol.*, 92, 107, 1966.
705. **Thornley, M. J. and Horne, R. W.**, Electron microscope observations on the structure of fimbriae, with particular reference to *Klebsiella* strains, by the use of the negative staining technique, *J. Gen. Microbiol.*, 28, 51, 1962.
706. **Hansen, A. J., Ingebritsen, A., and Weeks, O. B.**, Flagellation of *Flavobacterium piscicida, J. Bacteriol.*, 86, 602, 1963.
707. **Bradley, D. E.**, The structure and infective process of a *Pseudomonas aeruginosa* bacteriophage containing ribonucleic acid, *J. Gen. Microbiol.*, 45, 83, 1966.
708. **Fuerst, J. A. and Hayward, A. C.**, Surface appendages similar to fimbriae (pili) on *Pseudomonas* species, *J. Gen. Microbiol.*, 58, 227, 1969.
709. **Weiss, R. L.**, The structure and occurrence of pili (fimbriae) on *Pseudomonas aeruginosa, J. Gen. Microbiol.*, 67, 135, 1971.
710. **Kurosaka, R. and Horiuchi, S.**, Serological properties and amino acid composition of the fimbriae of *Aeromonas hydrophila, Bull. Tokyo Med. Dent. Univ.*, 21, 7, 1974.
711. **Hoeniger, J. F. M.**, Development of flagella by *Proteus mirabilis, J. Gen. Microbiol.*, 40, 29, 1965.
712. **Shedden, W. I. H.**, Fimbriae and haemagglutinating activity in strains of *Proteus hauseri, J. Gen. Microbiol.*, 28, 1, 1962.
713. **Coetzee, J. N., Pernet, G., and Theron, J. J.**, Fimbriae and haemagglutinating properties in strains of *Proteus, Nature (London)*, 196, 497, 1962.
714. **Schmidt, J. M.**, Observations on the adsorption of *Caulobacter* bacteriophages containing ribonucleic acid, *J. Gen. Microbiol.*, 45, 347, 1966.
715. **Schmidt, J. M.**, Prosthecate bacteria, *Annu. Rev. Microbiol.*, 25, 93, 1971.
716. **Hampp, E. G., Scott, D. B., and Wyckoff, R. W. G.**, Morphological characteristics of oral fusobacteria as revealed by the electron microscope, *J. Bacteriol.*, 79, 716, 1960.
717. **Bøvre, K. and Frøholm, L. O.**, Correlation between the fimbriated state and competence of genetic transformation in *Moraxella nonliquefaciens* strains, *Acta Pathol. Microbiol. Scand. Sect. B.*, 78, 526, 1970.
718. **Frøholm, L. O. and Sletten, K.**, Purification and N-terminal sequence of a fimbrial protein from *Moraxella nonliquefaciens, FEBS Lett.*, 73, 29, 1977.
719. **Frøholm, L. O.**, Antigenic characteristics of *Moraxella nonliquefaciens* fimbriae in double immunodiffusion, *Acta Pathol. Microbiol. Scand. Sect. B.*, 86, 169, 1978.

720. **Henriksen, S. D. and Frøholm, L. O.,** A fimbriated strain of *Pasteurella multocida* with spreading and corroding colonies, *Acta Pathol. Microbiol. Scand. Sect. B.*, 83, 129, 1975.
721. **Wistreich, G. A. and Baker, R. F.,** The presence of fimbriae (pili) in three species of *Neisseria, J. Gen. Microbiol.*, 65, 1971.
722. **Novotny, P., Short, J. A., and Walker, P. B.,** Electron-microscope study of naturally occurring and cultured cells of *Neisseria gonorrheae, J. Med. Microbiol.*, 8, 413, 1975.
723. **Punsalang, A. P., Jr. and Sawyer, W. D.,** Role of pili in the virulence of *Neisseria gonorrhoeae, Infect. Immun.*, 8, 255, 1973.
724. **Buchanan, T. M. and Pearce, W. A.,** Pili as mediator of the attachment of gonococci to human erythroyctes, *Infect. Immun.*, 13, 1483, 1976.
725. **Yanagawa, R., Otsuki, K., and Tokui, T.,** Electron microscopy of fine structure of *Corynebacterium renale* with special reference to pili, *Jpn. J. Vet. Res.*, 16, 31, 1968.
726. **Yanagawa, R. and Otsuki, K.,** Some properties of the pili of *Corynebacterium renale, J. Bacteriol.*, 101, 1063, 1970.
727. **Kumazawa, N. and Yanagawa, R.,** Chemical properties of the pili of *Corynebacterium renale, Infect. Immun.*, 5, 27, 1972.
728. **Henriksen, S. D. and Henrichsen, J.,** Twitching motility and possession of polar fimbriae in spreading *Streptococcus sanguis* isolates from the human throat, *Acta Pathol. Microbiol. Scand. Sect. B.*, 83, 133, 1975.
729. **Cole, R. M., Calandra, G. B., Huff, E., and Nugent, K. M.,** Attributes of potential utility in differentiating among "group H" streptococci or *Streptococcus sanguis, J. Dent. Res.*, 55, A142, 1976.
730. **Jones, G. W.,** The attachment of bacteria to the surfaces of animal cells, in *Microbial Interactions,* Reissig, J. L., Ed., Chapman and Hall, London, 1977, chap. 4.
731. **Duguid, J. P., Anderson, E. S., Alfredsson, G. A., Barker, R., and Old, D. C.,** A new biotyping scheme for *Salmonella typhimurium* and its phylogenetic significance, *J. Med. Microbiol.*, 8, 149, 1975.
732. **Old, D. C.,** Inhibition of the interaction between fimbrial haemagglutinins and erythrocytes by D-mannose and other carbohydrates, *J. Gen. Microbiol.*, 71, 149, 1972.
733. **Maccacaro, G. A. and Dettori, R.,** Studi sulle fimbrie batteriche. IV. Metabolismo ossidativo e fermentativo in cellule fimbriate e sfimbriate, *G. Microbiol.*, 7, 52, 1959.
734. **Brinton, C. C., Jr., Buzzell, A., and Laufer, M. A.,** Electrophoresis and phage susceptibility studies on a filament-producing variant of *E. coli* B bacterium, *Biochim. Biophys. Acta,* 15, 533, 1954.
735. **Maccacaro, G. A. and Turri, M.,** Studi sulle fimbrie batteriche. V. Sferoplasti fimbriati, *G. Microbiol.*, 7, 69, 1959.
736. **Maccacaro, G. A. and Hayes, W.,** The genetics of fimbriation in *Escherichia coli, Genet. Res. Camb.*, 2, 394, 1961.
737. **Maccacaro, G. A. and Hayes, W.,** Pairing interaction as a basis for negative interference, *Genet. Res. Camb.*, 2, 406, 1961.
738. **Iino, T.,** Genetics and chemistry of bacterial flagella, *Bacteriol. Rev.*, 33, 454, 1969.
739. **Zieg, J., Silverman, M., Hilmen, M., and Simon, M.,** Recombinational switch for gene expression, *Science,* 196, 170, 1977.
740. **Zieg, J., Hilmen, M., and Simon, M.,** Regulation of gene expression by site-specific inversion, *Cell,* 15, 237, 1978.
741. **Brinton, C. C., Jr. and Baron, L. S.,** Transfer of piliation from *Escherichia coli* to *Salmonella typhosa* by genetic recombination, *Biochim. Biophys. Acta,* 42, 298, 1960.
742. **Brinton, C. C., Jr., Gemski, P., Jr., Falkow, S., and Baron, L. S.,** Location of the piliation factor on the chromosome of *Escherichia coli, Biochem. Biophys. Res. Commun.*, 5, 293, 1961.
743. **Maccacaro, G. A., Colombo, C., and di Nardo, A.,** Studi sulle fimbrie batteriche. I. Lo studio genetico delle fimbrie, *G. Microbiol.*, 7, 1, 1959.
744. **Bachman, B. J., Low, K. B., and Taylor, A. L.,** Recalibrated linkage map of *Escherichia coli* K-12, *Bacteriol. Rev.*, 40, 116, 1976.
745. **Sanderson, K. E. and Hartman, P. E.,** Linkage map of *Salmonella typhimurium,* edition V., *Microbiol. Rev.*, 42, 471, 1978.
746. **Mulczyk, M. and Duguid, J. P.,** Influence of the state of fimbriation on transmission of the colicinogenic factor *coli* between strains of *Shigella flexneri, J. Gen. Microbiol.*, 45, 459, 1966.
747. **Meynell, G. G. and Lawn, A. M.,** Sex pili and common pili in conjugational transfer of colicin factor Ib by *Salmonella typhimurium, Genet. Res. Camb.*, 9, 359, 1967.
748. **Duguid, J. P.,** Functional anatomy of *Escherichia coli* with a special reference to enteropathogenic *E. coli, Rev. Latin-Am. Microbiol.*, 7 (Suppl. 13), 1, 1964.
749. **Duguid, J. P., Darekar, M. R., and Wheater, D. W. F.,** Fimbriae and infectivity in *Salmonella typhimurium, J. Med. Microbiol.*, 9, 459, 1976.
750. **Alexander, H. E. and Leidy, G.,** Transformation type specificity of *H. influenzae, Proc. Soc. Exp. Biol.*, 73, 485, 1950.

751. **Alexander, H. E. and Redman, W.,** Transformation of type specificity of meningococci, *J. Exp. Med.*, 97, 797, 1953.
752. **Balassa, R.,** Transformations mechanismen der Rhizobien, *Acta Microbiol. Acad. Sci. Hung.*, 2, 51, 1954.
753. **Spizizen, J.,** Transformation of biochemically deficient strains of *Bacillus subtilis* by deoxyribonucleate, *Proc. Natl. Acad. Sci. U.S.A.*, 44, 1072, 1958.
754. **Perry, D. and Slade, H. D.,** Transformation of streptococci to streptomycin resistance, *J. Bacteriol.*, 83, 443, 1962.
755. **Bøvre, K.,** Studies on transformation in *Moraxella* and organisms assumed to be related to *Moraxella*. I. A method for quantitative transformation in *Moraxella* and *Neisseria*, with streptomycin resistance as the genetic marker, *Acta Pathol. Microbiol. Scand.*, 61, 457, 1964.
756. **Moseley, B. E. B. and Setlow, J. K.,** Transformation in *Micrococcus radiodurans* and the ultraviolet sensitivity of its transforming DNA, *Proc. Natl. Acad. Sci. U.S.A.*, 61, 176, 1968.
757. **Cosloy, S. D. and Oishi, M.,** Genetic transformation in *Escherichia coli* K12, *Proc. Natl. Acad. Sci. U.S.A.*, 70, 84, 1973.
758. **Hinnen, A., Hicks, J. B., and Fink, G. R.,** Transformation in yeast, *Proc. Natl. Acad. Sci. U.S.A.*, 75, 1929, 1978.
759. **Hicks, J. B., Hinnen, A., and Fink, G. R.,** Properties of yeast transformation, *Cold Spring Harbor Symp. Quant. Biol.*, 43, 1305, 1979.
760. **Griffith, F.,** The significance of pneumococcal types, *J. Hyg. Camb.*, 27, 113, 1928.
761. **Avery, O. T., MacLeod, C. M., and McCarty, M.,** Studies on the chemical nature of the substance inducing transformation of pneumococcal types. I. Induction of transformation by a desoxyribonucleic acid fraction isolated from pneumococcus type III, *J. Exp. Med.*, 79, 137, 1944.
762. **Spizizen, J., Reilly, B. E., and Evans, A. H.,** Microbial transformation and transfection, *Annu. Rev. Microbiol.*, 20, 371, 1966.
763. **Tomasz, A.,** Some aspects of the competent state in genetic transformation, *Annu. Rev. Genet.*, 3, 217, 1969.
764. **Hotchkiss, R. D., and Gabor, M.,** Bacterial transformation, with special reference to recombination process, *Annu. Rev. Genet.*, 4, 193, 1970.
765. **Archer, L. J., Ed.,** *Bacterial Transformation*, Academic Press, London, 1973.
766. **Notani, N. K. and Setlow, J. K.,** Mechanism of bacterial transformation and transfection, *Prog. Nucl. Acid Res. Mol. Biol.*, 14, 39, 1974.
767. **Balassa, R.,** Einige Bedingungen der Ausbildung einer durch Desoxyribonukleinsäuren induzierten niedrigen Streptomycinresistenz bei *Rhizobium*, *Acta Microbiol. Acad. Sci. Hung.*, 4, 85, 1957.
768. **Goodgal, S. H. and Herriott, R. M.,** Studies on transformations of *Hemophilus influenzae*. I. Competence, *J. Gen. Physiol.*, 44, 1201, 1961.
769. **Spencer, H. T. and Herriott, R. M.,** Development of competence of *Haemophilus influenzae*, *J. Bacteriol.*, 90, 911, 1965.
770. **Catlin, B. W.,** Transformation of *Neisseria meningitidis* by deoxyribonucleates from cells and from culture slime, *J. Bacteriol.*, 79, 579, 1960.
771. **Lie, S.,** Studies on the phenotypic expression of competence in *Neisseria meningitidis*, *Acta Pathol. Microbiol. Scand.*, 64, 119, 1965.
772. **Hotchkiss, R. D.,** Cyclic behavior in pneumococcal growth and transformability occasioned by environmental changes, *Proc. Natl. Acad. Sci. U.S.A.*, 40, 49, 1954.
773. **Tomasz, A.,** Model for the mechanism controlling the expression of competent state in pneumococcus cultures, *J. Bacteriol.*, 91, 1050, 1966.
774. **Pakula, R. and Walczak, W.,** On the nature of competence of transformable streptococci, *J. Gen. Microbiol.*, 31, 125, 1963.
775. **Miller, D. H. and Huang, P. C.,** Identification of competence-repressing factors during log-phase growth of *Haemophilus influenzae*, *J. Bacteriol.*, 109, 560, 1972.
776. **Wise, E. M., Jr., Alexander, S. P., and Powers, M.,** Adenosine 3′:5′-cyclic monophosphate as a regulator of bacterial transformation, *Proc. Natl. Acad. Sci. U.S.A.*, 70, 471, 1973.
777. **Mosser, J. L. and Tomasz, A.,** Choline-containing teichoic acid as a structural component of pneumococcal cell wall and its role in sensitivity to lysis by an autolytic enzyme, *J. Biol. Chem.*, 245, 287, 1970.
778. **Ranhand, J. M., Leonard, C. G., and Cole, R. M.,** Autolytic activity associated with competent group H streptococci, *J. Bacteriol.*, 106, 257, 1971.
779. **Young, F. E., Spizizen, J., and Crawford, I. P.,** Biochemical aspects of competence in the *Bacillus subtilis* transformation system. I. Chemical composition of cell walls, *J. Biol. Chem.*, 238, 3119, 1963.
780. **Young, F. E. and Spizizen, J.,** Biochemical aspects of competence in the *Bacillus subtilis* transformation system. II. Autolytic enzyme activity of cell walls, *J. Biol. Chem.*, 238, 3126, 1963.
781. **Young, F. E., Tipper, D. J., and Strominger, J. L.,** Autolysis of cell walls of *Bacillus subtilis*. Mechanism and possible relationship to competence, *J. Biol. Chem.*, 239, 3600, 1964.

782. **Young, F. E.,** Autolytic enzyme associated with cell walls of *Bacillus subtilis, J. Biol. Chem.,* 241, 3462, 1966.

783. **Horváth, S.,** Competence in *Bacillus subtilis* transformation system, *J. Gen. Microbiol.,* 51, 85, 1968.

784. **Ottolenghi, E. and Hotchkiss, R. D.,** Release of genetic transforming DNA from pneumococcal cultures during growth and disintegration, *J. Exp. Med.,* 116, 491, 1962.

785. **Sinha, R. P. and Iyer, V. N.,** Competence for genetic transformation and the release of DNA from *Bacillus subtilis, Biochim. Biophys. Acta,* 232, 61, 1971.

786. **Jensen, R. A. and Haas, F. L.,** Electrokinetics and cell physiology. II. Relationship of surface charge to onset of bacterial competence for genetic transformation, *J. Bacteriol.,* 86, 79, 1963.

787. **Archer, L. J.,** Heat sensitivity of competent and pre-competent cells of *Bacillus subtilis,* in *Bacterial Transformation,* Archer, L. J., Ed., Academic Press, London, 1973, 43.

788. **Nester, E. W. and Stocker, B. A. D.,** Biosynthetic latency in early stages of deoxyribonucleic acid transformation in *Bacillus subtilis, J. Bacteriol.,* 86, 785, 1963.

789. **McCarthy, C. and Nester, E. W.,** Macromolecular synthesis in newly transformed cells of *Bacillus subtilis, J. Bacteriol.,* 94, 131, 1967.

790. **Laird, C. D., Wang, L., and Bodmer, W. F.,** Recombination and DNA replication in *Bacillus subtilis* transformation, *Mutat. Res.,* 6, 205, 1968.

791. **Erikson, R. J.,** New ideas and data on competence and DNA entry in transformation of *Bacillus subtilis, Curr. Top. Microbiol. Immunol.,* 53, 149, 1970.

792. **Dooley, D. C., Hadden, C. T., and Nester, E. W.,** Macromolecular synthesis in *Bacillus subtilis* during development of the competent state, *J. Bacteriol.,* 108, 668, 1971.

793. **Bøvre, K. and Frøholm, L. O.,** Competence in genetic transformation related to colony type and fimbriation in three species of *Moraxella, Acta Pathol. Microbiol. Scand. Sect. B.,* 80, 649, 1972.

794. **Bøvre, K. and Frøholm, L. O.,** Competence of genetic transformation correlated with the occurrence of fimbriae in three bacterial species, *Nature New Biol.,* 234, 151, 1971.

795. **Nava, G., Galis, A., and Beiser, S. M.,** Bacterial transformation: an antigen specific for 'competent' pneumococci, *Nature (London),* 197, 903, 1963.

796. **Tomasz, A. and Beiser, S. M.,** Relationship between the competence antigen and the competence-activator substance in pneumococci, *J. Bacteriol.,* 90, 1226, 1965.

797. **Pakula, R.,** Cellular sites for the competence-provoking factor of streptococci, *J. Bacteriol.,* 94, 75, 1967.

798. **Bingham, D. P. and Barnhart, B. J.,** Inhibition of transformation by antibodies against competent *Haemophilus influenzae, J. Gen. Microbiol.,* 75, 249, 1973.

799. **Tomasz, A.,** Interactions between cell surface components and DNA during genetic transformation of bacteria, in *Membrane Research,* Fox, C. F., Ed., Academic Press, New York, 1972, 311.

800. **Felkner, I. C. and Wyss, O.,** A substance produced by competent *Bacillus cereus* 569 cells that affects transformability, *Biochem. Biophys. Res. Commun.,* 16, 94, 1964.

801. **Charpak, M. and Dedonder, R.,** Production d'un "facteur de compétence" soluble par *Bacillus subtilis* Marburg ind^{-}_{168}, *C. R. Acad. Sci.,* 260, 5638, 1965.

802. **Barnhart, B. J.,** Competence-stimulating activity in sterile filtrates from *Hemophilus influenzae, Biochim. Biophys. Acta,* 142, 465, 1967.

803. **Akrigg, A., Ayad, S. R., and Barker, G. R.,** The nature of a competence-inducing factor in *Bacillus subtilis, Biochem. Biophys. Res. Commun.,* 28, 1062, 1967.

804. **Streips, U. N. and Young, F. E.,** Mode of action of the competence-inducing factor of *Bacillus stearothermophilus, J. Bacteriol.,* 106, 868, 1971.

805. **Joenje, H., Gruber, M., and Venema, G.,** Stimulation of the development of competence by culture fluids in *Bacillus subtilis* transformation, *Biochim. Biophys. Acta,* 262, 189, 1972.

806. **Goldsmith, M. R., Havas, S. W., Ma, R.-I., and Kallenbach, N. R.,** Intercellular effects on development of competence in *Bacillus subtilis, J. Bacteriol.,* 102, 774, 1970.

807. **Osoweicki, H., Nalecz, J., and Dobrzanski, W. T.,** The mechanism of competence in the transformation of streptococci of serological group H. Purification and some properties of the competence factor, *Mol. Gen. Genet.,* 105, 16, 1969.

808. **Leonard, C. G. and Cole, R. M.,** Purification and properties of streptococcal competence factor isolated from chemically defined medium, *J. Bacteriol.,* 110, 273, 1972.

809. **Leonard, C. G., Ranhand, J. M., and Cole, R. M.,** Competence factor production in chemically defined media by noncompetent cells of group H *Streptococcus* strain Challis, *J. Bacteriol.,* 104, 674, 1970.

810. **Nalecz, J. and Dobrzanski, W. T.,** Correlation between the occurrence of competence in the transformation of group H streptococci and the presence of the competence factor and the *in vitro* DNA-inactivating factor, *Mol. Gen. Genet.,* 114, 249, 1972.

811. **Ziegler, R. and Tomasz, A.,** Binding of the competence factor to receptors in the spheroplast membrane of pneumococci, *Biochem. Biophys. Res. Commun.,* 41, 1342, 1970.

812. **Javor, G. T. and Tomasz, A.,** An autoradiographic study of genetic transformation, *Proc. Natl. Acad. Sci. U.S.A.,* 60, 1216, 1968.

813. **Tomasz, A.,** Cell physiological aspects of DNA uptake during genetic transformation in bacteria, in *Informative Molecules in Biological Systems,* Ledoux, L., Ed., North-Holland, Amsterdam, 1971, 4.
814. **Tomasz, A., Zanati, E., and Ziegler, R.,** DNA uptake during genetic transformation and the growing zone of the cell envelope, *Proc. Natl. Acad. Sci. U.S.A.,* 68, 1848, 1971.
815. **Mandel, I. D.,** Dental plaque: nature, formation and effects, *J. Periodontol.,* 37, 357, 1966.
816. **Dawes, C.,** The nature of dental plaque, films, and calcareous deposits, *Ann. N. Y. Acad. Sci.,* 153, 102, 1968.
817. **Egelberg, J.,** A review of the development of dental plaque, in *Dental Plaque,* McHugh, W. D., Ed., E. & S. Livingstone, Edinburgh, 1970, 9.
818. **Schroeder, H. E. and de Boever, J.,** The structure of microbial dental plaque, in *Dental Plaque,* McHugh, W. D., Ed., E., & S. Livingstone, Edinburgh, 1970, 49.
819. **Leach, S. A.,** A review of the biochemistry of dental plaque, in *Dental Plaque,* McHugh, W. D., Ed., E. & S. Livingstone, Edinburgh, 1970, 143.
820. **Krasse, B.,** A review of the bacteriology of dental plaque, in *Dental Plaque,* McHugh, W. D., Ed., E. & S. Livingstone, Edinburgh, 1970, 199.
821. **Kleinberg, I.,** Biochemistry of the dental plaque, *Adv. Oral Biol.,* 4, 43, 1970.
822. **Scherp, H. W.,** Dental caries: prospects for prevention, *Science,* 173, 1199, 1971.
823. **Jenkins, G. N.,** Current concepts concerning the development of dental caries, *Int. Dent. J.,* 22 350, 1972.
824. **Gibbons, R. J.,** Microbial ecological models and dental diseases, *Va. Dent. J.,* 49, 7, 1972.
825. **Gibbons, R. J. and van Houte, J.,** On the formation of dental plaques, *J. Periodontol.,* 44, 347, 1973.
826. **Gibbons, R. J. and van Houte, J.,** Dental caries, *Annu. Rev. Med.,* 26, 121, 1975.
827. **van Houte, J.,** Oral bacterial colonization: mechanisms and implications, in *Microbial Aspects of Dental Caries,* Stiles, H. M., Loesche, W. J., and O'Brien, T. C., Eds., Information Retrieval, Washington, D. C., 1976, 3.
828. **Hardie, J. M. and Bowden, G. H.,** The microbial flora of dental plaque: bacterial succession and isolation considerations, in *Microbial Aspects of Dental Caries,* Stiles, H. M., Loesche, W. J., and O'Brien, T. C., Eds., Information Retrieval, Washington, D. C., 1976, 63.
829. **Guggenheim, B.,** Ultrastructure and some biochemical aspects of dental plaque: a review, in *Microbial Aspects of Dental Caries,* Stiles, H. M., Loesche, W. J., and O'Brien, T. C., Eds., Information Retrieval, Washington, D. C., 1976, 89.
830. **Gibbons, R. J. and van Houte, J.,** Oral bacterial ecology, in *Textbook of Oral Biology,* Shaw, J. H., Sweeney, E. A., Cappuccino, C. C., and Meller, S. M., Eds., W. B. Saunders, Philadelphia, 1978, 684.
831. **Bowden, G. H. W., Ellwood, D. C., and Hamilton, I. R.,** Microbiol ecology of the oral cavity, *Adv. Microb. Ecol.,* 3, 135, 1979.
832. **Gibbons, R. J., Socransky, S. S., de Araujo, W. C., and van Houte, J.,** Studies of the predominant cultivable microbiota of dental plaque, *Arch. Oral Biol.,* 9, 365, 1964.
833. **Socransky, S. S. and Manganiello, S. D.,** The oral microbiota of man from birth to senility, *J. Periodontol.,* 42, 485, 1971.
834. **Frank, R. M. and Brendel, A.,** Ultrastructure of the approximal dental plaque and the underlying normal and carious enamel, *Arch. Oral Biol.,* 11, 883, 1966.
835. **Carlsson, J. and Egelberg, J.,** Effect of diet on early plaque formation in man, *Odontol. Revy,* 16, 112, 1965.
836. **Theilade, E., Wright, W. H., Jensen, S. B., and Löe, H.,** Experimental gingivitis in man. II. A longitudinal clinical and bacteriological investigation, *J. Periodontal Res.,* 1, 1, 1966.
837. **Ritz, H. L.,** Microbial population shifts in developing human dental plaque, *Arch. Oral Biol.,*12, 1561, 1967.
838. **Clarke, J. K.,** On the bacterial factor in the aetiology of dental caries, *Br. J. Exp. Pathol.,* 5, 141, 1924.
839. **Zinner, D. D., Jablon, J. M., Aran, A. P., and Saslaw, M. S.,** Experimental caries induced in animals by streptococci of human origin, *Proc. Soc. Exp. Biol. Med.,* 118, 766, 1965.
840. **Krasse, B.,** The effect of caries-inducing streptococci in hamsters fed diets with sucrose or glucose, *Arch. Oral Biol.,* 10, 223, 1965.
841. **Newbrun, E.,** Sucrose, the arch criminal of dental caries, *Odontol. Revy,* 18, 373, 1967.
842. **Guggenheim, B. and Schroeder, H. E.,** Biochemical and morphological aspects of extracellular polysaccharides produced by cariogenic streptococci, *Helv. Odontol. Acta,* 11, 131, 1967.
843. **Wood, J. M. and Critchley, P.,** The extracellular polysaccharide produced from sucrose by a cariogenic streptococcus, *Arch. Oral Biol.,* 11, 1039, 1966.
844. **Gibbons, R. J. and Banghart, S. B.,** Synthesis of extracellular dextran by cariogenic bacteria and its presence in human dental plaque, *Arch. Oral Biol.,* 12, 11, 1967.
845. **Gibbons, R. J. and Nygaard, M.,** Synthesis of insoluble dextran and its significance in the formation of gelatinous deposits by plaque-forming streptococci, *Arch. Oral Biol.,* 13, 1249, 1968.
846. **Guggenheim, B. and Newbrun, E.,** Extracellular glucosyltransferase activity of an HS strain of *Streptococcus mutans, Helv. Odontol. Acta,* 13, 84, 1969.

847. **Kelstrup, J. and Funder-Nielsen, T. D.,** Molecular interactions between the extracellular polysaccharides of *Streptococcus mutans, Arch. Oral Biol.*, 17, 1659, 1972.
848. **de Stoppelaar, J. D., König, K. G., Plasschaert, A. J. M., and van der Hoeven, J. S.,** Decreased cariogenicity of a mutant of *Streptococcus mutans. Arch. Oral Biol.*, 16, 971, 1971.
849. **Johnson, M. C., Bozzola, J. J., and Schechmeister, I. L.,** Morphological study of *Streptococcus mutans* and two extracellular polysaccharide mutants, *J. Bacteriol.*, 118, 304, 1974.
850. **Johnson, M. C., Bozzola, J. J., Schechmeister, I. L., and Shklair, I. L.,** Biochemical study of the relationship of extracellular glucan to adherence and cariogenicity in *Streptococcus mutans* and an extracellular polysaccharide mutant, *J. Bacteriol.*, 129, 351, 1977.
851. **Kuramitsu, H. K.,** Properties of a mutant of *Streptococcus mutans* altered in glucosyltransferase activity, *Infect. Immun.*, 13, 345, 1976.
852. **Tanzer, J. M., Freedman, M. L., Fitzgerald, R. J., and Larson, R. H.,** Diminished virulence of glucan synthesis-defective mutants of *Streptococcus mutans, Infect. Immun.*, 10, 197, 1974.
853. **Nalbandian, J., Freedman, M. L., Tanzer, J. M., and Lovelace, S. M.,** Ultrastructure of mutants of *Streptococcus mutans* with reference to agglutination, adhesion, and extracellular polysaccharide, *Infect. Immun.*, 10, 1170, 1974.
854. **Gibbons, R. J., van Houte, J., and Liljemark, W. F.,** Parameters that affect the adherence of *Streptococcus salivarius* to oral epithelial surfaces, *J. Dent. Res.*, 51, 424, 1972.
855. **Lai, C., Listgarten, M., and Rosan, B.,** Serology of *Streptococcus sanguis:* localization of antigens with unlabeled antisera, *Infect. Immun.*, 8, 475, 1973.
856. **Liljemark, W. F. and Gibbons, R. J.,** Proportional distribution and relative adherence of *Streptococcus miteor (mitis)* on various surfaces in the human oral cavity, *Infect. Immun.*, 6, 852, 1972.
857. **Otake, S., McGhee, J. R., Hirasawa, M., Williams, K., Arnold, R. R., Babb, J. L., Kiyono, H., Cox, C., Michalek, S. M., Shiota, T., Ikeda, T., and Ochiai, K.,** Use of mutants in the elucidation of virulence of *Streptococcus mutans,* in *Secretory Immunity and Infection*, McGhee, J. R., Mestecky, J., and Babb, J. L., Eds., Plenum Press, New York, 1978, 673.
858. **Schachtele, C. F., Germaine, G. R., and Harlander, S. K.,** Production of elevated levels of dextransucrase by a mutant of *Streptococcus mutans, Infect. Immun.*, 12, 934, 1975.
859. **Janda, W. M. and Kuramitsu, H. K.,** Properties of a variant of *Streptococcus mutans* altered in its ability to interact with glucans, *Infect. Immun.*, 16, 575, 1977.
860. **Tanzer, J. M. and Freedman, M. L.,** Genetic alterations of *Streptococcus mutans* virulence, in *Secretory Immunity and Infection,* McGhee, J. R., Mestecky, J., and Babb, J. L., Eds., Plenum Press, New York, 1978, 661.
861. **Clark, W. B., Bammann, L. L., and Gibbons, R. J.,** Ability of *Streptococcus mutans* and a glucosyltransferase-defective mutant to colonize rodents and attach to hydroxyapatite surfaces, *Infect. Immun.*, 21, 681, 1978.
862. **Wu-Yuan, C. D., Tai, S., and Slade, H. D.,** Dextran/glucan binding by *Streptococcus mutans:* the role of molecular size and binding site in agglutination, in *Secretory Immunity and Infection,* McGhee, J. R., Mestecky, J., and Babb, J. L., Eds., Plenum Press, New York, 1978, 737.
863. **Olson, G. A., Guggenheim, B., and Small, P. A., Jr.,** Antibody-mediated inhibition of dextran/sucrose-induced agglutination of *Streptococcus mutans, Infect. Immun.*, 9, 273, 1974.
864. **McCabe, M. M., Keyes, P. H., and Howell, A., Jr.,** An *in vitro* method for assessing the plaque forming ability of oral bacteria, *Arch. Oral Biol.*, 12, 1653, 1967.
865. **Mukasa, H. and Slade, H. D.,** Mechanism of adherence of *Streptococcus mutans* to smooth surfaces. II. Nature of the binding site and the adsorption of dextran-levan synthetase enzymes on the cell-wall surface of the streptococcus, *Infect. Immun.*, 9, 419, 1974.
866. **Kuramitsu, H. K.,** Adherence of *Streptococcus mutans* to dextran synthesized in the presence of extracellular dextransucrase, *Infect. Immun.*, 9, 764, 1974.
867. **Spinell, D. M. and Gibbons, R. J.,** Influence of culture medium on the glucosyltransferase- and dextran-binding capacity of *Streptococcus mutans* 6715 cells, *Infect. Immun.*, 10, 1448, 1974.
868. **McCabe, M. M., Haynes, A. U., and Hamelik, R. M.,** Cell adherence of *Streptococcus mutans,* in *Microbial Aspects of Dental Caries,* Stiles, H. M., Loesche, W. J., and O'Brien, T. C., Eds., Information Retrieval, Washington, D. C., 1976, 413.
869. **McCabe, M. M. and Smith, E. E.,** Carbohydrate receptors of oral streptococci, in *Immunologic Aspects of Dental Caries,* Bowen, W. H., Genco, R. J., and O'Brien, T. C., Eds., Information Retrieval, Washington, D. C., 1976, 111.
870. **McCabe, M. M., Hamelik, R. M., and Smith, E. E.,** Purification of dextran-binding protein from cariogenic *Streptococcus mutans, Biochem. Biophys. Res. Commun.*, 78, 273, 1977.
871. **McCabe, M. M. and Hamelik, R. M.,** Multiple forms of dextran-binding proteins from *Streptococcus mutans,* in *Secretory Immunity and Infection,* McGhee, J. R., Mestecky, J., and Babb, J. L., Eds., Plenum Press, New York, 1978, 749.

872. **Wu-Yuan, C. D., Tai, S., and Slade, H. D.,** Properties of *Streptococcus mutans* grown in a synthetic medium: binding of glucosyltransferase and *in vitro* adherence, and binding of dextran/glucan and glycoprotein and agglutination, *Infect. Immun.*, 23, 600, 1979.
873. **McCabe, M. M. and Smith, E. E.,** Origin of the cell-associated dextransucrase of *Streptococcus mutans*, *Infect. Immun.*, 7, 829, 1973.
874. **Kuramitsu, H. K.,** Characterization of cell-associated dextransucrase activity from glucose-grown cells of *Streptococcus mutans*, *Infect. Immun.*, 10, 227, 1974.
875. **Janda, W. M. and Kuramitsu, H. K.,** Regulation of extracellular glucosyltransferase production and the relationship between extracellular and cell-associated activities in *Streptococcus mutans*, *Infect. Immun.*, 14, 191, 1976.
876. **Slade, H. D.,** Cell surface antigenic polymers of *Streptococcus mutans* and their role in adherence of the microorganism *in vitro*, in *Microbiology — 1977*, Schlessinger, D., Ed., American Society for Microbiology, Washington, D.C., 1977, 411.
877. **Bailey, R. W.,** Transglucosidase activity of rumen strains of *Streptococcus bovis*. II. Isolation and properties of dextransucrase, *Biochem. J.*, 72, 42, 1959.
878. **Carlsson, J. and Elander, B.,** Regulaton of dextransucrase formation by *Streptococcus sanguis*, *Caries Res.*, 7, 89, 1973.
879. **Wittenberger, C. L., Beaman, A. J., and Lee, L. N.,** Tween 80 effect on glucosyltransferase synthesis by *Streptococcus salivarius*, *J. Bacteriol.*, 133, 231, 1978.
880. **Gibbons, R. J.,** Formation and significance of bacterial polysaccharides in caries etiology, *Caries Res.*, 2, 164, 1968.
881. **Guggenheim, B.,** Extracellular polysaccharides and microbial plaque, *Int. Dent. J.*, 20, 657, 1970.
882. **Guggenheim, B.,** Enzymatic hydrolysis and structure of water-insoluble glucan produced by glucosyltransferase from a strain of *Streptococcus mutans*, *Helv. Odontol. Acta*, 14, 89, 1970.
883. **Fitzgerald, R. J., Keyes, P. H., Stoudt, T. H., and Spinell, D. M.,** The effects of a dextranase preparation on plaque and caries in hamsters, a preliminary report, *J. Am. Dent. Assoc.*, 76, 301, 1968.
884. **Gibbons, R. J. and Keyes, P. H.,** Inhibition of insoluble dextran synthesis, plaque formation and dental caries in hamsters by low molecular weight dextran, *Arch. Oral Biol.*, 14, 721, 1969.
885. **Guggenheim, B., Regolati, B., and Mühlemann, H. R.,** Caries and plaque inhibition by mutanase in rats, *Caries Res.*, 6, 289, 1972.
886. **Kelstrup, J., Funder-Nielsen, T. D., and Møller, E. N.,** Enzymatic reduction of the colonization of *Streptococcus mutans* in human dental plaque, *Acta Odontol. Scand.*, 31, 249, 1973.
887. **Ciardi, J. E.,** Glycosyltransferases of *Streptococcus mutans*, in *Immunologic Aspects of Dental Caries*, Bowen, W. H., Genco, R. J., and O'Brien, T. C., Eds., Information Retrieval, Washington, D.C., 1976, 101.
888. **Chludzinski, A. M., Germaine, G. R., and Schachtele, C. F.,** Purification and properties of dextransucrase from *Streptococcus mutans*, *J. Bacteriol.*, 118, 1, 1974.
889. **Germaine, G. R., Chludzinski, A. M., and Schachtele, C. F.,** *Streptococcus mutans* dextransucrase: requirement for primer dextran, *J. Bacteriol.*, 120, 287, 1974.
890. **Fukui, K., Fukui, Y., and Moriyama, T.,** Purification and properties of dextransucrase and invertase from *Streptococcus mutans*, *J. Bacteriol.*, 118, 796, 1974.
891. **Mukasa, H. and Slade, H. D.,** Mechanism of the adherence of *Streptococcus mutans* to smooth surfaces. III. Purification and properties of the enzyme complex responsible for adherence, *Infect. Immun.*, 10, 1135, 1974.
892. **Kuramitsu, H. K.,** Characterization of extracellular glucosyltransferase activity of *Streptococcus mutans*, *Infect. Immun.*, 12, 738, 1975.
893. **Ciardi, J. E., Hageage, G. J., Jr., and Wittenberger, C. L.,** Multicomponent nature of the glucosyltransferase system of *Streptococcus mutans*, *J. Dent. Res.*, 55, C87, 1976.
894. **Ciardi, J. E., Beaman, A. J., and Wittenberger, C. L.,** Purification, resolution, and interaction of the glucosyltransferases of *Streptococcus mutans* 6715, *Infect. Immun.*, 18, 237, 1977.
895. **Hillman, J. D., van Houte, J., and Gibbons, R. J.,** Sorption of bacteria to human enamel powder, *Arch. Oral Biol.*, 15, 899, 1970.
896. **Gibbons, R. J. and Spinell, D. M.,** Salivary-induced aggregation of plaque bacteria, in *Dental Plaque*, McHugh, W. D., Ed., E. & S. Livingstone, Edinburgh, 1970, 207.
897. **Hay, D. I., Gibbons, R. J., and Spinell, D. M.,** Characteristics of some high molecular weight constituents with bacterial aggregating activity from whole saliva and dental plaque, *Caries Res.*, 5, 111, 1971.
898. **Kashket, S. and Donaldson, C. G.,** Saliva-induced aggregation of oral streptococci, *J. Bacteriol.*, 112, 1127, 1972.
899. **Ørstavik, D., Kraus, F. W., and Henshaw, L. C.,** In vitro attachment of streptococci to the tooth surface, *Infect. Immun.*, 9, 794, 1974.
900. **Ericson, T., Pruitt, K., and Wedel, H.,** The reaction of salivary substances with bacteria, *J. Oral Pathol.*, 4, 307, 1975.

901. **Ericson, T., Carlen, A., and Dagerskog, E.,** Salivary aggregation factors, in *Microbial Aspects of Dental Caries,* Stiles, H. M., Loesche, W. J., and O'Brien, T. C., Eds., Information Retrieval, Washington, D.C., 1976, 151.
902. **Williams, R. C. and Gibbons, R. J.,** Inhibition of bacterial adherence by secretory immunoglobulin A: a mechanism of antigen disposal, *Science,* 177, 697, 1972.
903. **Williams, R. C. and Gibbons, R. J.,** Inhibition of streptococcal attachment to receptors on human buccal epithelial cells by antigenically similar salivary glycoproteins, *Infect. Immun.,* 11, 711, 1975.
904. **Gibbons, R. J. and Qureshi, J. V.,** Interactions of *Streptococcus mutans* and other oral bacteria with blood group reactive substances, in *Microbial Aspects of Dental Caries,* Stiles, H. M., Loesche, W. J., and O'Brien, T. C., Eds., Information Retrieval, Washington, D. C., 1976, 163.
905. **Gibbons, R. J. and Qureshi, J. V.,** Selective binding of blood group-reactive salivary mucins by *Streptococcus mutans* and other oral organisms, *Infect. Immun.,* 22, 665, 1978.
906. **Sirisinha, S.,** Reactions of human salivary immunoglobulins with indigenous bacteria, *Arch. Oral Biol.,* 15, 551, 1970.
907. **Kashket, S. and Hankin, S. R.,** Interaction of streptococcal aggregating factors with thiol- and disulphide-reactive compounds, *Arch. Oral Biol.,* 22, 49, 1977.
908. **Friedberg, I. and Avigad, G.,** High lysozyme concentration and lysis of *Micrococcus lysodeikticus, Biochim. Biophys. Acta,* 127, 532, 1966.
909. **Kamaya, T.,** Flocculation phenomenon of *Candida albicans* by lysozyme, *Mycopathol. Mycol. Appl.,* 37, 320, 1969.
910. **Metcalf, R. H. and Deibel, R. H.,** Differential lytic response of enterococci associated with addition order of lysozyme and anions, *J. Bacteriol.,* 99, 674, 1969.
911. **Kruse, H. and Hurst, A.,** Preparation of spheroplasts from *Streptococcus lactis, Can. J. Microbiol.,* 18, 825, 1972.
912. **Pollock, J. J., Iacono, V. J., Bicker, H. G., MacKay, B. J., Katona, L. I., Taichman, L. B., and Thomas, E.,** The binding, aggregation and lytic properties of lysozyme, in *Microbial Aspects of Dental Caries,* Stiles, H. M., Loesche, W. J., and O'Brien, T. C., Eds., Information Retrieval, Washington, D.C., 1976, 325.
913. **McBride, B. C. and Gisslow, M. T.,** Role of sialic acid in saliva-induced aggregation of *Streptococcus sanguis, Infect. Immun.,* 18, 35, 1977.
914. **Howell, A., Jr. and Jordan, H. V.,** Production of an extracellular levan by *Odontomyces viscosus, Arch. Oral Biol.,* 12, 571, 1967.
915. **Rosan, B. and Hammond, B. F.,** Extracellular polysaccharides of *Actinomyces viscosus, Infect. Immun.,* 10, 304, 1974.
916. **Pabst, M. J.,** Levan and levansucrase of *Actinomyces viscosus, Infect. Immun.,* 15, 518, 1977.
917. **McBride, B. C. and Bourgeau, G.,** Dextran-induced aggregation of *Actinomyces viscosus, Arch. Oral Biol.,* 20, 837, 1975.
918. **Bourgeau, G. and McBride, B. C.,** Dextran-mediated interbacterial aggregation between dextran-synthesizing streptococci and *Actinomyces viscosus, Infect. Immun.,* 13, 1228, 1976.
919. **Jones, S. J.,** Natural plaque on tooth surfaces: a scanning electron microscopy study, *Apex J. Univ. Coll. Hosp. Dent. Soc.,* 5, 93, 1971.
920. **Jones, S. J.,** A special relationship between spherical and filamentous microorganisms in mature human dental plaque, *Arch. Oral Biol.,* 17, 613, 1972.
921. **Listgarten, M. A., Mayo, H., and Amsterdam, M.,** Ultrastructure of the attachment devise between coccal and filamentous microorganisms in "corn cob" formations of dental plaque, *Arch. Oral Biol.,* 18, 651, 1973.
922. **Takazoe, I., Matsukubo, T., and Kato, T.,** Experimental formation of "corn cob" *in vitro, J. Dent. Res.,* 57, 384, 1978.
923. **Gibbons, R. J. and Nygaard, M.,** Interbacterial aggregation of plaque bacteria, *Arch. Oral Biol.,* 15, 1397, 1970.
924. **Ellen, R. P. and Balcerzak-Raczkowski, I. B.,** Interbacterial aggregation of *Actinomyces naeslundii* and dental plaque streptococci, *J. Periodontal Res.,* 12, 11, 1977.
925. **McIntire, F. C., Vatter, A. E., Baros, J., and Arnold, J.,** Mechanism of coaggregation betwen *Actinomyces viscosus* T14V and *Streptococcus sanguis* 34, *Infect. Immun.,* 21, 978, 1978.
926. **Girard, A. E. and Jacius, B. H.,** Ultrastructure of *Actinomyces viscosus* and *Actinomyces naeslundii, Arch. Oral Biol.,* 19, 71, 1974.
927. **Hammond, B. F., Steel, C. F., and Peindl, K. S.,** Antigens and surface components associated with virulence of *Actinomyces viscosus, J. Dent. Res.,* 55, A19, 1976.
928. **Cisar, J. O., McIntire, F. C., and Vatter, A. E.,** Fimbriae of *Actinomyces viscosus* T14V: their relationship to the virulence-associated antigen and to coaggregation with *Streptococcus sanguis* 34, in *Secretory Immunity and Infection,* McGhee, J. R., Mastecky, J., and Babb, J. L., Eds., Plenum Press, New York, 1978, 695.

929. **Cisar, J. O., Kolenbrander, P. E., and McIntire, F. C.,** Specificity of coaggregation reactions between human oral streptococci and strains of *Actinomyces viscosus* or *Actinomyces naeslundii, Infect. Immun.*, 24, 742, 1979.
930. **Bladen, H., Hageage, G., Pollock, F., and Harr, R.,** Plaque formation *in vitro* on wires by Gram-negative oral microorganisms *(Veillonella), Arch. Oral Biol.*, 15, 127, 1970.
931. **Silverman, G. and Kleinberg, I.,** Studies on factors affecting the aggregation of the microorganisms in human dental plaque, *Arch. Oral Biol.*, 12, 1407, 1967.
932. **Miller, C. H., Palenik, C. J., and Stamper, K. E.,** Factors affecting the aggregation of *Actinomyces naeslundii* during growth and in washed cell suspensions, *Infect. Immun.*, 21, 1003, 1978.
933. **Olsson, J. and Glantz, P.-O.,** Effect of pH and counter ions on the zeta-potential of oral streptococci, *Arch. Oral Biol.*, 22, 461, 1977.
934. **Lubick, H. A.,** Effects of cyclic nucleotides on *Streptococcus mutans, J. Dent. Res.*, 58 (Spec. Issue A), 94, 1979.
935. **Liljemark, W. F. and Bloomquist, C. G.,** Preparation of a cell wall protein from *S. sanguis* which mediates its adherence to hydroxyapatitie, *J. Dent. Res.*, 58 (Spec. Issue A), 201, 1979.
936. **Fives-Taylor, P. and Fachon, S.,** Study of the surface structures of *Streptococcus sanguis, J. Dent. Res.*, 58 (Spec. Issue A), 202, 1979.
937. **Kolenbrander, P. E., Cisar, J. O., and McIntire, F. C.,** Specific coaggregations between oral streptococci and actinomycetes, *J. Dent. Res.*, 58 (Spec. Issue A), 203, 1979.
938. **MacIntyre, B. A., Spector, P. C., and Handelman, S. L.,** A method to determine the effect of trace elements on bacterial aggregation, *J. Dent. Res.*, 58 (Spec. Issue A), 203, 1979.
939. **Rundegren, J., Ericson, T., and Sjöström, R.,** Ca-dependence of a *Streptococcus mutans* agglutinin, *J. Dent. Res.*, 58 (Spec. Issue A), 252, 1979.
940. **Palenik, C. J., Miller C. H., and Sze, D.,** Lectin-induced aggregation of *Actinomyces naeslundii, J. Dent. Res.*, 58 (Spec. Issue A), 340, 1979.
941. **Dunny, G. M., Brown, B. L., and Clewell, D. B.,** Induced cell aggregation and mating in *Streptococcus faecalis:* evidence for a bacterial sex pheromone, *Proc. Natl. Acad. Sci. U.S.A.*, 75, 3479, 1978.
942. **Courvalin, P. M., Carlier, C., and Chabbert, Y. A.,** Plasmid-linked tetracycline and erythromycin resistance in group D "streptococcus", *Ann. Inst. Pasteur Paris*, 123, 755, 1972.
943. **Tomura, T., Hirano, T., Ito, T., and Yoshioka, M.,** Transmission of bacteriocinogenicity by conjugation in group D streptococci, *Jpn. J. Microbiol.*, 17, 445, 1973.
944. **Clewell, D. B., Yagi, Y., Dunny, G. M., and Schultz, S. K.,** Characterization of three plasmid deoxyribonucleic acid molecules in a strain of *Streptococcus faecalis:* identification of a plasmid determining erythromycin resistance, *J. Bacteriol.*, 117, 283, 1974.
945. **Jacob, A. E. and Hobbs, S. J.,** Conjugal transfer of plasmid-borne multiple antibiotic resistance in *Streptococcus faecalis* var. *zymogenes, J. Bacteriol.*, 117, 360, 1974.
946. **Jacob, A. E., Douglas, G. J., and Hobbs, S. J.,** Self-transferable plasmids determining the hemolysin and bacteriocin of *Streptococcus faecalis* var. *zymogenes, J. Bacteriol.*, 121, 863, 1975.
947. **Dunny, G. M. and Clewell, D. B.,** Transmissible toxin (hemolysin) plasmid in *Streptococcus faecalis* and its mobilization of a noninfectious drug resistance plasmid, *J. Bacteriol.*, 124, 784, 1975.
948. **Oliver, D. R., Brown, B. L., and Clewell, D. B.,** Analysis of plasmid deoxyribonucleic acid in a cariogenic strain of *Streptococcus faecalis:* an approach to identifying genetic determinants on cryptic plasmids, *J. Bacteriol.*, 130, 759, 1977.
949. **Oliver, D. R., Brown, B. L., and Clewell, D. B.,** Characterization of plasmids determining hemolysin and bacteriocin production in *Streptococcus faecalis* 5952, *J. Bacteriol.*, 130, 948, 1977.
950. **van Embden, J. D. A., Engel, H. W. B., and van Klingeren, B.,** Drug resistance in group D streptococci of clinical and nonclinical origin: prevalence, transferability, and plasmid properties, *Antimicrob. Agents Chemother.*, 11, 925, 1977.
951. **Marder, H. P. and Kayser, F. H.,** Transferable plasmids mediating multiple-antibiotic resistance in *Streptococcus faecalis* subsp. *liquefaciens, Antimicrob. Agents Chemother.*, 12, 261, 1977.
952. **Franke, A. E., Dunny, G. M., Brown, B. L., An, F., Oliver, D. R., Damle, S. P., and Clewell, D. B.,** Gene transfer in *Streptococcus faecalis:* evidence for the mobilization of chromosomal determinants by transmissible plasmids, in *Microbiology — 1978*, Schlessinger, D., Ed., American Society for Microbiology, Washington, D.C., 1978, 45.
953. **Tomich, P. K., An, F. Y., Damle, S. P., and Clewell, D. B.,** Plasmid-related transmissibility and multiple drug resistance in *Streptococcus faecalis* subsp. *zymogenes* strain DS16, *Antimicrob. Agents Chemother.*, 15, 828, 1979.
954. **Dunny, G. M., Craig, R. A., Carron, R. L., and Clewell, D. B.,** Plasmid transfer in *Streptococcus faecalis:* production of multiple sex pheromones by recipients, *Plasmid*, 2, 454, 1979.
955. **Yagi, Y. and Clewell, D. B.,** Conjugation in *Streptococcus faecalis:* the nature of the donor aggregation response, *J. Dent. Res.*, 58 (Spec. Issue A), 336, 1979.

956. **Craig, R. A. and Clewell, D. B.,** On the nature of bacterial sex pheromones in *Streptococcus faecalis, Abstr. Annu. Meet. Am. Soc. Microbiol.,* 1979, 135, 1979.
957. **Clewell, D. B., Dunny, G. M., Brown, B. L., Craig, R. A., and Yagi, Y.,** Sex pheromones in *Streptococcus faecalis,* in *Pathogenic Streptococci,* Parker, M. T., Ed., Reedbooks, Chertsey, Surrey, 1979, 284.
958. **Campbell, A., Starlinger, P., Berg, D. E., Botstein, D., Lederberg, E. M., Novick, R. P., and Szybalski, W.,** Nomenclature of transposable elements in prokaryotes, *Plasmid,* 2, 466, 1979.
959. **Tomich, P. K., An, F. Y., and Clewell, D. B.,** A transposon (Tn *917*) in *Streptococcus faecalis* which exhibits enhanced transposition during induction of drug resistance, *Cold Spring Harbor Symp. Quant. Biol.,* 43, 1217, 1978.
960. **Franke, A. E. and Clewell, D. B.,** Transposon specific cell aggregation response and enhanced mating in *Streptococcus faecalis, Abstr. Annu. Meet. Am. Soc. Microbiol.,* 1979, 126, 1979.
961. **Link, H. F.,** Observationes in ordines plantarum naturales. Dissertatio prima complectens anandrarum ordines epiphytes, mucedines, gastromycos et fungos, *Mag. Neuesten Entdeckungen Gesammten Naturkinde, Gessellschaft Naturforschender Freunde, Berlin,* 3, 1, 1809.
962. **Stanier, R. Y.,** Order VIII. *Myxobacterales* Jahn, 1915, in *Bergey's Manual of Determinative Bacteriology,* 7th ed., Breed, R. S., Murray, E. G. D., and Smith, N. R., Eds., Williams & Wilkins, Baltimore, 1957, 854.
963. **Quinlan, M. S. and Raper, K. B.,** Development of the myxobacteria, in *Differenzierung und Entwicklung, Handbuch der Pflanzenphysiologie,* Vol. 15, Part 1, Lang, A., Ed., Springer-Verlag, Berlin, 1965, 596.
964. **Starr, M. P. and Skerman, V. B. D.,** Bacterial diversity: the natural history of selected morphologically unusual bacteria, *Annu. Rev. Microbiol.,* 19, 407, 1965.
965. **Dworkin, M.,** Biology of the myxobacteria, *Annu. Rev. Microbiol.,* 20, 75, 1966.
966. **Larpent, J. P.,** *De la cellule a l'organisme (Acrasiales, Myxomycètes, Myxobactériales),* Masson et Cie, Paris, 1970, 75.
967. **Dworkin, M.,** The myxobacteria: new directions in studies of procaryotic development, *Crit. Rev. Microbiol.,* 1, 435, 1972.
968. **Dworkin, M.,** The *Myxobacterales* [(fruiting) *Myxobacterales*], in *CRC Handbook of Microbiology,* Vol. 1, Laskin, A. I. and Lechavalier, H. A., Eds., CRC Press, Boca Raton, Fla., 1973, 191; in 2nd ed., 1977, p. 131.
969. **Dworkin, M.,** Cell-cell interactions in the myxobacteria, in *Microbial Differentiation,* Ashworth, J. M. and Smith, J. E., Eds., Cambridge University Press, Cambridge, U. K., 1973, 125.
970. **McCurdy, H. D.,** The fruiting myxobacteria, in *Bergey's Manual of Determinative Bacteriology,* 8th ed., Buchanan, R. E., and Gibbons, N. E., Eds., Williams & Wilkins, Baltimore, 1974, 76.
971. **Wireman, J. W. and Dworkin, M.,** Morphogenesis and developmental interactions in myxobacteria, *Science,* 189, 516, 1975.
972. **Kaiser, D., Manoil, C., and Dworkin, M.,** Myxobacteria: cell interactions, genetics, and development, *Annu. Rev. Microbiol.,* 33, 595, 1979.
973. **Zusman, D. R.,** Genetic approaches to the study of development in the myxobacteria, in *The Molecular Genetics of Development: an Introduction to Recent Research in Experimental Systems,* Leighton, T. and Loomis, W. F., Eds., Academic Press, New York, 1980, 41.
974. **Reichenbach, H. and Dworkin, M.,** Introduction to the gliding bacteria, in *The Prokaryotes: a Handbook of Habitats, Isolation, and Identification of Bacteria,* Vol. 1, Starr, M. P., Stolp, H., Trüper, H. G., Balows, A., and Schlegel, H. G., Eds., Springer-Verlag, Berlin, 1981, 315.
975. **Singh, B. N.,** Myxobacteria in soils and composts; their distribution, number and lytic action on bacteria, *J. Gen. Microbiol.,* 1, 1, 1947.
976. **McCurdy, H. D., Jr.,** A method for the isolation of myxobacteria in pure culture, *Can. J. Microbiol.,* 9, 282, 1963.
977. **Kühlwein, H. and Reichenbach, H.,** Anreicherung und Isolierung von Myxobakterien, *Zentralbl. Bakteriol. Parasitenk. Infektionskr. Hyg. Abt. 1,* Suppl. 1, 57, 1966.
978. **McCurdy, H. D.,** Studies on the taxonomy of the Myxobacterales. I. Record of Canadian isolates and survey of methods, *Can. J. Microbiol.,* 15, 1453, 1969.
979. **Peterson, J. E.,** Isolation, cultivation and maintenance of the myxobacteria, *Methods in Microbiology,* Vol. 3B, Norris, J. R. and Ribbons, D. W., Eds., Academic Press, London, 1969, 185.
980. **Reichenbach, H.,** *Myxococcus* spp. (Myxobacteriales) Schwarmentwicklung und Bildung von Protocysten, Film E778/1965, *Encyclopedia Cinematographica,* Wolf, G., Ed., Inst. Wiss. Film, Gottingen, 1966, 557.
981. **Reichenbach, H.,** Schwarmentwicklung und Bildung von Protocysten, Film E777/1965, *Encyclopedia Cinematographica,* Wolf, G., Ed., Inst. Wiss. Film, Gottingen, 1968, 317.
982. **Kühlwein, H. and Reichenbach, H.,** Schwarmentwicklung und Morphogenese bei Myxobakterien: *Archangium, Myxococcus, Chondrococcus, Chondromyces,* Film C893/1965, *Encyclopedia Cinematographica,* Wolf, G., Ed., Inst. Wiss. Film, Gottingen, 1968, 335.

983. **Kühlwein, H.,** Film E1583/1971, *Encyclopedia Cinematographica,* Wolf, G., Ed., Inst. Wiss. Film, Gottingen, 1971, 3.

984. **Grimm, K.,** Film E1588/1971, *Encyclopedia Cinematographica,* Wolf, G., Ed., Inst. Wiss. Film, Gottingen, 1971, 3.

985. **Reichenbach, H.,** Film E779/1965, *Encyclopedia Cinematographica,* Wolf, G., Ed., Inst. Wiss. Film, Gottingen, 1974, 3.

986. **Zahler, S. A. and McCurdy, H. D.,** Family I. *Myxococcaceae* Jahn 1924, 84, in *Bergey's Manual of Determinative Bacteriology,* 8th ed., Buchanan, R. E. and Gibbons, N. E., Eds., Williams & Wilkins, Baltimore, 1974, 79.

987. **McCurdy, H. D.,** Family II. *Archangiaceae* Jahn 1924, 66, in *Bergey's Manual of Determinative Bacteriology,* 8th ed., Buchanan, R. E. and Gibbons, N. E., Eds., Williams & Wilkins, Baltimore, 1974, 83.

988. **McCurdy, H. D.,** Family III. *Cystobacteraceae* McCurdy 1970, 286, in *Bergey's Manual of Determinative Bacteriology,* 8th ed., Buchanan, R. E. and Gibbons, N. E., Eds., Williams & Wilkins, Baltimore, 1974, 86.

989. **McCurdy, H. D. and Peterson, J. E.,** Family IV. *Polyangiaceae* Jahn 1924, 75, in *Bergey's Manual of Determinitive Bacteriology,* 8th ed., Buchanan, R. E. and Gibbons, N. E., Williams & Wilkins, Baltimore, 1974, 92.

990. **Leadbetter, E. R.,** Order II. *Cytophagales nomen novum,* in *Bergey's Manual of Determinative Bacteriology,* 8th ed., Buchanan, R. E. and Gibbons, N. E., Eds., Williams & Wilkins, Baltimore, 1974, 99.

991. **Soriano, S. and Lewin, R. A.,** Gliding microbes: some taxonomic reconsiderations, *Antonie van Leeuwenhoek, J. Microbiol. Serol.,* 31, 66, 1965.

992. **Mandel, M. and Leadbetter, E.R.,** Deoxyribonucleic acid base composition of myxobacteria, *J. Bacteriol.,* 90, 1795, 1965.

993. **McCurdy, H. D. and Wolf, S.,** Deoxyribonucleic acid base compositions of fruiting *Myxobacterales, Can. J. Microbiol.,* 13, 1707, 1967.

994. **Mandel, M. and Lewin, R. A.,** Deoxyribonucleic acid base composition of flexibacteria, *J. Gen. Microbiol.,* 58, 171, 1969.

995. **McCurdy, H. D. and Wolf, S.,** Studies on the taxonomy of fruiting *Myxobacterales, Bacteriol. Proc.,* p. 39, 1967.

996. **Lewin, R. A.,** A classification of flexibacteria, *J. Gen. Microbiol.,* 58, 189, 1969.

997. **Colwell, R. R.,** Numerical taxonomy of the flexibacteria, *J. Gen. Microbiol.,* 58, 207, 1969.

998. **Geitler, L.,** Über *Polyangium parasiticum* n. sp., eine submerse, parasitische Myxobacteriacee, *Arch. Protistenk.,* 50, 67, 1925.

999. **Jeffers, E. E.,** Myxobacters of a freshwater lake and its environs, *Int. Bull. Bacteriol. Nomencl. Taxon.,* 14, 115, 1964.

1000. **Shilo, M.,** Lysis of blue-green algae by myxobacter, *J. Bacteriol.,* 104, 453, 1970.

1001. **Reichenbach, H. and Kleinig, H.,** The carotenoids of *Myxococcus fulvus* (Myxobacterales), *Arch. Mikrobiol.,* 76, 364, 1971.

1002. **Kleinig, H.,** Membranes from *Myxococcus fulvus (Myxobacterales)* containing carotenoid glucosides. I. Isolation and composition, *Biochim. Biophys. Acta,* 274, 489, 1972.

1003. **Dworkin, M.,** Sensitivity of gliding bacteria to actinomycin D, *J. Bacteriol.,* 98, 851, 1969.

1004. **White, D., Dworkin, M., and Tipper, D. J.,** Peptidoglycan of *Myxococcus xanthus:* structure and relation to morphogenesis, *J. Bacteriol.,* 95, 2186, 1968.

1005. **Reichenbach, H.,** Rhythmische Vorgänge bei der Schwarmentfaltung von Myxobakterien, *Ber. Deutsch. Bot. Ges.,* 78, 102, 1965.

1006. **Sutherland, I. W. and Thomson, S.,** Comparison of polysaccharides produced by myxococcus strains, *J. Gen. Microbiol.,* 89, 124, 1975.

1007. **Gnosspelius, G.,** Myxobacterial slime and proteolytic activity, *Arch. Microbiol.,* 116, 51, 1978.

1008. **Adye, J. C. and Powelson, D. M.,** Microcyst of *Myxococcus xanthus:* chemical composition of the wall, *J. Bacteriol.,* 81, 780, 1961.

1009. **Dworkin, M. and Gibson, S. M.,** A system for studying microbial morphogenesis: rapid formation of microcysts in *Myxococcus xanthus, Science,* 146, 243, 1964.

1010. **Reichenbach, H., Voelz, H., and Dworkin, M.,** Structural changes in *Stigmatella aurantiaca* during myxospore induction, *J. Bacteriol.,* 97, 905, 1969.

1011. **McCurdy, H. D. and Khouw, B. T.,** Studies on *Stigmatella brunnea, Can. J. Microbiol.,* 15, 731, 1969.

1012. **Sadler, W. and Dworkin, M.,** Induction of cellular morphogenesis in *Myxococcus xanthus.* II. Macromolecular synthesis and mechanism of inducer action, *J. Bacteriol.,* 91, 1520, 1966.

1013. **Reichenbach, H. and Dworkin, M.,** Induction of myxospore formation in *Stigmatella aurantiaca (Myxobacterales)* by monovalent cations, *J. Bacteriol.,* 101, 325, 1970.

1014. **Witkin, S. S. and Rosenberg, E.,** Induction of morphogenesis by methionine starvation in *Myxococcus xanthus:* polyamine control, *J. Bacteriol.,* 103, 641, 1970.

1015. **Gerth, K. and Reichenbach, H.**, Induction of myxospore formation in *Stigmatella aurantiaca* (Myxobacterales). I. General characterization of the system, *Arch. Microbiol.*, 117, 173, 1978.

1016. **Sudo, S. Z. and Dworkin, M.**, Resistance of vegetative cells and microcysts of *Myxococcus xanthus*, *J. Bacteriol.*, 98, 883, 1969.

1017. **Sudo, S. Z. and Dworkin, M.**, Comparative biology of prokaryotic resting cells, *Adv. Microb. Physiol.*, 9, 153, 1973.

1018. **Burchard, R. P.**, Myxospore induction in a nondispersed growing mutant of *Myxococcus xanthus*, *J. Bacteriol.*, 122, 302, 1975.

1019. **Burchard, R. P.**, Gliding motility mutants of *Myxococcus xanthus*, *J. Bacteriol.*, 104, 940, 1970.

1020. **Qualls, G. T., Stephens, K., and White, D.**, Morphogenetic movements and multicellular development in the fruiting myxobacterium, *Stigmatella aurantiaca*, *Dev. Biol.*, 66, 270, 1978.

1021. **Doetsch, R. N. and Hageage, G. H.**, Motility in procaryotic organisms: problems, points of view, and perspectives, *Biol. Rev. Camb. Philos. Soc.*, 43, 317, 1968.

1022. **Henrichsen, J.**, Bacterial surface translocation: a survey and a classification, *Bacteriol. Rev.*, 36, 478, 1972.

1023. **Grimm, K. and Kühlwein, H.**, Untersuchungen an spontanen Mutanten von *Archangium violaceum* (Myxobacterales). I. Bewegliche und unbewegliche Zellen von *A. violaceum*, *Arch. Mikrobiol.*, 89, 105, 1973.

1024. **Dawid, W.**, *Myxococcus virescens* — ein Schleimbakterium mit grünen Fruchtkörpern. Beobachtungen und Untersuchungen zur Kultur und Biologie eines Schleimbakteriums, *Mikrokosmos*, 63, 1, 1974.

1025. **Hodgkin, J. and Kaiser, D.**, Genetics of gliding motility in *Myxococcus xanthus* (Myxobacterales): genes controlling movement of single cells, *Mol. Gen. Genet.*, 171, 167, 1979.

1026. **Graf, W.**, Bewegungsorganellen bei Myxobakterien, *Arch. Hyg. Bakteriol.*, 149, 518, 1965.

1027. **MacRae, T. H. and McCurdy, H. D.**, Ultrastructural studies of *Chondromyces crocatus* vegetative cells, *Can. J. Microbiol.*, 21, 1815, 1975.

1028. **MacRae, T. H. and McCurdy, H. D.**, The isolation and characterization of gliding motility mutants of *Myxococcus xanthus*, *Can. J. Microbiol.*, 22, 1282, 1976.

1029. **MacRae, T. H. and McCurdy, H. D.**, Evidence for motility-related fimbriae in the gliding microorganism *Myxococcus xanthus*, *Can. J. Microbiol.*, 22, 1589, 1976.

1030. **MacRae, T. H., Dobson, W. J., and McCurdy, H. D.**, Fimbriation in gliding bacteria, *Can. J. Microbiol.*, 23, 1096, 1977.

1031. **Hodgkin, J. and Kaiser, D.**, Genetics of gliding motility in *Myxococcus xanthus* (Myxobacterales): two gene systems control movement, *Mol. Gen. Genet.*, 171, 177, 1979.

1032. **Schmidt-Lorenz, W. and Kühlwein, H.**, Intracelluläre Bewegungsorganellen der Myxobakterien, *Arch. Mikrobiol.*, 60, 95, 1968.

1033. **Burchard, A. C., Burchard, R. P., and Kloetzel, J. A.**, Intracellular, periodic structures in the gliding bacterium *Myxococcus xanthus*, *J. Bacteriol.*, 132, 666, 1977.

1034. **Jahn, E.**, *Beiträge zur Botanischen Protistologie. I. Die Polyangiden*, Verlag von Gebrüder Borntraeger, Leipzig, 1924.

1035. **Ridgway, H. F. and Lewin, R. A.**, Goblet-shaped sub-units from the wall of a marine gliding microbe, *J. Gen. Microbiol.*, 79, 119, 1973.

1036. **Pate, J. L. and Chang, L.-Y. E.**, Evidence that gliding motility in prokaryotic cells is driven by rotary assemblies in the cell envelopes, *Curr. Microbiol.*, 2, 59, 1979.

1037. **Berg, H. C. and Anderson, R. A.**, Bacteria swim by rotating their flagellar filaments, *Nature (London)*, 245, 380, 1973.

1038. **Silverman, M. and Simon, M.**, Flagellar rotation and the mechanism of bacterial motility, *Nature (London)*, 249, 73, 1974.

1039. **Larsen, S. H., Adler, J., Gargus, J. J., and Hogg, R. W.**, Chemomechanical coupling without ATP: the source of energy for motility and chemotaxis in bacteria, *Proc. Natl. Acad. Sci. U.S.A.*, 71, 1239, 1974.

1040. **Manson, M. D., Tedesco, P., Berg, H. C., Harold, F. M., and van der Drift, C.**, A protonmotive force drives bacterial flagella, *Proc. Natl. Acad. Sci. U.S.A.*, 74, 3060, 1977.

1041. **Ridgway, H. F.**, Source of energy for gliding motility in *Flexibacter polymorphus:* effects of metabolic and respiratory inhibitors on gliding movement, *J. Bacteriol.*, 131, 544, 1977.

1042. **Stanier, R. Y.**, A note on elasticotaxis in myxobacteria, *J. Bacteriol.*, 44, 405, 1942.

1043. **Bonner, J. T.**, *Morphogenesis: an Essay on Development*, Princeton University Press, Princeton, N.J., 1952, 165.

1044. **Ho, J. and McCurdy, H. D.**, Demonstration of positive chemotaxis to cyclic GMP and 5′-AMP in *Myxococcus xanthus* by means of a simple apparatus for generating practically stable concentration gradients, *Can. J. Microbiol.*, 25, 1214, 1979.

1045. **Rosenberg, E., Filer, D., Zafriti, D., and Kindler, S. H.**, Aspartokinase activity and the developmental cycle of *Myxococcus xanthus*, *J. Bacteriol.*, 115, 29, 1973.

1046. **Bretscher, A. P. and Kaiser, D.,** Nutrition of *Myxococcus xanthus,* a fruiting myxobacterium, *J. Bacteriol.*, 133, 763, 1978.
1047. **Filer, D., Rosenberg, E., and Kindler, S. H.,** Aspartokinase of *Myxococcus xanthus:* "feedback stimulation" by required amino acids, *J. Bacteriol.*, 115, 23, 1973.
1048. **McCurdy, H. D., Ho, J., and Dobson, W. J.,** Cyclic nucleotides, cyclic nucleotide phosphodiesterase, and development in *Myxococcus xanthus, Can. J. Microbiol.*, 24, 1475, 1978.
1049. **Campos, J. M. and Zusman, D. R.,** Regulation of development in *Myxococcus xanthus:* effect of 3′:5′-cyclic AMP, ADP, and nutrition, *Proc. Natl. Acad. Sci. U.S.A.*, 72, 518, 1975.
1050. **Yajko, D. M. and Zusman, D. R.,** Changes in cyclic AMP levels during development in *Myxococcus xanthus, J. Bacteriol.*, 133, 1540, 1978.
1051. **Parish, J. H., Wedgwood, K. R., and Herries, D. G.,** Morphogenesis in *Myxococcus xanthus* and *Myxococcus virescens* (Myxobacterales), *Arch. Microbiol.*, 107, 343, 1976.
1052. **Lazzarini, R. A. and Cashel, M.,** On the regulation of guanosine tetraphosphate levels in stringent and relaxed strains of *Escherichia coli, J. Biol. Chem.*, 246, 4381, 1971.
1053. **Cashel, M. and Gallant, J.,** Cellular regulation of guanosine tetraphosphate and guanosine pentaphosphate, in *Ribosomes,* Nomura, M., Tissières, A., and Lengyel, P., Eds., Cold Spring Harbor Laboratory, Cold Spring Harbor, N. Y., 1974, 733.
1054. **Ishiguro, E. E. and Ramey, W. D.,** Stringent control of peptidoglycan biosynthesis in *Escherichia coli* K-12, *J. Bacteriol.*, 127, 1119, 1976.
1055. **Hecht, S. M., Faulkner, R. D., and Hawrelak, S. D.,** Competitive inhibition of beef heart cyclic AMP phosphodiesterase by cytokinins and related compounds, *Proc. Natl. Acad. Sci. U.S.A.*, 71, 4670, 1974.
1056. **Coppola, S., Zoina, A., and Marino, P.,** Interactions of N^6-(Δ^2-isopentenyl)adenine with cyclic AMP on the regulation of growth and β-galactosidase synthesis in *Escherichia coli, J. Gen. Microbiol.*, 94, 436, 1976.
1057. **Jones, M. V.,** Inhibition of fruiting body formation in *Myxococcus xanthus* by cytokinins, *FEMS Microbiol. Lett.*, 3, 147, 1978.
1058. **Lev, M.,** Demonstration of a diffusible fruiting factor in myxobacteria, *Nature (London),* 173, 501, 1954.
1059. **Fluegel, W.,** Myxobacterial chemotaxis, *Bacteriol. Proc.*, p. 48, 1963.
1060. **Bonner, J. T.,** Evidence for the formation of cell aggregates by chemotaxis in the development of the slime mold *Dictyostelium discoideum, J. Exp. Zool.*, 106, 1, 1947.
1061. **Lonski, J., Heromin, R., and Ingraham, D.,** A quantitative assay to study cell movement in the myxobacteria, *J. Cell Sci.*, 25, 173, 1977.
1062. **Burchard, R. P.,** Growth of surface colonies of the gliding bacterium *Myxococcus xanthus, Arch. Microbiol.*, 96, 247, 1974.
1063. **Fluegel, W.,** Fruiting body populations of *Myxococcus fulvus* (Myxobacterales), *Growth,* 29, 183, 1965.
1064. **McVittie, A., Messik, F., and Zahler, S. A.,** Developmental biology of *Myxococcus, J. Bacteriol.*, 84, 546, 1962.
1065. **Coucke, P.,** Morphology and morphogenesis of *Sorangium compositum, J. Appl. Bacteriol.*, 32, 24, 1969.
1066. **Qualls, G. T., Stephens, K., and White, D.,** Light-stimulated morphogenesis in the fruiting myxobacterium *Stigmatella aurantiaca, Science,* 201, 444, 1978.
1067. **Baur, E.,** Myxobakterien-Studien, *Arch. Protistenk.*, 5, 92, 1905.
1068. **Aschner, M. and Chorin-Kirsh, I.,** Light-oriented locomotion in certain myxobacter species, *Arch. Mikrobiol.*, 74, 308, 1970.
1069. **Inouye, M., Inouye, S., and Zusman, D. R.,** Gene expression during development of *Myxococcus xanthus:* pattern of protein synthesis, *Dev. Biol.*, 68, 579, 1979.
1070. **Inouye, M., Inouye, S., and Zusman, D. R.,** Biosynthesis and self-assembly of protein S, a development-specific protein of *Myxococcus xanthus, Proc. Natl. Acad. Sci. U.S.A.*, 76, 209, 1979.
1071. **Cumsky, M. and Zusman, D. R.,** Myxobacterial hemagglutinin: a development-specific lectin of *Myxococcus xanthus, Proc. Natl. Acad. Sci. U.S.A.*, 76, 5505, 1979.
1072. **Dobson, W. J. and McCurdy, H. D.,** The function of fimbriae in *Myxococcus xanthus.* I. Purification and properties of *M. xanthus* fimbriae, *Can. J. Microbiol.*, 25, 1152, 1979.
1073. **Dobson, W. J., McCurdy, H. D., and MacCrae, T. H.,** The function of fimbriae in *Myxococcus xanthus.* II. The role of fimbriae in cell-cell interaction, *Can. J. Mcrobiol.*, 25, 1359, 1979.
1074. **Rudd, K. and Zusman, D. R.,** Rifampin-resistant mutants of *Myxococcus xanthus* defective in development, *J. Bacteriol.*, 137, 295, 1979.
1075. **Bolle, A., Epstein, R. H., Salser, W., and Geiduschek, E. P.,** Transcription during bacteriophage T4 development: synthesis and relative stability of early and late RNA, *J. Mol. Biol.*, 31, 325, 1968.
1076. **Pero, J., Nelson, J., and Losick, R.,** In vitro and in vivo transcription by vegetative and sporulating *Bacillus subtilis,* in *Spores VI,* Gerhardt, P., Costilow, R. N., and Sadoff, H. L., Eds., American Society for Microbiology, Washington, D. C., 1975, 202.
1077. **Talkington, C. and Pero, C.,** Restriction fragment analysis of the temporal program of bacteriophage SP01 transcription and its control by phage-modified RNA polymerases, *Virology,* 83, 365, 1977.

1078. **Hagen, D. C., Bretscher, A. P., and Kaiser, D.,** Synergism between morphogenetic mutants of *Myxococcus xanthus, Dev. Biol.*, 64, 284, 1978.
1079. **Hodgkin, J. and Kaiser, D.,** Cell-to-cell stimulation of movement in nonmotile mutants of *Myxococcus, Proc. Natl. Acad. Sci.*, U.S.A., 74, 2938, 1977.
1080. **Burchard, R. P. and Dworkin, M.,** Light-induced lysis and carotenogenesis in *Myxococcus xanthus, J. Bacteriol.*, 91, 535, 1966.
1081. **Burchard, R. P., Burchard, A. C., and Parish, J. H.,** Pigmentation phenotype instability in *Myxococcus xanthus, Can. J. Microbiol.*, 23, 1657, 1977.
1082. **Grimm, K.,** Klonkulturen von *Archangium violaceum, Arch. Mikrobiol.*, 57, 283, 1967.
1083. **Grimm, K. and Kühlwein, H.,** Untersuchungen an spontanen Mutanten von *Archangium violaceum* (Myxobacterales). II. Über den Einfluss des Schleims auf die Bewegung der Zellen und die Entstehung stabiler Suspensionskulturen, *Arch. Mikrobiol.*, 89, 121, 1973.
1084. **Grimm, K. and Kühlwein, H.,** Untersuchungen an spontanen Mutanten von *Archangium violaceum* (Myxobacterales). III. Über weitere Eigenschaften der *K*- und *S-Zellen. Arch. Mikrobiol.*, 89, 133, 1973.
1085. **Mason, D. J. and Powelson, D.,** Lysis of *Myxococcus xanthus, J. Gen. Microbiol.*, 19, 65, 1958.
1086. **Wireman, J. W. and Dworkin, M.,** Developmentally induced autolysis during fruiting body formation by *Myxococcus xanthus, J. Bacteriol.*, 129, 796, 1977.
1087. **Kühlwein, H.,** Beiträge zur Biologie und Entwicklungsgeschichte der *Myxobakterien, Arch. Mikrobiol.*, 14, 678, 1950.
1088. **Voelz, H. and Dworkin, M.,** Fine structure of *Myxococcus xanthus* during morphogenesis, *J. Bacteriol.*, 84, 943, 1962.
1089. **McCurdy, H. D., Jr.,** Light and electron microscope studies on the fruiting bodies of *Chondromyces crocatus, Arch. Mikrobiol.*, 65, 380, 1969.
1090. **Voelz, H. and Reichenbach, H.,** Fine structure of fruiting bodies of *Stigmatella aurantiaca (Myxobacterales), J. Bacteriol.*, 99, 856, 1969.
1091. **Smit, M. and Clark, A. G.,** The observation of myxobacterial fruiting bodies, *J. Appl. Bacteriol.*, 34, 399, 1971.
1092. **McNeil, K. E. and Skerman, V. B. D.,** Examination of myxobacteria by scanning electron microscopy, *Int. J. Syst. Bacteriol.*, 22, 243, 1972.
1093. **Brockman, E. R. and Todd, R. L.,** Fruiting myxobacters as viewed with a scanning electron microscope, *Int. J. Syst. Bacteriol.*, 24, 118, 1974.
1094. **Shimkets, L. and Seale, T. W.,** Fruiting-body formation and myxospore differentiation and germination in *Myxococcus xanthus* viewed by scanning electron microscopy, *J. Bacteriol.*, 121, 711, 1975.
1095. **Grilione, P. L. and Pangborn, J.,** Scanning electron microscopy of fruiting body formation by myxobacteria, *J. Bacteriol.*, 124, 1558, 1975.
1096. **Barber, R. T.,** Interaction of bubbles and bacteria in the formation of organic aggregates in sea-water, *Nature (London)*, 211, 257, 1966.
1097. **Marshall, K. C.,** *Interfaces in Microbial Ecology*, Harvard University Press, Cambridge, Mass., 1976.
1098. **Marshall, K. C.,** Theoretical and practical significance of bacteria at interfaces, *Dev. Ind. Microbiol.*, 20, 1, 1979.
1099. **Much, H.,** Über eine Vorstufe des Fibrinfermentes in Kulturen von Staphylokokkus aureus, *Biochem. Z.*, 14, 143, 1908.
1100. **Birch-Hirschfeld, L.,** Über die Agglutination von Staphylokokken durch Bestandteile des Säugetierblutplasmas, *Klin. Wochenschr.*, 13, 331, 1934.
1101. **Berger, F. M.,** Clumping of pathogenic staphylococci in plasma, *J. Pathol. Bacteriol.*, 55, 435, 1943.
1102. **Duthie, E. S.,** The action of fibrinogen on certain pathogenic cocci, *J. Gen. Microbiol.*, 13, 383, 1955.
1103. **Elkins, K. L., Hyde, R. M., and Kelly, F. C.,** Immunofluorescence examination of typical *Staphylococcus aureus* variants and L-forms for clumping factor, *J. Gen. Microbiol.*, 60, 1, 1970.
1104. **Flegel, T. W.,** Let's call a yeast a yeast, *Can. J. Microbiol.*, 23, 945, 1977.
1105. **Lodder, J.,** General classification of the yeasts, in *The Yeasts: a Taxonomic Study*, 2nd ed., Lodder, J., Ed., North-Holland, Amsterdam, 1970, chap. 1.
1106. **Rose, A. H. and Harrison, J. S., Eds.,** *The Yeasts*, Vols. 1 to 3, Academic Press, London, 1969, 1970, 1971.
1107. **Reed, G. and Peppler, H. J.,** *Yeast Technology*, AVI Publishing, Westport, Conn., 1973.
1108. **Prescott, D. M., Ed.,** *Yeast Cells, Methods in Cell Biology*, Vol. 12, Academic Press, New York, 1975.
1109. **Phaff, H. J., Miller, M. W., and Mrak, E. M.,** *The Life of Yeasts*, 2nd ed., Harvard University Press, Cambridge, Mass., 1978.
1110. **Crandall, M.,** Mechanisms of fusion in yeast cells, in *Microbial and Plant Protoplasts*, Peberdy, J. F., Rose, A. H., Rogers, H. J., and Cocking, E. C., Eds., Academic Press, London, 1976, 161.
1111. **Taylor, N. W.,** Inactivation of sexual agglutination in *Hansenula wingei* and *Saccharomyces kluyveri* by disulfide-cleaving agents, *J. Bacteriol.*, 88, 929, 1964.

1112. **Brock, T. D.,** Protein as a specific cell surface component in the mating reaction of *Hansenula wingei, J. Bacteriol.*, 76, 334, 1958.
1113. **Brock, T. D.,** Mating reaction in *Hansenula*. Relation of cell surface properties to agglutination, *J. Bacteriol.*, 78, 59, 1959.
1114. **Brookbank, J. W. and Heisler, M. R.,** Immunology of the yeast *Hansenula wingei, J. Bacteriol.*, 85, 509, 1963.
1115. **Crandall, M. A. and Brock, T. D.,** Molecular aspects of specific cell contact, *Science*, 161, 473, 1968.
1116. **Taylor, N. W.,** Purification of sexual agglutination factor from the yeast *Hansenula wingei* by chromatography and gradient sedimentation, *Arch. Biochem. Biophys.*, 111, 181, 1965.
1117. **Taylor, N. W. and Tobin, R.,** Sexual agglutination in yeast. IV. Minimum particle count per cell for fast-sedimenting 5-agglutinin, *Arch. Biochem. Biophys.*, 115, 271, 1966.
1118. **Taylor, N. W. and Orton, W. L.,** Sexual agglutination in yeast. V. Small particles of 5-agglutinin, *Arch. Biochem. Biophys.*, 120, 602, 1967.
1119. **Taylor, N. W., Orton, W. L., and Babcock, G. E.,** Sexual agglutination in yeast. VI. Role of disulfide bonds in 5-agglutinin, *Arch. Biochem. Biophys.*, 123, 265, 1968.
1120. **Karush, F.,** Immunologic specificity and molecular structure, *Adv. Immunol.*, 2, 1, 1962.
1121. **Herman, A. I., Wickerham, L. J., and Griffin, P.,** Allelic determinants of agglutination in the yeast *Hansenula wingei, Genetics*, 54, 339, 1966.
1122. **Crandall, M.,** Microtiter plate assay for sexual agglutination in the yeast *Hansenula wingei, J. Bacteriol.*, 137, 1051, 1979.
1123. **Wickerham, L. J. and Burton, K. A.,** Phylogeny of phosphomannan-producing yeasts. I. The genera, *J. Bacteriol.*, 82, 265, 1961.
1124. **Crandall, M. and Robinow, C. F.,** The chromosome number of the yeast *Hansenula wingei* observed in mitosis and meiosis, *Can. J. Genet. Cytol.*, 15, 874, 1973.
1125. **Crandall, M. and Richter, R. H.,** Genetics of resistance to ethidium bromide in the petite-negative yeast *Hansenula wingei, Mol. Gen. Genet.*, 125, 279, 1973.
1126. **Herman, A. I.,** Sex-specific growth responses in yeasts, *Antonie van Leeuwenhoek J. Microbiol. Serol.*, 37, 379, 1971.
1127. **Conti, S. F. and Brock, T. D.,** Electron microscopy of cell fusion in conjugating *Hansenula wingei, J. Bacteriol.*, 90, 524, 1965.
1128. **Slooff, W. C.,** Genus 19. *Schizosaccharomyces* Lindner, in *The Yeasts: a Taxonomic Study,* 2nd ed., Lodder, J., Ed., North-Holland, Amsterdam, 1970, 733.
1129. **Lindner, P.,** *Schizosaccharomyces Pombe* n. sp., ein neuer Gährungserreger, *Wochenschr. Brau.*, 10, 1298, 1893.
1130. **Leupold, U.,** Genetical methods for *Schizosaccharomyces pombe*, in *Methods in Cell Physiology*, Vol. 4, Prescott, D. M., Ed., Academic Press, New York, 1970, 169.
1131. **Mitchison, J. M.,** The growth of single cells. I. *Schizosaccharomyces pombe, Exp. Cell Res.*, 13, 244, 1957.
1132. **Mitchison, J. M.,** Physiological and cytological methods for *Schizosaccharomyces pombe*, in *Methods in Cell Physiology*, Vol. 4, Prescott, D. M., Ed., Academic Press, New York, 1970, 131.
1133. **Robinow, C. F.,** The number of chromosomes in *Schizosaccharomyces pombe:* light microscopy of stained preparations, *Genetics*, 87, 491, 1977.
1134. **Robinow, C. F.,** Further to the number of chromosomes in *Schizosaccharomyces pombe, Genetics*, 90, 207, 1978.
1135. **Schopfer, W. H., Wustenfeld, D., and Turian, G.,** La division nucleaire chez *Schizosaccharomyces pombe, Arch. Mikrobiol.*, 45, 304, 1963.
1136. **Mitchison, J. M.,** *The Biology of the Cell Cycle*, Cambridge University Press, Cambridge, U.K., 1971.
1137. **Mitchison, J. M.,** The cell cycle of a eukaryote, in *Microbial Differentiation*, Ashworth, J. M. and Smith, J. E., Eds., Cambridge University Press, Cambridge, U.K., 1973, 189.
1138. **Johnson, B. F., Calleja, G. B., Yoo, B. Y., Zuker, M., and McDonald, I. J.,** Cell division: key to cellular morphogenesis in the fission yeast, *Schizosaccharomyces pombe, Int. Rev. Cytol.*, 75, 167, 1982.
1139. **Egel, R.,** Commitment to meiosis in fission yeast, *Mol. Gen. Genet.*, 121, 277, 1973.
1140. **Suminoe, K. and Dukmo, H.,** The life cycles of *Schizosaccharomyces* with reference to the mode of conjugation and ascospore formation, *J. Gen. Appl. Microbiol.*, 9, 243, 1963.
1141. **Conti, S. F. and Naylor, H. B.,** Electron microscopy of ultrathin sections of *Schizosaccharomyces octosporus*. II. Morphological and cytological changes preceding ascospore formation, *J. Bacteriol.*, 79, 331, 1960.
1142. **Tsuboi, M., Ohashi, K., Takahara, M., and Hayashibe, M.,** Sexual process in eight-spored ascus-forming yeast, *Schizosaccharomyces japonicus*. I. Culture medium for the synchronous sexual process, *Plant Cell Physiol.*, 19, 1327, 1978.
1143. **Calleja, G. B., Johnson, B. F., and Yoo, B. Y.,** Macromolecular changes and commitment to sporulation in the fission yeast *Schizosaccharomyces pombe, Plant Cell Physiol.*, 21, 613, 1980.

1144. **Fowell, R. R.,** Factors controlling sporulation of *Schizosaccharomyces*, *Antonie van Leeuwenhoek J. Microbiol. Serol.*, 35, (Suppl. Yeast Symp.), H21, 1969.
1145. **Fraser, R. S. S., Creanor, J., and Mitchison, J. M.,** Rapid and selective inhibition of the synthesis of high molecular weight RNA in yeast by lomofungin, *Nature (London)*, 244, 222, 1973.
1146. **Calleja, G. B.,** Studies on the sexuality of *Schizosaccharomyces pombe* NCYC 132 and its derivatives, *Genetics*, 71, s9, 1972.
1147. **Calleja, G. B., Zuker, M., and Johnson, B. F.,** Temporal asymmetry of sex interconversion in a strain of the homothallic fission yeast *Schizosaccharomyces pombe*, *Curr. Microbiol.*, 6, 225, 1981.
1148. **Egel, R.,** The genetic instabilities of the mating type locus in fission yeast, *Mol. Gen. Genet.*, 145, 281, 1976.
1149. **Egel, R.,** Frequency of mating-type switching in homothallic fission yeast, *Nature (London)*, 266, 172, 1977.
1150. **Gutz, H. and Doe, J. F.,** Two different h^- mating types in *Schizosaccharomyces pombe*, *Genetics*, 74, 563, 1973.
1151. **Gutz, H. and Doe, J. F.,** On homo- and heterothallism in *Schizosaccharomyces pombe*, *Mycologia*, 67, 748, 1975.
1152. **Leupold, U.,** Some data on polyploid inheritance in *Schizosaccharomyces pombe*, *C. R. Trav. Lab. Carlsberg Ser. Physiol.*, 26, 221, 1956.
1153. **Egel, R.,** Rearrangements at the mating type locus in fission yeast, *Mol. Gen. Genet.*, 148, 149, 1976.
1154. **Fincham, J. R. S. and Day, P. R.,** *Fungal Genetics*, 3rd ed., Blackwell Scientific, Oxford, 1971.
1155. **Egel, R.,** Orientation of "*plus*" genes at the mating-type locus in homothallic fission yeast, *Mol. Gen. Genet.*, 161, 305, 1978.
1156. **Gutz, H., Meade, J. H., and Walker, S.,** Mating type modifier genes in *Schizosaccharomyces pombe*, *Genetics*, 80, s38, 1975.
1157. **Bresch, C., Müller, G., and Egel, R.,** Genes involved in meiosis and sporulation of a yeast, *Mol. Gen. Genet.*, 102, 301, 1968.
1158. **Egel, R.,** Genes involved in mating type expression of fission yeast, *Mol. Gen. Genet.*, 122, 339, 1973.
1159. **Kawanabe, Y., Yoshida, K., and Yanagishima, N.,** Sexual cell agglutination in relation to the formation of zygotes in *Saccharomyces cerevisiae*, *Plant Cell Physiol.*, 20, 423, 1979.
1160. **Megnet, R.,** Effect of 2-deoxyglucose on *Schizosaccharomyces pombe*, *J. Bacteriol.*, 90, 1032, 1965.
1161. **Johnson, B. F. and Rupert, C. M.,** Cellular growth rates of the fission yeast, *Schizosaccharomyces pombe*, and variable sensitivity to 2-deoxyglucose, *Exp. Cell Res.*, 48, 618, 1967.
1162. **Poon, N. H., Martin, J., and Day, A. W.,** Conjugation in *Ustilago violacea*. I. Morphology, *Can. J. Microbiol.*, 20, 187, 1974.
1163. **van der Walt, J. P.,** Genus 16. *Saccharomyces* emend. Reess, in *The Yeasts: a Taxonomic Study*, 2nd ed., Lodder, J., Ed., North-Holland, Amsterdam, 1970, 555.
1164. **Winge, Ö.** On haplophase and diplophase in some Saccharomycetes, *C. R. Trav. Lab. Carlsberg Ser. Physiol.*, 21, 77, 1935.
1165. **Ahmad, M.,** The mating system in *Saccharomyces*, *Ann. Bot.*, 17, 329, 1953.
1166. **Mortimer, R. K. and Hawthorne, D. C.,** Yeast genetics, in *The Yeasts*, Vol. 1, Rose, A. H. and Harrison, J. S., Eds., Academic Press, London, 1969, 386.
1167. **Fowell, R. R.,** Life cycles in yeasts, in *The Yeasts*, Vol. 1, Rose, A. H. and Harrison, J. S., Eds., Academic Press, London, 1969, 461.
1168. **Biliński, T., Litwińska, J., Zuk, J., and Gajewski, W.,** Synchronous zygote formation in yeasts, in *Methods of Cell Biology*, Vol. 11, Prescott, D. M., Ed., Academic Press, New York, 1975, 89.
1169. **Yanagishima, N.,** Physiology of the mating reaction in yeast, in *Commentaries in Plant Science*, Smith, H., Ed., Pergamon Press, Oxford, 1976, 63.
1170. **Yanagishima, N., Yoshida, K., Hagiya, M., Kawanabe, Y., Shimoda, C., Sakurai, A., Tamura, S., and Osumi, M.,** Sexual cell agglutination in *Saccharomyces cerevisiae*, in *Growth and Differentiation in Microorganisms*, Ishikawa, T., Maruyama, Y., and Matsumiya, H., Eds., University of Tokyo Press, Tokyo, 1977, 193.
1171. **MacKay, V. L.,** Mating-type specific pheromones as mediators of sexual conjugation in yeast, in *Molecular Control of Proliferation and Differentiation*, Papaconstantinou, J. and Rutter, W. J., Eds., Academic Press, New York, 1978, 243.
1172. **Sakai, K. and Yanagishima, N.,** Mating reaction in *Saccharomyces cervisiae*. II. Hormonal regulation of agglutinability of *a* type cells, *Arch. Mikrobiol.*, 84, 191, 1972.
1173. **MacKay, V. and Manney, T. R.,** Mutations affecting sexual conjugation and related processes in *Saccharomyces cerevisiae*. I. Isolation and phenotypic characterization of nonmating mutants, *Genetics*, 76, 255, 1974.
1174. **Throm, E. and Duntze, W.,** Mating-type-dependent inhibition of deoxyribonucleic acid synthesis in *Saccharomyces cerevisiae*, *J. Bacteriol.*, 104, 1388, 1970.

1175. **Shimoda, C. and Yanagishima, N.,** Mating reaction in *Saccharomyces cerevisiae*. IV. Retardation of deoxyribonucleic acid synthesis, *Physiol. Plant.*, 29, 54, 1973.

1176. **Biliński, T., Jachymczyk, W., Litwińska, J., Żuk, J., and Gajewski, W.,** Mutual inhibition of DNA synthesis in *a*- and α-cells of *Saccharomyces cerevisiae* during conjugation, *J. Gen. Microbiol.*, 82, 97, 1974.

1177. **Żuk, J., Zaborowska, D., Litwińska, J., Chlebowicz, E., and Biliński, T.,** Macromolecular synthesis during conjugation in yeast, *Acta Microbiol. Polon. Ser. A.*, 7, 67, 1975.

1178. **Petes, T. D. and Fangman, W. L.,** Preferential synthesis of yeast mitochondrial DNA in α factor-arrested cells, *Biochem. Biophys. Res. Commun.*, 55, 603, 1973.

1179. **Lipke, P. N., Taylor, A., and Ballou, C. E.,** Morphogenic effects of α-factor on *Saccharomyces cerevisiae a* cells, *J. Bacteriol.*, 127, 610, 1976.

1180. **Tkacz, J. S. and MacKay, V. L.,** Sexual conjugation in yeast. Cell surface changes in response to the action of mating hormones, *J. Cell Biol.*, 80, 326, 1979.

1181. **Yanagishima, N., Shimoda, C., Tsuboi, M., Hagiya, M., Takao, N., Sakurai, A., and Tamura, S.,** Hormonal regulation in the life cycle of *Saccharomyces cerevisiae*, in *Plant Growth Substances 1973*, Proc. 8th Int. Conf. Plant Growth Substances, 1973, 173.

1182. **Sakurai, A., Tamura, S., Yanagishima, N., and Shimoda, C.,** Isolation and chemical characterization of the peptidyl factor inducing sexual agglutination in *Saccharomyces cerevisiae*, *Agric. Biol. Chem.*, 40, 255, 1976.

1183. **Doi, S., Suzuki, Y., and Yoshimura, M.,** Induction of sexual cell agglutinability of *a* mating type cells by α factor in *Saccharomyces cerevisiae*, *Biochem. Biophys. Res. Commun.*, 91, 849, 1979.

1184. **Duntze, W.,** Control of intercellular cooperation in yeast by hormone-like substances, *Postepy Mikrobiol.*, 13, 41, 1974.

1185. **Tanaka, T., Kita, H., Murakami, T., and Narita, K.,** Purification and amino acid sequence of mating factor from *Saccharomyces cerevisiae*, *J. Biochem.*, 82, 1681, 1977.

1186. **Tanaka, T., Kita, H., Murakami, T., and Narita, K.,** Purification and its structure activity relationship of mating factor of *Saccharomyces cerevisiae*, *Proc. Jpn. Acad.*, 53, 67, 1977.

1187. **Masui, Y., Chino, N., Sakakibara, S., Tanaka, T., Murakami, T., and Kita, H.,** Synthesis of the mating factor of *Saccharomyces cerevisiae* and its truncated peptides: the structure-activity relationship, *Biochem. Biophys. Res. Commun.*, 78, 534, 1977.

1188. **Ciejek, E., Thorner, J., and Geier, M.,** Solid phase peptide synthesis of α-factor, a yeast mating pheromone, *Biochem. Biophys. Res. Commun.*, 78, 952, 1977.

1189. **Masui, Y., Tanaka, T., Chino, N., Kita, H., and Sakakibara, S.,** Amino acid substitution of mating factor of *Saccharomyces cerevisiae:* structure-activity relationship, *Biochem. Biophys. Res. Commun.*, 86, 982, 1979.

1190. **Yanagishima, N., Shimizu, T., Hagiya, M., Yoshida, K., Sakurai, A., and Tamura, S.,** Physiological detection of a binding substance for the agglutinability-inducing pheromone, α substance-I in *Saccharomyces cerevisiae*, *Plant Cell Physiol.*, 18, 1181, 1977.

1191. **Chan, R. K.,** Recovery of *Saccharomyces cerevisiae* mating-type *a* cells from G1 arrest by α factor, *J. Bacteriol.*, 130, 766, 1977.

1192. **Tanaka, T. and Kita, H.,** Site of action of mating factor in *a*-mating type cell of *Saccharomyces cerevisiae*, *Biochem. Biophys. Res. Commun.*, 83, 1319, 1978.

1193. **Finkelstein, D. B. and Strausberg, S.,** Metabolism of α-factor by *a* mating type cells of *Saccharomyces cerevisiae*, *J. Biol. Chem.*, 254, 796, 1979.

1194. **Scherer, G., Haag, G., and Duntze, W.,** Mechanism of α factor biosynthesis in *Saccharomyces cerevisiae*, *J. Bacteriol.*, 119, 386, 1974.

1195. **Tanaka, T. and Kita, H.,** Degradation of mating factor by α-mating type cells of *Saccharomyces cerevisiae*, *J. Biochem.*, 82, 1689, 1977.

1196. **Fehrenbacher, G., Perry, K., and Thorner, J.,** Cell-cell recognition in *Saccharomyces cerevisiae:* regulation of mating-specific adhesion, *J. Bacteriol.*, 134, 893, 1978.

1197. **Betz, R., Duntze, W., and Manney, T. R.,** Mating-factor-mediated sexual agglutination in *Saccharomyces cerevisiae*, *FEMS Microbiol. Lett.*, 4, 107, 1978.

1198. **Shimoda, C., Yanagishima, N., Sakurai, A., and Tamura, S.,** Induction of sexual agglutinability of *a* mating-type cells as the primary action of the peptidyl sex factor from α mating-type cells in *Saccharomyces cerevisiae*, *Plant Cell Physiol.*, 19, 513, 1978.

1199. **MacKay, V. and Manney, T. R.,** Mutations affecting sexual conjugation and related processes in *Saccharomyces cerevisiae*. II. Genetic analysis of nonmating mutants, *Genetics*, 76, 273, 1974.

1200. **Wilkinson, L. E. and Pringle, J. R.,** Transient G1 arrest of *S. cerevisiae* cells of mating type α by a factor produced by cells of mating type *a*, *Exp. Cell Res.*, 89, 175, 1974.

1201. **Lee, E.-H. and Johnson, B. F.,** Volume-related mitochondrial deoxyribonucleic acid synthesis in zygotes and vegetative cells of *Saccharomyces cerevisiae*, *J. Bacteriol.*, 129, 1066, 1977.

1202. **Betz, R., MacKay, V. L., and Duntze, W.,** *a*-factor from *Saccharomyces cerevisiae:* partial characterization of a mating hormone produced by cells of mating type *a*, *J. Bacteriol.*, 132, 462, 1977.

1203. **Tohoyama, H., Hagiya, M., Yoshida, K., and Yanagishima, N.,** Regulation of the production of the agglutination substances responsible for sexual agglutination in *Saccharomyces cerevisiae:* changes associated with conjugation and temperature shift, *Mol. Gen. Genet.*, 174, 269, 1979.

1204. **Shimoda, C., Matsushima, Y., and Yanagishima, N.,** Mating reaction in *Saccharomyces cerevisiae*. X. Agglutinability-inactivating factor: a factor which destroys sexual agglutinability of *a* mating-type cells, *Antonie van Leeuwenhoek J. Microbiol. Serol.*, 42, 511, 1976.

1205. **Campbell, D. A.,** Kinetics of the mating-specific aggregation in *Saccharomyces cerevisiae*, *J. Bacteriol.*, 116, 323, 1973.

1206. **Rogers, D. and Bussey, H.,** Fidelity of conjugation in *Saccharomyces cerevisiae*, *Mol. Gen. Genet.*, 162, 173, 1978.

1207. **Manney, T. R. and Woods, V.,** Mutants of *Saccharomyces cerevisiae* resistant to the α mating-type factor, *Genetics*, 82, 639, 1976.

1208. **Matsushima, Y., Shimoda, C., and Yanagishima, N.,** A factor in cell-free extracts inactivating sexual agglutinability of *a* mating type in *Saccharomyces cerevisiae*, *Plant Cell Physiol.*, 17, 621, 1976.

1209. **Hicks, J. B. and Herskowitz, I.,** Evidence for a new diffusible element of mating pheromones in yeast, *Nature (London)*, 260, 246, 1976.

1210. **Barksdale, A. W.,** The uptake of exogenous hormone A by certain strains of *Achlya*, *Mycologia*, 55, 164, 1963.

1211. **Machlis, L.,** The chemotactic activity of various sirenins and analogues and the uptake of sirenin by the sperm of *Allomyces*, *Plant Physiol.*, 52, 527, 1973.

1212. **Musgrave, A. and Nieuwenhuis, D.,** Metabolism of radioactive antheridiol by *Achlya* species, *Arch. Microbiol.*, 105, 313, 1975.

1213. **Tkacz, J. S., Cybulska, E. B., and Lampen, J. O.,** Specific staining of wall mannan in yeast cells with fluorescein-conjugated concanavalin A, *J. Bacteriol.*, 105, 1, 1971.

1214. **Kawanabe, Y., Hagiya, M., Yoshida, K., and Yanagishima, N.,** Effect of concanavalin A on the mating reaction in *Saccharomyces cerevisiae*, *Plant Cell Physiol.*, 19, 1207, 1978.

1215. **Firnhaber, J. and Windisch, S.,** Über das Kopulationsverhalten von *Saccharomyces*-Hefen, *Arch. Mikrobiol.*, 65, 329, 1969.

1216. **Haefner, K.,** A rapid method for obtaining zygotes and to determine the mating type in *Saccharomyces*, *Z. Allg. Mikrobiol.*, 5, 77, 1965.

1217. **Iguti, S. and Onobu, T.,** Conjugation in *Saccharomyces cerevisiae*, *Bot. Mag.*, 77, 181, 1964.

1218. **Ephrussi, B. and Hottinguer, H.,** Cytoplasmic constituents of heredity: on an unstable cell state in yeast, *Cold Spring Harbor Symp. Quant. Biol.*, 16, 75, 1951.

1219. **Shimoda, C. and Yanagishima, N.,** Mating reaction in *Saccharomyces cerevisiae*. III. Changes in autolytic activity, *Arch. Mikrobiol.*, 85, 310, 1972.

1220. **del Rey, F., Santos, T., García-Acha, I., and Nombela, C.,** Synthesis of 1,3-β-glucanases in *Saccharomyces cerevisiae* during the mitotic cycle, mating, and sporulation, *J. Bacteriol.*, 139, 924, 1979.

1221. **Svoboda, A.,** Mating reaction in yeast protoplasts, *Arch. Microbiol.*, 110, 313, 1976.

1222. **Hawthorne, D. C.,** A deletion in yeast and its bearing on the structure of the mating type locus, *Genetics*, 48, 1727, 1963.

1223. **Hicks, J. B. and Herskowitz, I.,** Interconversion of yeast mating types. I. Direct observations of the action of the homothallism (*HO*) gene, *Genetics*, 83, 245, 1976.

1224. **Takano, I. and Oshima, Y.,** An allele specific and a complementary determinant controlling homothallism in *Saccharomyces oviformis*, *Genetics*, 57, 875, 1967.

1225. **Strathern, J. N. and Herskowitz, I.,** Asymmetry and directionality in production of new cell types during clonal growth: the switching pattern of homothallic yeast, *Cell*, 17, 371, 1979.

1226. **Naumov, G. I. and Tolstorukov, I. I.,** Comparative genetics of yeast. X. Reidentification of mutators of mating types in *Saccharomyces*, *Genetika*, 9, 82, 1973.

1227. **Harashima, S. and Oshima, Y.,** Mapping of the homothallic genes, *HM*α and *HMa*, in Saccharomyces yeasts, *Genetics*, 84, 437, 1976.

1228. **Hicks, J., Strathern, J. N., and Klar, A. J. S.,** Transposable mating type genes in *Saccharomyces cerevisiae*, *Nature (London)*, 282, 478, 1979.

1229. **Klar, A. J. S. and Fogel, S.,** Activation of mating type genes by transposition in *Saccharomyces cerevisiae*, *Proc. Natl. Acad. Sci. U.S.A.*, 76, 4539, 1979.

1230. **Hopper, A. K. and Hall, B. D.,** Mutation of a heterothallic strain to homothallism, *Genetics*, 80, 77, 1975.

1231. **Takahashi, T. and Ikeda, Y.,** Bisexual mating reaction in *Saccharomyces chevalieri*, *Genetics*, 44, 375, 1959.

1232. **Herman, A. and Roman, H.,** Allele specific determinants of homothallism in *Saccharomyces lactis*, *Genetics*, 53, 727, 1966.

1233. **Santa Maria, J. and Vidal, D.**, Segregación anormal del "mating type" en Saccharomyces, *Instituto Nacional de Investigaciones Agronómicas (Madrid) Conferencias*, 30, 1, 1970.
1234. **Oshima, Y. and Takano, I.**, Genetic controlling system for homothallism and a novel method for breeding triploid cells in *Saccharomyces*, in *Fermentation Technology Today*, Proc. 4th Int. Fermentation Symp., Terui, G., Ed., Society of Fermentation Technology, Osaka, 1972, 847.
1235. **Takano, I., Kusumi, T., and Oshima, Y.**, An α mating-type allele insensitive to the mutagenic action of the homothallic gene system in *Saccharomyces diastaticus*, *Mol. Gen. Genet.*, 126, 19, 1973.
1236. **Takano, I. and Oshima, Y.**, Allelism tests among various homothallism-controlling genes and gene systems in *Saccharomyces*, *Genetics*, 64, 229, 1970.
1237. **Herskowitz, I., Strathern, J. N., Hicks, J. B., and Rine, J.**, Mating type interconversion in yeast and its relationship to development in higher eucaryotes, in *Molecular Approaches to Eucaryotic Genetic Systems*, Wilcox, G., Abelson, J., and Fox, C. F., Eds., Academic Press, New York, 1977, 193.
1238. **Hicks, J. B. and Herskowitz, I.**, Interconversion of yeast mating types. II. Restoration of mating ability to sterile mutants in homothallic and heterothallic strains, *Genetics*, 85, 373, 1977.
1239. **Strathern, J. N., Blair, L. C., and Herskowitz, I.**, Healing of *mat* mutations and control of mating type interconversion by the mating type locus in *Saccharomyces cerevisiae*, *Proc. Natl. Acad. Sci. U.S.A.*, 76, 3425, 1979.
1240. **Blair, L. C., Kushner, P. J., and Herskowitz, I.**, Mutations of the *HMa* and *HMα* loci and their bearing on the cassette model of mating type interconversion in yeast, in *Eucaryotic Gene Regulation*, Axel, R., Maniatis, T., and Fox, C. F., Eds., Academic Press, New York, 1979, 13.
1241. **Hicks, J., Klar, A., Strathern, J. N., and Newlon, C. S.**, Control of mating type in yeast, in *Eucaryotic Gene Regulation*, Axel, R., Maniatis, T., and Fox, C. F., Eds., Academic Press, New York, 1979, 27.
1242. **Haber, J. E. and Garvik, B.**, A new gene affecting the efficiency of mating-type interconversions in homothallic strains of *Saccharomyces cerevisiae*, *Genetics*, 87, 33, 1977.
1243. **Hicks, J. B., Strathern, J. N., and Herskowitz, I.**, Interconversion of yeast mating types. III. Action of the homothallism *(HO)* gene in cells homozygous for the mating locus, *Genetics*, 85, 395, 1977.
1244. **Klar, A. J. S. and Fogel, S.**, The action of homothallism genes in *Saccharomyces* diploids during vegetative growth and the equivalence of *hma* and *HMα* loci functions, *Genetics*, 85, 407, 1977.
1245. **Takano, I., Oshima, T., Harashima, S., and Oshima, Y.**, Tetraploid formation through the conversion of the mating-type alleles by the action of homothallic genes in the diploid cells of *Saccharomyces* yeasts, *J. Ferment. Technol.*, 55, 1, 1977.
1246. **Mortimer, R. K.**, Radiobiological and genetic studies on a polyploid series (haploid to hexaploid) of *Saccharomyces cerevisiae*, *Radiat. Res.*, 9, 312, 1958.
1247. **Roman, H. and Sands, S. M.**, Heterogeneity of clones of *Saccharomyces* derived from haploid ascospores, *Proc. Natl. Acad. Sci. U.S.A.*, 39, 171, 1953.
1248. **Gilliland, R. B.**, The flocculation of brewing yeast, *Wallerstein Lab. Commun.*, 20, 41, 1957.
1249. **Bishop, L. R. and Whitley, W. A.**, On the function of wort flocculum in affecting attenuation and yeast behaviour, *J. Inst. Brew. London*, 44, 73, 1938.
1250. **Curtis, N. S.**, Wort attenuation and yeast flocculation, *Brew. Guild J.*, 42, 175, 1956.
1251. **Jansen, H. E.**, Flocculation of yeasts, in *The Chemistry and Biology of Yeasts*, Cook, A. H., Ed., Academic Press, New York, 1958, 635.
1252. **Van Laer, H.**, Sur quelques phénomènes de coagulation produits par les borates (Agglutination de la levure), *Bull. Soc. Chim. Belg.*, 19, 31, 1905.
1253. **Hayduck, F. and Schüking, K.**, Über die Bedeutung der Kalksalze im Brauwasser auf Hefe und Gärung, *Wochenschr. Brau.*, 25, 241, 1908.
1254. **Schönfeld, F.**, Die chemische Zusammensetzung der Hefe in Beziehung zu ihrem Verhalten bei der Gärung, *Wochenschr. Brau.*, 29, 393, 1912.
1255. **Moufang, E.**, Zur Frage der Wirkung eines reduzierten Säuregrades beim Maischprozess, *Wochenschr. Brau.*, 29, 721, 1912.
1256. **Schönfeld, F. and Schönfelder, G.**, Die Mineralbestandteile der Hefe und ihre Bedeutung für den Lebenszustand derselben, *Wochenschr. Brau.*, 31, 245, 1914.
1257. **Schönfeld, F. and Krumhaar, H.**, Die Bruch- und Staubform der Hefe — ihre Ursachen, *Wochenschr. Brau.*, 35, 302, 1918.
1258. **Beijerinck, M. W.**, Die Erscheinung der Flockenbildung oder Agglutination bei Alkoholhefen, *Centralbl. Bakteriol. Parasitenk. Abt. 2*, 20, 641, 1908.
1259. **Comrie, A. A. D.**, Yeast flocculation, *Wallerstein Lab. Commun.*, 15, 339, 1952.
1260. **Thorne, R. S. W.**, The problem of flocculence in brewery yeasts, *Wallerstein Lab. Commun.*, 15, 201, 1952.
1261. **Lüers, H.**, Betrachtungen über die Bruchbildung der Hefe, *Brauwissenschaft*, 4, 49, 1953.
1262. **Gilliland, R. B.**, The flocculation of brewing yeast, *Brew. Guild J.*, 41, 246, 1955.
1263. **St. Johnston, J. H.**, Yeast flocculation: a review, *Brew. Dig.*, 33, 62, 1958.

1264. **Silbereisen, K.,** Allgemeine physikalische Eigenschaften, in *Die Hefen,* Vol. 1, Reiff, F., Kautzmann, R., Lüers, H., and Lindemann, M., Eds., Verlag Hans Carl, Nürnberg, 1960, 323.
1265. **Bunker, H. J.,** Recent research on the yeasts, in *Progress in Industrial Microbiology,* Vol. 3, Hockenhull, D. J. D., Ed., Heywood & Co., London, 1961, 2.
1266. **Rainbow, C.,** Flocculation of brewer's yeast, *Process Biochem.,* 1, 489, 1966.
1267. **Rainbow, C.,** Brewer's yeasts, in *The Yeasts,* Vol. 3, Rose, A. H. and Harrison, J. S., Eds., Academic Press, London, 1970, 147.
1268. **Kirsop, B. H.,** Yeast separation, *Brew. Guardian,* 100, 75, 1971.
1269. **Geilenkotten, I. and Nyns, E. J.,** The biochemistry of yeast flocculence, *Brew. Dig.,* 46, 64, 1971.
1270. **Stewart, G. G. and Russell, I.,** Yeast flocculation, in *Brewing Science,* Vol. 2, Pollock, J. R. A., Ed., Academic Press, London, 1981, 61.
1271. **Gaeng, F. E.,** *Petit J. Brass.,* 62, 260, 1954.
1272. **Burns, J. A.,** Brewing of top fermentation beers. II. Practical aspects of top fermentation, *Brew Dig.,* 23, 58, 1948.
1273. **Masschelein, C. A. and Jeunehomme, C.,** Floculence et dégénérescence des levures basses, *Brass. Malt. Belg.,* 9, 184, 1959.
1274. **Hough, J. S.,** Flocculence in relation to continuous fermentation, *J. Inst. Brew. London,* 64, 368, 1958.
1275. **Lange, H.,** Über den physiologischen Zustand der Hefe, *Wochenschr. Brau.,* 24, 445, 1907.
1276. **Ranken, C.,** The surface of yeast as a factor in fermentation, *J. Inst. Brew. London,* 33, 76, 1927.
1277. **Ranken, C. and Bell, J. R.,** The surface of yeast as a factor in fermentation. II, *J. Inst. Brew. London,* 34, 265, 1928.
1278. **Lange, H.,** Über den physiologischen Zustand der Hefe, *Wochenschr. Brau.,* 24, 417, 1907.
1279. **Eddy, A. A.,** The mechanism of yeast flocculation, Eur. Brew. Conv. Proc. Congr., Baden-Baden, 1955, 65.
1280. **Oldfield, A. I., St. Johnston, J. H., and Taylor, A. E.,** Some factors affecting the formation of the yeast head in top fermentation, Eur. Brew. Conv. Proc. Congr., Brighton, 1951, 98.
1281. **Miki, B. L. A., Poon, N. H., James, A. P., and Seligy, V. L.,** Flocculation in *Saccharomyces cerevisiae:* mechanism of cell-cell interactions, in *Current Developments in Yeast Research,* Stewart, G. G. and Russell, I., Eds., Pergamon Press, Toronto, 1981, 193.
1282. **Jansen, H. E. and Mendlik, F.,** A study on yeast flocculation. II, Eur. Brew. Conv. Proc. Congr., Nice, 1953, 143.
1283. **Lyons, T. P. and Hough, J. S.,** Cation-binding by yeast cell walls, *Biochem. J.,* 119, 10p, 1970.
1284. **Lyons, T. P. and Hough, J. S.,** Flocculation of brewer's yeast, *J. Inst. Brew. London,* 76, 564, 1970.
1285. **Lyons, T. P. and Hough, J. S.,** Further evidence for the cross-bridging hypothesis for flocculation of brewer's yeast, *J. Inst. Brew. London,* 77, 300, 1971.
1286. **MacWilliam, I. C.,** The structure, synthesis and functions of the yeast cell wall — a review, *J. Inst. Brew. London,* 76, 524, 1970.
1287. **Phaff, H. J.,** Cell wall of yeasts, *Annu. Rev. Microbiol.,* 17, 15, 1963.
1288. **Ballou, C. E.,** Some aspects of the structure, immunochemistry, and genetic control of yeast mannans, in *Advances in Enzymology and Related Areas of Molecular Biology,* Vol. 40, Meister, A., Ed., John Wiley & Sons, New York, 1974, 239.
1289. **Griffin, S. R. and MacWilliam, I. C.,** Variation of cell wall content in flocculent and non-flocculent yeast strains, *J. Inst. Brew. London,* 75, 355, 1969.
1290. **Stewart, G. G. and Garrison, I. F.,** Some observations on co-flocculation in *Saccharomyces cerevisiae, Am. Soc. Brew. Chem. Proc.,* p. 118, 1972.
1291. **Isaeva, V. S., Zhvirblyanskaya, A. E., and Kritskova, G. M.,** Relation of flocculation to the content of polysaccharides in the cell walls of yeast used in the continuous brewing process, *Kvasny Prum.,* 20, 98, 1974.
1292. **Masschelein, C. A., Jeunehomme-Ramos, C., Castiau, C., and Devreux, A.,** Mechanism of phenotypic variations in the flocculence character of yeast, *J. Inst. Brew. London,* 69, 332, 1963.
1293. **Cawley, T. N. and Ballou, C. E.,** Identification of two *Saccharomyces cerevisiae* cell wall mannan chemotypes, *J. Bacteriol.,* 111, 690, 1972.
1294. **Marfey, P., Sørensen, S. B., and Ottesen, M.,** Studies on yeast flocculation. Comparison of enzymatic digests of flocculent and non-flocculent cells of *Saccharomyces carlsbergensis, Carlsberg Res. Commun.,* 42, 353, 1977.
1295. **Friis, J. and Ottolenghi, P.,** The genetically determined binding of Alcian blue by a minor fraction of yeast cell walls, *C. R. Trav. Lab. Carlsberg Ser. Physiol.,* 37, 327, 1970.
1296. **Davies, R. and Wayman, F. J.,** The effect of thiols on *Saccharomyces fragilis, Antonie van Leeuwenhoek J. Microbiol. Serol.,* 41, 33, 1975.
1297. **Russell, I., Garrison, I. F., and Stewart, G. G.,** Studies on the formation of spheroplasts from stationary phase cells of *Saccharomyces cerevisiae, J. Inst. Brew. London,* 79, 48, 1973.

1298. **Nishihara, H., Toraya, T., and Fukui, S.,** Effect of chemical modification of cell surface components of a brewer's yeast on the floc-forming ability, *Arch. Microbiol.*, 115, 19, 1977.
1299. **St. Johnston, J. H.,** The flocculation of proteins in wort and beer, *Eur. Brew. Conv. Proc. Congr.*, 1, 62, 1949.
1300. **Rohrer, E.,** Yeast flocculation, *Brew. Dig.*, 25, 4, 1950.
1301. **Wiles, A. E.,** The action of finings and its relation to the electrokinetic properties of the yeast cell, Eur. Brew. Conv. Proc. Congr., Brighton, 1951, 84.
1302. **Taylor, N. W. and Orton, W. L.,** Aromatic compounds and sugars in flocculation of *Saccharomyces cerevisiae, J. Inst. Brew. London,* 84, 113, 1978.
1303. **Miki, B. L. A., Poon, N. H., James, A. P., and Seligy, V. L.,** Possible mechanism for flocculation interactions governed by gene *FL01* in *Saccharomyces cerevisiae, J. Bacteriol.*, 150, 878, 1982.
1304. **Lands, W. E. M. and Graff, G.,** Effects of fatty acid structure on yeast cell physiology, in *Current Developments in Yeast Research*, Stewart, G. G. and Russell, I., Eds., Pergamon Press, Toronto, 1981, 331.
1305. **Schönfeld, F.,** Die Beeinflussung der Eigenschaften obergäriger Brauereihefen, *Wochenschr. Brau.*, 27, 541, 1910.
1306. **Stockhausen, F. and Silbereisen, K.,** Ueber Flockung und Bruch verursachende Stoffe aus Braureihefen, *Wochenschr. Brau.*, 50, 349, 1933.
1307. **Stockhausen, F. and Silbereisen, K.,** Hefegummistudien I. Die Bestimmung des Hefegummis in der Hefe, *Wochenschr. Brau.*, 52, 145, 1935.
1308. **Kalyuzhnyi, M. Ya.,** Flocculation and sorption of yeast on media on non-alimentary raw materials, *Microbiology*, 31, 586, 1963.
1309. **Patel, G. B. and Ingledew, W. M.,** The relationship of acid-soluble glycogen to yeast flocculation, *Can. J. Microbiol.*, 21, 1608, 1975.
1310. **Williams, N. J. and Wiseman, A.,** Brewers' yeast flocculation, *J. Sci. Food Agric.*, 25, 1046, 1974.
1311. **Williams, N. J. and Wiseman, A.,** Metabolic regulation of flocculation in yeast, *J. Appl. Chem. Biotechnol.* 25, 557, 1975.
1312. **Beavan, M. J., Belk, D. M., Stewart, G. G., and Rose, A. H.,** Changes in electrophoretic mobility and lytic enzyme activity associated with development of flocculating ability in *Saccharomyces cerevisiae, Can. J. Microbiol.*, 25, 888, 1979.
1313. **Hayduck, F.,** Über die Bedeutung des Eiweiss im Hefenleben, *Wochenschr. Brau.*, 22, 661, 1905.
1314. **Lange, H.,** Bericht der Abteilung: Versuchsanstalt des Vereins der Kornbrennereibesitzer und der Presshefefabrikanten Deutschlands, *Jahrb. Ver. Spiritusfabr. Dtsch.*, 4, 135, 1904.
1315. **Heron, H.,** Oxygen as a factor in the production of sound beer, *J. Inst. Brew. London,* 32, 261, 1926.
1316. **Miki, B. L. A., Poon, N. H., and Seligy, V. L.,** Repression and induction of flocculation interactions in *Saccharomyces cerevisiae. J. Bacteriol.*, 150, 890, 1982.
1317. **Devreux, A.,** Definition, characteristics, causes and mechanisms of (yeast) degeneration, *J. Inst. Brew. London*, 69, 266, 1963.
1318. **Wilkie, D. and Nudd, R. C.,** Aspects of mitochondrial control of cell surface characteristics in *Saccharomyces cerevisiae*, in *Current Developments in Yeast Research,* Stewart, G. G. and Russell, I., Eds., Pergamon Press, Toronto, 1981, 345.
1319. **Jeffery, E. J.,** Brewery yeast propagation with special reference to flocculence in its relationship to attenuation, *J. Inst. Brew. London,* 62, 309, 1956.
1320. **Curtis, N. S.,** Factors affecting the control of wort attenuation, *J. Inst. Brew. London,* 69, 426, 1963.
1321. **Norstedt, C. and Bengtsson, A.,** Changes in the behaviour of yeast, *Eur. Brew. Conv. Proc. Congr.*, 12, 263, 1969.
1322. **Thompson, C. C.,** Conical fermentors, *J. Inst. Brew. London,* 76, 423, 1970.
1323. **Pomper, S. and Burkholder, P. R.,** Studies on the biochemical genetics of yeast, *Proc. Natl. Acad. Sci. U.S.A.*, 35, 456, 1949.
1324. **Thorne, R. S. W.,** The genetics of flocculence in *Saccharomyces cerevisiae, C. R. Trav. Lab. Carlsberg, Ser. Physiol.*, 25, 101, 1951.
1325. **Roman, H., Hawthorne, D. C., and Douglas, H. C.,** Polyploidy in yeast and its bearing on the occurrence of irregular genetic ratios, *Proc. Natl. Acad. Sci. U.S.A.*, 37, 79, 1951.
1326. **Gilmore, R. A. and Murphy, J.,** Flocculence in *Saccharomyces cerevisiae, Genetics,* 71, s19, 1972.
1327. **Farris, J. S. and Gilmore, R. A.,** Genetics of flocculence in *Saccharomyces cerevisiae, Genetics,* 77, s21, 1974.
1328. **Lewis, C. W. and Johnston, J. R.,** Genes controlling flocculation in yeast, *Proc. Soc. Gen. Microbiol.*, 1, 73, 1974.
1329. **Johnston, J. R. and Lewis, C. W.,** Genetic analysis of flocculation in *Saccharomyces cerevisiae* and tetrad analysis of commercial brewing and baking yeasts, in *2nd Int. Symp. Genet. Ind. Microorganisms,* Macdonald, K. D., Ed., Academic Press, London, 1976, 339.

1330. **Rothstein, R. J. and Sherman, F.,** Genes affecting the expression of cytochrome *c* in yeast: genetic mapping and genetic interactions, *Genetics*, 94, 871, 1980.

1331. **Russell, I., Stewart, G. G., Reader, H. P., Johnston, J. R., and Martin, P. A.,** Revised nomenclature of genes that control yeast flocculation, *J. Inst. Brew. London*, 86, 120, 1980.

1332. **Holmberg, S.,** Isolation and characterization of a polypeptide absent from non-flocculent mutants of *Saccharomyces cerevisiae, Carlsberg Res. Commun.*, 43, 401, 1978.

1333. **Šilhánková, L., Šavel, J., and Moštek, J.,** Respiratory deficient mutants of bottom brewer's yeast. I. Frequencies and types of mutant in various strains, *J. Inst. Brew. London*, 76, 280, 1970.

1334. **Gyllang, H. and Martinson, E.,** Properties of some spontaneous mutants from brewers yeast with increased sedimentation rate, *Eur. Brew. Conv. Proc. Congr.*, 13, 265, 1971.

1335. **Cowan, W. D., Hoggan, J., and Smith, J. E.,** Induction of respiratory deficient mutants in brewery yeast by formaldehyde and its technological significance, in *Proc. 4th Int. Symp. Yeasts*, Part 1, Klaushofer, H. and Sleytr, U. B., Eds., Hochschule f. Bodenkultur, Wien, 1974, 153.

1336. **Egilsson, V., Evans, I. H., and Wilkie, D.,** Toxic and mutagenic effects of carcinogens on the mitochondria of *Saccharomyces cerevisiae, Mol. Gen. Genet.*, 174, 39, 1979.

1337. **Spencer, J. F. T., Miller, R., Spencer, D. M., and Wilkie, D.,** Flocculation and starch utilization in some antibiotic-resistant mitochondrial mutants of *Saccharomyces diastaticus*, in *Current Developments in Yeast Research*, Stewart, G. G. and Russell, I., Eds., Pergamon Press, Toronto, 1981, 33.

1338. **Nunokawa, Y., Toba, H., and Ouchi, K.,** Froth flotation of yeast cells (I) Correlation between high flotability of cells and their froth forming ability in sake mash, *J. Ferment. Technol.*, 49, 959, 1971.

1339. **Ouchi, K., Takahashi, K., Suzuki, S., and Nunokawa, Y.,** Non-foaming mutants of *saké* yeast: comparison of the cell wall composition between the parent and mutants, *J. Gen. Appl. Microbiol.*, 19, 429, 1973.

1340. **Curtis, N. S. and Wenham S.,** Mutual flocculation as a factor influencing wort attenuation in the brewery, *J. Inst. Brew. London*, 64, 368, 1958.

1341. **Eddy, A. A. and Rudin, A. D.,** The structure of the yeast cell wall. I. Identification of charged groups at the surface, *Proc. R. Soc. London Ser. B.*, 148, 419, 1958.

1342. **Stewart, G. G.,** Recent developments in the characterization of brewery yeast strains, *Tech. Q. Master Brew. Assoc. Am.*, 9, 183, 1972.

1343. **Stewart, G. G.,** Some observations on maltose, fructose, and glucose metabolism in *Saccharomyces cerevisiae, Am. Soc. Brew. Chem. Proc.*, p. 86, 1973.

1344. **Momose, H., Iwano, K., and Tonoike, R.,** Studies on the aggregation of yeasts caused by lactobacilli. III. Effects of various conditions on the aggregation, *J. Agric. Chem. Soc. Jpn.*, 43, 119, 1969.

1345. **Momose, H., Iwano, K., and Tonoike, R.,** Studies on the aggregation of yeast caused by lactobacilli. IV. Force responsible for aggregation, *J. Gen. Appl. Microbiol.*, 15, 19, 1969.

1346. **Freiberg, K. J. and Cruess, W. V.,** A study of certain factors affecting the growth of flor yeast, *Appl. Microbiol.*, 3, 208, 1955.

1347. **Castor, J. G. B. and Archer, T. E.,** Nutrient requirements for growth of the sherry flor yeast, *Saccharomyces beticus, Appl. Microbiol.*, 5, 56, 1957.

1348. **Gotô, S.,** Film-forming wine yeasts. I. Film formation and distribution, *J. Ferment. Technol.*, 37, 384, 1959.

1349. **Gotô, S.,** Film-forming wine yeasts. II. Classification and taxonomic characteristics, *J. Ferment. Technol.*, 37, 390, 1959.

1350. **Gotô, S.,** Film-forming wine yeasts. III. Characteristics and cultural conditions, *J. Ferment. Technol.*, 37, 412, 1959.

1351. **Gotô, S.,** Effect of vitamins and amino acids upon the film formation of pseudo-film forming yeasts, *J. Ferment. Technol.*, 39, 709, 1961.

1352. **Iimura, Y., Hara, S., and Otsuka, K.,** Cell surface hydrophobicity as a pellicle formation factor in film strain of *Saccharomyces, Agric. Biol. Chem.*, 44, 1215, 1980.

1353. **Iimura, Y., Hara, S., and Otsuka, K.,** Fatty acids as hydrophobic substance on cell surface of film strain of *Saccharomyces, Agric. Biol. Chem.*, 44, 1223, 1980.

1354. **Wickerham, L. J. and Stodola, F. H.,** Formation of extracellular sphingolipids by microorganisms. I. Tetraacetylphytosphingosine from *Hansenula ciferri, J. Bacteriol.*, 80, 484, 1960.

1355. **Stodola, F. H. and Wickerham, L. J.,** Formation of extracellular sphingolipids by microorganisms. II. Structural studies on tetraacetylphytosphingosine from the yeast *Hansenula ciferrii, J. Biol. Chem.*, 235, 2584, 1960.

1356. **Maister, H. G., Rogovin, S. P., Stodola, F. H., and Wickerham, L. J.,** Formation of extracellular sphingolipids by microorganisms. IV. Pilot-plant production of tetraacetylphytosphingosine by *Hansenula ciferrii, Appl. Microbiol.*, 10, 401, 1962.

1357. **van der Walt, J. P.,** Genus 8. *Kluyveromyces* van der Walt emend. van der Walt, in *The Yeasts: a Taxonomic Study*, 2nd ed., Lodder, J., Ed., North-Holland, Amsterdam, 1970, 316.

1358. **van Uden, N. and Vidal-Leiria, M.,** Genus 10. *Torulopsis* Berlese, in *The Yeasts: a Taxonomic Study*, 2nd ed., Lodder, J., Ed., North-Holland, Amsterdam, 1970, 1235.
1359. **McCullough, J. and Herskowitz, I.,** Mating pheromones of *Saccharomyces kluyveri:* pheromone interactions between *Saccharomyces kluyveri* and *Saccharomyces cerevisiae*, *J. Bacteriol.*, 138, 146, 1979.
1360. **Sakurai, A., Sato, Y., Park, K. H., Takahashi, N., Yanagishima, N., and Banno, I.,** Isolation and chemical characterization of a mating pheromone produced by *Saccharomyces kluyveri*, *Agric. Biol. Chem.* 44, 1451, 1980.
1361. **Burke, D., Mendonça-Previato, L., and Ballou, C. E.,** Cell-cell recognition in yeast: purification of *Hansenula wingei* 21-cell sexual agglutination factor and comparison of the factors from three genera, *Proc. Natl. Acad. Sci. U.S.A.*, 77, 318, 1980.
1362. **Barker, E. R. and Miller, M. W.,** Some properties of *Saccharomyces kluyveri*, *Antonie van Leeuwenhoek J. Microbiol. Serol.*, 35, 159, 1969.
1363. **Starmer, W. T., Phaff, H. J., Miranda, M., and Miller, M. W.,** *Pichia amethionina*, a new heterothallic yeast associated with the decaying stems of cereoid cacti, *Int. J. Syst. Bacteriol.*, 28, 433, 1978.
1364. **van der Walt, J. P. and Liebenberg, N. V. D. W.,** Agglutinative mating types of *Saccharomyces transvaalensis*, *Antonie van Leeuwenhoek J. Microbiol. Serol.*, 39, 629, 1973.
1365. **Thom, R.,** *Structural Stability and Morphogenesis*, W. A. Benjamin, Reading, Mass., 1975.
1366. **Chung, B.,** personal communication.
1367. **Crandall, M.,** personal communication.

[illegible]

[illegible]

[illegible]

[illegible]

[illegible]

[illegible]

[illegible]

[illegible]

[illegible]

APPENDIX: SYMBOLS AND ABBREVIATIONS

(The following symbols and abbreviations are either undefined in the text or, if defined, also found in places far removed from where they are first mentioned and defined.)

A.	*Actinomyces*
a	Mating type of *Saccharomyces cerevisiae, Sacch. kluyveri,* and *Pichia amethionina;* complementary to **α** *(vide infra)*
α	Mating type of *Saccharomyces cerevisiae, Sacch. kluyveri,* and *Pichia amethionina;* complementary to **a** *(vide supra)*
ADP	Adenosine-5′-diphosphate
AMP	Adenosine-5′-monophosphate
ATP	Adenosine-5′-triphosphate
B.	*Bacillus*
C.	*Chlamydomonas*
CF	Competence factor for bacterial transformation
CIA	Clumping-inducing agent in *Streptococcus faecalis*
Cyclic AMP	Adenosine-3′:5′-cyclic monophosphate
Cyclic GMP	Guanosine-3′:5′-cyclic monophosphate
D.	*Dictyostelium*
d	Cell or colony type of a homothallic strain of *Schizosaccharomyces pombe;* complementary to P *(vide infra)*
DEAE-	Diethylaminoethyl-
E.	*Escherichia*
EDTA	Ethylenediaminetetraacetic acid; [ethylenedinitrilo] tetraacetic acid
F	Fertility plasmid in *Escherichia coli*
5f	Agglutination factor from strain 5 of *Hansenula wingei;* complementary to 21f *(vide infra)*
21f	Agglutination factor from strain 21 of *Hansenula wingei;* complementary to 5f *(vide supra)*
Fab	Fragment of antibody
GMP	Guanosine-5′-monophosphate
H.	*Hansenula*
K.	*Klebsiella*
M.	*Myxococcus*
MAT	Mating-type locus in several systems; dominant allele of *mat (vide infra)*
mat	Mating-type locus in several systems; recessive allele of *MAT (vide supra)*
NSI	Nonspecific inhibitor of sexual agglutination in *Hansenula wingei*
P	Cell or colony type of a homothallic strain of *schizosaccharomyces pombe;* complementary to d *(vide supra)*
Q_{10}	K_{T+10}/K_T, where K is some value at temperature T
R.	*Rhizobium*
S.	*Streptococcus, Saccharomyces,* or *Schizosaccharomyces*
$S_{20,W}$	Sedimentation coefficient corrected to 20°C, water
SDS	Sodium dodecylsulfate; SLS; sodium lauryl sulfate
S-S	Disulfide bond
UV	Ultraviolet light
v/v	Volume of solute in volume of solvent
w/v	Weight of solute in volume of solvent
X	X-ray

[illegible]

The following [illegible]

[illegible]

INDEX

A

B

C

F

G

H

I

J

K

L

O

P

Q

R

S

T

U

V

W

Y

Z